PROCEEDINGS OF THE FIRST INTERNATIONAL SYMPOSIUM ON THE DEVELOPMENT OF NATURAL SCIENCE MUSEUMS UNDER THE BELT AND ROAD INITIATIVE

"一带一路"科普场馆发展国际研讨会

论文集

中国科学技术馆◎编

科学普及出版社

·北 京·

图书在版编目（CIP）数据

"一带一路"科普场馆发展国际研讨会论文集：汉文、英文 / 中国科学技术馆编 . —北京：科学普及出版社，2019.9

ISBN 978-7-110-09993-3

Ⅰ.①一… Ⅱ.①中… Ⅲ.①科学技术—展览馆—世界—文集—汉、英 Ⅳ.① N281-53

中国版本图书馆 CIP 数据核字（2019）第 186859 号

策划编辑	郑洪炜 李 洁	
责任编辑	李 洁 史朋飞	
封面设计	中文天地	
内文设计	中文天地	
责任校对	杨京华	
责任印制	马宇晨	

出 版	科学普及出版社
发 行	中国科学技术出版社有限公司发行部
地 址	北京市海淀区中关村南大街16号
邮 编	100081
发行电话	010-62173865
传 真	010-62173081
网 址	http://www.cspbooks.com.cn

开 本	787mm×1092mm 1/16
字 数	740千字
印 张	33.5
版 次	2019年9月第1版
印 次	2019年9月第1次印刷
印 刷	北京九州迅驰传媒文化有限公司
书 号	ISBN 978-7-110-09993-3 / N·243
定 价	98.00元

Proceedings of the First International Symposium on the Development of Natural Science Museums under The Belt and Road Initiative

CONTENTS

International Symposium on the Development of Natural Science Museums Under The Belt and Road Initiative - Keynote Speeches

International Symposium on the Development of Natural Science Museums under The Belt and Road Initiative - BRISMIS Forum

International Symposium on the Development of Natural Science Museums Under The Belt and Road Initiative-Keynote Speeches

Beijing Declaration

1. Background and Origin

In order to promote the scientific and cultural communication between all the nations along The Belt and Road Initiative (The Silk Road Economic Belt and the 21st Century Maritime Silk Road) as well as to strengthen the cooperation between national science museums in countries along The Belt and Road Initiative, the First International Symposium on the Development of Natural Science Museums under The Belt and Road Initiative (BRISMIS) is held in Beijing and Shanghai from November 27th to 30th, 2017. It is hosted by Chinese Association of Natural Science Museums and jointly organized by China Science and Technology Museum and Shanghai Science and Technology Museum. Altogether there are 28 science popularization venues/organizations from 22 countries along the routes of The Belt and Road Initiative who have confirmed that they will attend the symposium.

Through our contact with those who are about to participate in the symposium, we have learned that some delegates proposed that this symposium should become a good beginning for further cooperation between science and technology museums in countries along The Belt and Road Initiative. What is more important is that a long-term effective cooperation mechanism should be established so that all science and technology museums can share information in many fields such as cultural exchange, educational cooperation and scientific development; construct a unified platform with combined efforts; share achievements so as to promote social progress and the integrated development of all humanity.

Some delegates even suggested to the chairman of the symposium that before its opening ceremony, all delegates should come together to discuss how to consolidate and enrich the possible results of the coming symposium, proposing to have a declaration so as to enhance the impact of this first symposium, setting the tone for sustainable development in the future. And this proposal has been fully affirmed and warmly supported by the chairman.

Thereby, altogether 12 members from 12 different science popularization institutions in countries along the routes of The Belt and Road Initiative have jointly made up a core team drafting the *Beijing Declaration*. The institutions span different fields such as nature, science and astronomy. After several rounds of group discussion and revision, a common prospect for further cooperation and development in the future is settled, hence, *the Beijing Declaration*.

2. *Beijing Declaration*

The current world is a mixture of opportunities and challenges, in which every nation seeks peace, development and cooperation. The Leaders' Roundtable Summit at The Belt and Road Initiative Forum for International Cooperation was wrapped up in China on May 15th, 2017, giving the strong impetus to building a community of shared future.

We, the delegates to the First International Symposium on the Development of Natural Science Museums under The Belt and Road Initiative, gathered in Beijing of China from November 27th to 28th, 2017. Committed to the principle enshrined in The Belt and Road Initiative Forum, we hold it to be self-evident that our aim is to discuss cooperation, establish a platform and share results. Having made in-depth analysis and detailed discussions based on the theme of "Collaborative Sharing and Reciprocation: Towards the Co-construction of a Silk Road for Science Communication", as well as in consideration of the unique role of science-related museums in supporting the realization of the Sustainable Development Goals of the United Nations, we hereby issue a declaration named *Beijing Declaration*.

We believe it is imperative for science-related museums in countries along the routes of The Belt and Road Initiative to cooperate and exchange insights in this era of information explosion and interdisciplinary integration. Frequent and effective exchanges between science-related museums along the routes of The Belt and Road Initiative are vital to realizing cross-border policy coordination, facilities connectivity, cultural interaction and people-to-people bonds; to enhance mutual understanding of the involved museums' education, exhibition, collection and research in different cultural contexts; to further promote technological development and innovation, which ultimately spurs sustainable social-economic growth and human progress.

We believe building trust among science-related museums in countries along the routes of The Belt and Road Initiative is crucial for us to play our role as a reliable link and an invaluable transmitter, as well as to encourage tolerance and critical thinking. It is also the basis for sharing the advances of science, which is also an important part of The Belt and Road Initiative Initiative. The concept of "science improves our lives" will develop into a "new Silk Road" for science popularization through the cooperation between science-related museums in countries involved in The Belt and Road Initiative, which helps remove the barrier to spreading scientific knowledge and together build a community of shared future that communicates science, culture and spirit.

We believe building The Belt and Road Initiative requires scientific attitude and spirit in addition to scientific knowledge and culture, and those science-related museums play an indispensable role in science popularization. The science-related museums along the routes of The Belt and Road Initiative should seek various cooperative and communicative methods and take into account to combine traditional object-oriented exhibitions with cutting-edge technologies to promote public engagement of science.

In view of what is said above: We are willing to establish an alliance for more widespread science communication in countries along The Belt and Road Initiative; to work to develop and tour travelling exhibitions, tap into museum education resources, make regular visits to museum partners, conduct in-service training for museum staff, cooperate in museum research and collection, jointly hold science festivals and youth competition of scientific knowledge. All the contributions mentioned above will be made to promote innovation and growth, make resources more enriching, develop a win-win relationship between museums and improve our ability to communicate science.

We must make use of the major resources of these science-related museums, build a platform for active and effective communication, and develop a cooperative framework and a network of collaboration, as a means of achieving more extensive and in-depth cultural and technological exchange and making contributions to the community of shared future.

We suggest setting up the "New Silk Road Academy of Science-related Museums," a non-profit academic organization, in an attempt to have better knowledge of each other, to promote innovation and to pursue common progress.

We suggest setting up a foundation especially for the cooperation and development of science-related museums in countries along The Belt and Road Initiative, and release a newsletter including updated information of science-related museums on a regular basis, etc. to support and promote sustained mutual exchanges among museums. As this year's forum has been successfully wrapped up and led to numerous academic outcomes, we suggest we remain committed to being "equal, voluntary and reciprocal" and hold the forum every two years on the basis of cooperation, consultation, sharing and being win-win, as a way to build a close partnership among countries along The Belt and Road Initiative.

In the spirit of openness and inclusiveness, we welcome the involvement of relevant organizations in other countries who are prepared to support the goals of this *Beijing Declaration*.

Roles of Natural Science Museums in the Achievement of *2030 Agenda* under The Belt and Road Initiative

Yoslan Nur[①]

Abstract The Belt and Road Initiative and *UNESCO's 2030 Agenda for Sustainable Development* (*2030 Agenda*) are different in their nature and scope, but both have sustainable development as the overarching objective. Thus, The Belt and Road Initiative has immense potential to be used as the implementation tool of the *2030 Agenda*. The paper focuses on how the development of natural sciences museums, as science infrastructure, could contribute in the achievement of *2030 Agenda*, especially in five priority areas of The Belt and Road Initiative including policy coordination, infrastructure connectivity, unimpeded trade, financial integration and people-to-people linkage. The paper specifically discusses the four following issues: the potential role of natural sciences museums in promoting science, technology and innovation cooperation among The Belt and Road Initiative countries; how to create the interconnectivity among natural sciences museums in The Belt and Road Initiative countries; the potential of overseas development assistance in the development of natural sciences museums within the context of The Belt and Road Initiative; the role of natural sciences museums in strengthening cultural understanding between people.

Keywords *2030 Agenda for Sustainable Development* The Belt and Road Initiative natural science museums

First of all, on behave of UNESCO, I would like to congratulate Chinese Association of

---

① Yoslan Nur: UNESCO Programme Specialist working at the Division of Science Policy and Capacity Building, Natural Sciences Sector. Email: y.nur@unesco.org.

Natural Science Museum for this excellent initiative in organizing this International Symposium on the Development of Natural Science Museums under The Belt and Road Initiative. With the theme of "Collaborative Sharing and Reciprocation: Towards the Co-construction of a Silk Road for Science Communication", this symposium will discuss how the development of the natural science museums contributes to the realization of the United Nations *2030 Agenda for Sustainable Development* and calls on the natural science museums to help the technology innovation cooperation among The Belt and Road Initiative countries and construction of a community with a shared future for mankind. The following section will focus on the roles and responsibilities of museums of natural science in the *2030 Agenda for Sustainable Development* and in The Belt and Road Initiative.

1. *2030 Agenda for Sustainable Development*

The United Nations' *2030 Agenda for Sustainable Development* was unanimously adopted by the United Nations' 193 member states at the historic summit in September 2015. The 2030 Agenda is integral and indivisible, and takes into account the three aspects of sustainable development: economy, society and environment, involving more than 7 billion people in both developing and developed countries. Former United Nation Secretary-General Ban Ki-moon pointed out: "These goals address the needs of the people in both developed and developing countries and are the shared vision of mankind. They are also the social contract reached between the leaders and peoples of various countries in the world. It is a list of actions benefiting the humanity and the earth, and a blueprint for success as well." The *2030 Agenda* includes 17 goals, 169 targets with 230 indicators, that allow all people to enjoy human rights, achieve gender equality and empower all women and girls, stressing "Leave no one behind".

Eliminating poverty and hunger and allowing all people to enjoy the fruits of economic and social development on an equal footing remains the highest priority for sustainable development. With the development of economy and science and technology, the problem of food shortage has been greatly alleviated. However, the problems of malnutrition and overnutrition that have long plagued developed countries have become increasingly prominent in developing countries. Diabetes, obesity and cardiovascular diseases are increasingly threatening the health and future of mankind. Due to lack of fresh water resources, uneven distribution, and deterioration of water quality, about 1/5 of the world's population does not have access to safe drinking water.

Moreover, some areas are suffering from flooding and exposed to cholera, measles, malaria and other infectious diseases and nutritional deficiencies caused by the lack of clean, safe water and sanitation systems. The key to solving these problems is the high global involvement. It will bring together the governments, private sectors, society, the United Nation system and other parties and mobilize all available resources to help implement 17 goals listed in the *2030 Agenda*. The long-term practice of promoting sustainable development proves that science, technology and innovation are fundamental approaches. The United Nations Development Summit held in September 2015 strengthened science, technology and innovation as an important means of implementation for putting into practice *2030 Agenda*. On the United Nations' first "Multi-stakeholder Forum on Science, Technology and Innovation for the Sustainable Development Goals" held in June 2016 and the United Nations' High-level Political Forum for Sustainable Development" held in July, the delegates unanimously stressed the support of scientific and technological innovation and the importance and irreplaceability of sustainable development, emphasizing that science, technology and innovation should run through all sustainable development goals.

The phenomena of inequality, discrimination, xenophobia and related intolerance in today's world have become major threats to human security, peace and development. Vulnerable groups suffer from deprivation of resources and are more prone to widespread environmental degradation. "Leave no one behind" is the core commitment of the *2030 Agenda* to ensure that everyone enjoys dignity, reflecting the importance attached to inclusiveness. This commitment has prompted the global rethinking of the definitions, characteristics and causes of vulnerable groups, social, economic, political and environmental constraints and other factors affecting vulnerable groups. Furthermore, the *2030 Agenda* is used to strengthen the rights and interests of vulnerable groups, bring benefit to the most vulnerable groups, and ensure social justice, democracy, gender equality, sustainability and inclusiveness.

The implementation of the *2030 Agenda* needs to promote sustainable development in a mutually beneficial manner at all levels, including individuals, communities, countries and the international community. The first is to strengthen institutional development, enhance capabilities, reform policies, and strengthen democratic governance. This requires the nation and the community to extensively reflect on what role they play and how to identify and seize new opportunities. The second is to provide global public goods based on the above principles and to reduce power asymmetries in the global governance of monetary, finance, trade technological and environmental matter. The third is the need for coordination within and between the countries. The fourth is to formulate national development plan, budget and business models needed for sustainable development, such as the use of carbon-containing products and the formation of a good low-carbon industrial cluster, and the industrial chain to achieve a low-carbon economy.

Countries face different situations and diversified needs, capabilities and resources. First, it is necessary to formulate new strategic planning, policies, plans and investment methods related to the sustainable development goals in accordance with the *2030 Agenda*. The second is to set new indicators, establish a comprehensive monitoring and evaluation system, clearly set goals and specific roadmaps, track implementation and measure impact. The third is to form a set of powerful tools on finance, science and technology, free trade, and information acquisition needed to achieve sustainable development goals. The fourth is to urge governments and their partners to cooperate in various sectors and fields with new models to promote comprehensive implementation.

However, a considerable number of government officials, policymakers, scientists and economic players are not fully aware of Sustainable development goals and believe that sustainable development is more about environmental business. Therefore, policymakers, scientists, economic players and the public need to fully understand the Sustainable development goals, discover the opportunities and responsibilities that they have, deepen the relationships among stakeholders, keep pace with policies, and reduce the negative effects on sustainable development agenda, and make more positive contributions.

In achieving sustainable development goals, the natural science museums play a prominent role, and two cases can evidence: First, November 10, 2016 is the first "International Science Center and Science Museum Day", where the public was invited to review the activities held for the United Nations sustainable development goals, and to explore the practice and philosophy so that science centers and science museums demonstrate their impact in encouraging public participation in science and technology topics on the international stage; second, with the theme of "Connecting the World and Achieving a Sustainable Future", the second Science Center World Summit held on November 15–17, 2017, released the Tokyo Protocol signed by the leaders of 6 regional science center associations in the world. The Tokyo Protocol calls for science centers to give priority consideration to the importance and urgency of expanding public awareness of, and engagement in actions that help achieve the Sustainable Development Goals. We should undertake actions relevant and appropriate to local communities with consideration for the Sustainable Development Goals; serve as platforms for discourse and exchange among all diverse actors in society.

The natural science museum aims at 5 groups of people to carry out targeted activities to promote the realization of sustainable development goals. First, policymakers are the best candidates to complement science and technology and policies and promote the integration of science and technology. They promote the building of a knowledge-based society by encouraging governments to formulate new science, technology, and innovation strategies. The natural science museum can participate in the development, evaluation, and reform of existing scientific and technological innovation systems and assist in the formulation of strategies and action plans. Second, scientists are an important force in promoting sustainable development goals. The natural science museum actively cooperates with scientists and encourages scientists to actively participate in achieving sustainable development goals. Third, the production sector (including industry and agriculture) uses scientific knowledge to increase productivity and economic efficiency. Fourth, the general public is the beneficiaries of the sustainable development goals. At the same time, public participation is also a necessary guarantee for achieving sustainable development. The goals and actions for sustainable development must rely on the recognition, support, and participation of the public and social groups to the maximum. Fifth, young people are the future, and only they receive a good education and master scientific knowledge, can they achieve ultimate success in sustainable development.

2. The Belt and Road Initiative

More than 2000 years ago, the industrious and brave people on Eurasia explored a number of trade and humanities exchange paths linking Asia, Europe, and Africa, which were collectively referred to as the "Silk Road". In September and October 2013, during the visit to the Central Asian and Southeast Asian countries, President Xi Jinping proposed the building of the Silk Road Economic Belt and the 21st Century Maritime Silk Road Initiative (referred to as The Belt and Road Initiative), arousing worldwide attention. The Belt and Road Initiative aims at linking different geographical areas physically with the big infrastructure, with telecommunication, with information digitally. During his speech in the Forum for International Cooperation of The Belt and Road Initiative in Beijing on 14 and 15 May this year, President Xi Jinping explained 5 priorities of The Belt and Road Initiative.

First priority is deepening policy connectivity between China and the neighboring countries. More than 120 agreements have already been signed between China and other countries within this context. These agreements not only cover "hardware" foundations such as transportation, infrastructure and energy; but also include "software" foundations such as telecommunications, customs and quarantine. The agreements also include plans and projects for economic and trade, industry, e-commerce, marine and green economic cooperation.

The second priority is enhancing infrastructure connectivity. China and relevant countries jointly accelerate the advancement of the Jakarta-Bandung high-speed rail, China-Laos railway, Addis Ababa-Djibouti railway, Hungary-Serbia railway and other projects, the construction of Gwadar Port (Pakistan), Piraeus Port (Greece) and other ports, planning to implement a large number of interconnection projects. At present, led by China-Pakistan, China-Mongolia-Russia, New Eurasian Landbridge and other economic corridors, a complex infrastructure network is being formed, with land, sea and air channels, and information highways as the backbone, and by relying on major projects of railways, ports, and pipe networks.

Third, strengthening trade connectivity. From 2014 to 2016, China's total trade volume with The Belt and Road Initiative countries exceeded US$ 3 trillion, and China's total investment in The Belt and Road Initiative countries exceeded US$ 50 billion. Chinese enterprises have built 56 economic and trade cooperation zones in more than 20 countries, creating nearly US$ 1.1 billion in tax revenue and 180,000 jobs for the countries concerned.

Fourth, strengthening financial connectivity. The Asian Infrastructure Investment Bank provided US$ 1.7 billion in loans for 9 projects in The Belt and Road Initiative countries. The Silk Road Fund has invested US$ 4 billion, and China and Central and Eastern Europe 16+1 Financial Holding Company were formally established. These new financial mechanisms and the traditional multilateral financial institutions such as the World Bank have different focuses and are complementary to each other, forming a "Belt and Road" financial cooperation network with a clear hierarchy and scale.

Fifth, strengthening private connectivity. In this context, China has carried out cooperation in the fields of science, education, culture, health, and humanities. Each year the Chinese government provides 10,000 government scholarships to the countries concerned. Chinese local governments also set up special Silk Road scholarships to encourage international cultural and educational exchanges.

What is the connection between Sustainable Development Goals and The Belt and Road

Initiative? Both The Belt and Road Initiative and the *2030 Agenda* have sustainable development goals as the overarching objective. During The Belt and Road Initiative International Forum in 2017, The United Nation secretary Antonio Guterres pointed out that the *2030 Agenda* and The Belt and Road Initiative share the same macro goals. Both are aimed at creating opportunities for bringing public goods for the benefit of the globe, and promote global linkages in various aspects, including infrastructure construction, trade, finance, policies, and cultural exchanges, bringing new markets and opportunities. Therefore, The Belt and Road Initiative has great potential for being used as an implementation tool for the United Nation *2030 Agenda*.

3. Natural Science Museums as a Platform for Sustainable Development

How should the development of natural science museums be under the background of The Belt and Road Initiative? What talent and intellectual support can the natural science museums provide?

First, the Natural Science Museum is a bridge and link between countries along The Belt and Road Initiative. There were more than 40 foreign delegates of natural science museums that came from 17 countries to participate in this symposium and signed a number of cooperation agreements to strengthen the dialogue and exchange between the natural science museums of The Belt and Road Initiative countries. The dialogue platform helps the natural science museums of The Belt and Road Initiative countries deepen understanding of all parties, identify common ground, and jointly develop cooperative projects.

Second, under the background of The Belt and Road Initiative, natural science museums explore the potential of their own international assistance. The Natural Science Museum of China provides appropriate assistance to museums in countries along The Belt and Road Initiative. Assistance includes three types: First, technical assistance - By sharing a lot of knowledge and experience of China in the development of natural science museums, to achieve respective advantages, sound cooperation, and integration and development between China and natural science museums in The Belt and Road Initiative countries; second, capacity building assistance - through the establishment of training workshops, we cultivate high-level talents for natural science museums and future core competitiveness; third, financial aid - it's possible to have grant from China in the development of natural science museums in The Belt and Road Initiative countries. UNESCO is very willing to facilitate the development of these activities and actively promote the strategic docking and cooperation between China, The Belt and Road Initiative countries and non-The Belt and Road Initiative countries.

Third, science centers and science museums should act as a platform for the communication of Sustainable Development Goals. What should we do about this? Refer to the Tokyo Protocol mentioned a little bit before.

Fourth, it is to use Natural Science Museum to strengthen the people-to-people connection. As a platform for informal science education, the Natural Science Museum provides a different learning environment from formal education. Essentially, learning has become a voluntary, autonomous action. The exhibition method of the Natural Science Museum emphasizes participation experience, with its topics involving the development of science and technology, and the influence of science and technology on society. The scientific principles are embedded in the exhibition exhibits. Through interaction with the exhibits, the science is closely linked with the public's life and the public's interest in science and technology is stimulated. Second, it promotes dialogue between the scientific community, society and policymakers. As a platform for scientific and technological innovation policy dialogues, the Natural Science Museum raises the awareness of the importance of "science for policy" and "policy for science". There is not enough budget to conduct activities. Still in many developing countries, they don't have science policy. If they have a science policy in their national development plan, it is just a kind of decoration because there is no budget. They are there just to be politically correct. The most important target group in science communication is the decision-makers, who provide corresponding budgets for technological innovation and science policies. The Museum of Natural Science has the ability to influence the decision-making process, but it is by no means easy. Again, under The Belt and Road Initiative, the Museum of Natural Science is a platform for intercultural understanding and peacebuilding. The history of science is a topic that the Natural Science Museum has always paid attention to and explored. It can raise public awareness of the rich history of science and technology, let the public know the historical roots of science and provide them with an opportunity to understand the multicultural dimensions of science and technology development, as we know today. For thousands of years, civilizations such as Islam, China, South Korea, Japan, Greece, and India have communicated with each other through the Silk Road. By understanding the cultures of others, we can have respect to other cultures. Science and technology can be a peaceful way of communication, and it is a way by which science centers and science museums can make contributions to peacebuilding around the world. In 2010, with Dr. Cheng

Donghong here, we organized the International Symposium on History of China of Science and its interaction among the civilizations and discussed how science has been used as a vehicle of exchange among the civilizations. The last thing is that Natural Science Museums are not technological innovators, however, they can play a role in introducing new technologies and the latest research to the public. Help the public understand the process of innovation-the skills to design, build, and use technology-and the impact of science and technology. In the past decade, science centers and science museums have played an increasingly important role in introducing new innovations to the public, especially technological innovations related to the current global challenges.

Museums as Cutting-edge Laboratories of Our Shared Humanity

An Laishun [①]

Abstract After more than 250 years of development and evolution, contemporary public museums are undergoing significant changes in their social, cultural, professional and even economic environment. The new trend will undoubtedly bring new opportunities, new possibilities and new challenges to the museum. One of the questions that museums need to answer today is always at the core: whether our museums are important to the society, no matter in the immediate future or in the long run? The relevant discussion around this area makes the contemporary museum's new culture view becoming clearer, so that the museum has a new coordinate for the strategic choice to adapt to the new situation, and finally leads to the possible reconstruction of the museum's value system and new operating rules. Therefore, in terms of theory and practice, the relationship between museums and our society is becoming more dynamic, becoming an important witness and active participant in the development of society.

Keywords museum trends strategy prospects

1. Introduction

Public museums have undergone evolution and development for more than 250 years. Today's museums, whether in terms of epistemology, typology or science of function and methodology, are vastly different from those classical museums in Europe at the end of the 19th century. One of the most obvious trends is that more and more science popularization

---

①　An Laishun: the Vice President of the International Council of Museums, ICOM and Secretary General of Chinese Museums Association, CMA Email: an_laishun@vip.163.com.

and promotion agencies share the same social responsibilities with those traditional museums of art history, anthropology and archeology, follow the same professional ethics, and devote themselves to the same purpose of "serving the society and social development". As a professional organization of global museums and their practitioners, International Council of Museums defined those public institutions engaged in science popularization as the extremely important type of museum in its Statutes of 1961. Among International Council of Museums' 30 international special committees, the International Committee for Museums and Collections of Science and Technology has been showing great vitality and potential since it was formally established in 1972. The committee, comprised of museum professionals from science and technology circles in various countries, serves not only science and technology museums in the traditional sense, but also plays an increasingly important role in promoting the development of science centers and other institutions. China's science and technology museum industry is deeply involved in the work of this Committee and plays an important role. As one of the most important platforms for communication and cooperation, International Council of Museum has joined together museums and museum professionals from different subject areas throughout the world with the same goals and responsibilities. The following is intended to conduct a discussion from three aspects of "situation", "strategy" and "foresight" by centering on the active relation between contemporary museums and social development (i.e. how does the modern society affect the development of museums? How can museums act on today's society).

2. Situation: The Social Environment of Museum Development has Undergone Changes

The changes of the social environment in the 21st century have had the extremely profound impact on the development of modern museums.

First, from a broad sense of cultural perspective. Changes are occurring in the connotation and denotation of four major types of culture, i.e. the material culture on which mankind depends to survive, the health culture featured with mankind's pursuit for the life quality, the social & political culture featured with mankind's joint compliance, and the aesthetic culture featured with mankind's pursuit for beauty in leisure-time lives. These changes that have emerged for more than 40 years are neither linear nor unidirectional. Instead, they are interrelated and spiral. The reason for museums, regardless of which country, what kind of disciplines, and what size of museums, can not be drifting out of the impact of these cultural changes from both contents to forms of expression, lies in the final analysis that the four types of culture above are the main objects of museum preservation, research, exhibition and dissemination. In the face of these new trends, museums and their practitioners can not do their job effectively just by standing on a single discipline, while interdisciplinary or even multidisciplinary integration has become an inevitable solution.

Second, from the perspective of the participatory society. The public's in-depth participation is an important feature of the modern society. The impact of the characteristics of participatory society on the traditional concept of the museum is enormous and profound. In fact, public participation in its true sense has gone far beyond the limited intervention of public institutions (such as museums) in terms of attracting public participation within the framework of their own business functions. Rather, these institutions have realized spiritual and ethical democratization and socialization, requiring that public institutions should recognize the equal status supposed to be possessed by the public and also deem the public as the main body of their own existences and developments, rather than insignificant objects, requiring public authorities to respect the public's right to share the resources of public institutions, requiring public institutions to encourage the public to participate in institutions as subjects with independent awareness and rights, and requiring public institutions to be more inclusive and allowing the public to express their opinions in institutions while maintaining their rights of expression in their own disciplinary positions. The above characteristics will inevitably bring about a huge impact on the traditional concept of museums, because in the traditional museum concept, museums always regard themselves as the only authority in the subject, of whose authoritativeness is not allowed to be challenged by the public.

Third, from the perspective of the economic environment under which public institutions run. The uncertain and unstable factors in the worldwide financial situation are gradually increasing, thus causing the situation of public institutions to become more and more complicated. Although the economic situation is not necessarily transmitted to the field of the museums in the most direct and quick manner, However its indirect and long-term influence cannot be ignored, and strong vigilance should be aroused among policy makers, managers and operators of museums. It is precisely because of such vigilance or the pressure experienced, that some museums have shifted their paths of survival and development to the community and educational activities and taken the shift as a strategy for adjustment, in order to prove their values and attract more local financial support. This phenomenon, known as the "third-generation museum" or even the "fourth-generation museum" in the field of museology, is indeed worth studying. The realities are much more than this, and a topic of concern is that the changing economic environment has forced museums into the economic sphere relatively unfamiliar to them. There are indeed opportunities here, but the challenges are also everywhere. What's more, even the arguments have been directed at such essential rather than simply operational issues like, "how museums stick to the attributes as public cultural institutions" and "what are the ethical standards of museums in economic activities".

Fourth, from the fields that are relatively close to museums. The ever-changing digital technology is unexceptionally infiltrating all social cells, and leisure industries that can provide multiple alternatives are competing with museums for culture consumption among the public. In this case, museums must offer the public sufficient reasons to frequently visit museums and the knowledge economy has enabled people today to no longer acquire knowledge and skills through one-time learning, thus arousing the enthusiasm for lifelong learning among the public. The museum's "Third Space", the tourism industry and local socio-economic development, intangible heritage, social hot topics, application of new technologies, and the provision of more extensive and multidisciplinary information have become six new focuses of museums in the 21st century.

3. Strategy: Functional Improvement and Transformation of Contemporary Museums

Contemporary museums have to answer a fundamental question like this, are our museums important to the society, regardless of at this stage or in the future? Selecting functional upgrading and transformation is the museums' strategy to answer the above questions.

First of all, the cultural perspective of the museum with contemporary characteristics has gradually formed. Since the 21st century, the evolving cultural perspective of the museum is provided with the following new value orientation, that is, museums are no longer just for the sake of preserving traces of the past. Instead, they are committed to preserving the soul of culture which are of great significance to the contemporary society, the country and the nation. Museums are no longer merely entrusted by our ancestors to extend the life of heritage, but entrusted by contemporary and future generations to play their roles better. Museums are no longer advocates of some groups' interests or discipline systems. Instead, they possess unique social identities and values, and become advocates of the country and national conscience. The transformation of the above value orientation is undoubtedly one of the most important progress of contemporary museum culture.

Secondly, the museum strategy accommodates to social reforms, while making adjustments and optimizations. The challenges of social reforms are objective, yet they also provide new opportunities for museums, i.e. opportunities for preserving, recording, witnessing and participation of reforms. Museums have been exploring strategic options that meet the social expectations. This includes the museums' "must dos", that is, museums must assume the duty of cultural memory bank (broad cultural memory: material culture, health culture, socio-political culture, aesthetic culture) because the memory function serves as one of the most original and central functions of a museum. Museums must enable history and culture to be understood by the present-day people, and build a bridge between the public and the most direct and authentic museum resource. Museums must reflect the issues of social concern, at least be sensitive as they face the social issues, and make themselves the channel calling for action for social progress. Museums must become an integral part of the public cultural system in their communities, extending the museums' services to all and serving as fusion agents of existing and different cultures. In this perspective, there also include what museums should "do", that is, museums should promote more in-depth understandings among different nations and countries, lead a country's civilization to a wider world, enable the free flows of knowledge of all ethnic groups and exchange of cultural information. Museums should interpret traditional cultures as well as new cultures, let the people access to information with tolerant attitude, enable them to keep curious attitude and respect to other cultures while respecting their own achievements and promote cultural inheritance and innovation. Museums should become catalysts for new cultures. Through displays, exhibitions and other means, new knowledge and cultures can be generated and expedited in the conjunction of museums with their collections or exhibits. In the practices of different museums, the "musts" and "shoulds" metioned above are respectively given different priorities.

It is especially pointed out here that science and technology museums should assume more important responsibilities in the field of the sustainable environment through strategic adjustments. Urbanization is an important trend in social development, of which pace is accelerating. At the same time, it puts biodiversity and other sustainable environment issues before the mankind. They are no longer distant and abstract issues; Instead, they are current, concrete and closely related to people's daily life. As one of the strategies in the aspect of the sustainable environment, museums aim to encourage the whole society to closely link issues that appear to belong to science and technology with those prominent social issues, strive to help people understand the past while drawing lessons from today and devoting themselves to the creation of sustainable future.

4. Foresight: Reconstruction of Museum Value System is Becoming Possible

Since its founding in 1946, the International Council of Museums has led eight rounds of revisions to the definition of the museum, and the currently ongoing ninth revision attempts to respond to the ever-changing development situation of the museum. The current authoritative definition of the museum was formed in 1974 and revised in 2007, that is, "A museum is a non-profit, permanent institution in the service of society and its development, open to the public, which acquires, conserves, researches, communicates and exhibits the tangible and intangible heritage of humanity and its environment for the purposes of education, study and enjoyment." In this definition, "in the service of society and its development" is a statement extremely prone to trigger discussion. It can be expected that a new definition of the museum will be formally proposed at the 2019 General Assembly of the International Council of Museums, and it should be a statement that reflects the new trends, new conditions, new missions and new possibilities of museums in today's social conditions.

There is no doubt that a matured museum industry must have a set of commonly-shared values and spiritual self-discipline system, as well as effective fundamental behavior standards, i.e. the code of professional ethics. In 1986, the International Council of Museums passed the *International Council of Museums Code of Ethics*, which was universally respected and adhered to by nearly 37,000 museum agencies and practitioners throughout 150 countries. It set out the codes of conduct concerning museum organizers, museum collections, museums and the public, the operation of museums in a legal and professional manner, and individual conducts of museum practitioners. After four rounds of subsequent revisions, this code better reflects the new situation and new requirements for the development of contemporary museums, and becomes a "soft law" of museums in many countries. Therefore, these countries no longer formulate specific laws.

In November 2015, UNESCO adopted the *International Council of Museums-drafted Recommendation on the Protection and Promotion of Museums and Collections, Their Diversity and Their Role in Society* (hereinafter referred to as *Recommendation 2015*). As for museums, *Recommendation 2015* is a powerful guidance document. Of particular concern is the fact that *Recommendation 2015* places the statement of the museum for the society and social development in a series of specific modern value systems, such as equity, freedom, solidarity, social inclusion, social integration, sustainable development, etc. Compared with the existing definition defined by the International Council of Museums in 2007, UNESCO requires museums to play a more active role in aspects related to human being and humanity, and extend the action areas of museums from the collection, historic sites and other aspects to a broader field of humanities.

In November 2016, UNESCO held a High Level Forum on Museums in Shenzhen, for the implementation of *Recommendation 2015*. On the opening ceremony, President Xi Jinping sent a congratulatory letter, pointing out that "Museums serve as bridges connecting with the past, the present and the future. They are not only preserves and recorders of history, but also witness and participants of the contemporary Chinese peoples' struggle to realize the Chinese Dream of national rejuvenation." It is no coincidence that Irene Bokova, then Director-General of UNESCO, stressed at the closing ceremony of the forum that "museums are cutting-edge laboratories of our shared humanity, which protect our heritage, inspire new creativity and help us to capture the complexities of the world." This may indicate that a museum era geared to the needs of today and tomorrow is coming to us. Correspondingly, the construction of the new museum value system will become possible in the near future.

Opening Up a Bright Future for the Development of Natural Science Museums under The Belt and Road Initiative

Cheng Donghong [1]

Abstract Natural science museums are defined as socially-involved institutions dedicated to science communication and education. This speech firstly reviews the history and development of natural science museums, as well as introduces the participating museums of the First International Symposium on the Development of Natural Science Museums under The Belt and Road Initiative. Secondly, this speech analyzes the major impacts exerted on natural science museums by the sustainable development goal, the uneven regional development, and the interaction between science and technology on the one hand and society on the other hand. It also analyzes the actions taken by the international community and the field of natural science museums in order to solve the above-mentioned issues. As pointed out in the speech, the proposal of The Belt and Road Initiative has provided natural science museums with a key platform whereby to meet such challenges. In view of this, the speech puts forward multiple proposals that are aimed at promoting reciprocal and resources-sharing partnerships among natural science museums in countries involved in The Belt and Road Initiative: setting up platforms for exchanges, conducting cooperative projects, establishing long-term mechanisms, and strengthening capacity building,... so as to work with concerted efforts to build a community of shared future for all natural science museums in countries involved in The Belt and Road Initiative.

Keywords Natural science museums   The Belt and Road Initiative   science communication   co-operation and resources-sharing

---

① Cheng Donghong: the President of the Chinese Association of Natural Science Museums (CANSM). Email: dhcheng@cast.org.cn.

Natural science museums function as a platform for the science culture globally, nationally, at city level and down to the township level. Even a mini-science museum plays a very important role in bringing science and applied technology to the local people, especially the kids, to enrich their daily life and their future. The Belt and Road Initiative has provided natural science museums with new opportunities and a new pathway to development.

1. History & Development of Natural Science Museums

In a constantly changing society, the definition of museums also changes with the times. Set up in 1946, The International Council of Museum has revised for eight times in its statute the definition of museum. The 8th revision made in 2007 points out that: "A museum is a non-profit, permanent institution in the service of society and its development, open to the public, which acquires, conserves, researches, communicates and exhibits the tangible and intangible heritage of humanity and its environment for the purposes of education, study and enjoyment". According to Dr. An Laishun, Vice President of The International Council of Museum, the 9th revision of the definition of museum is already under discussions.

Bernard Schiele, professor of the University of Quebec, Canada and chairman of the International Scientific Advisory Committee especially set up for the content development of the new venue China Science and Technology Museum, has for a long time focused his researches on natural science museums. According to his definition, natural science museum is a socially-involved institution dedicated to making the general public aware of the latest science discoveries and development of technology application. [1] Historically, the development of natural science museums has undergone 4 phases. In every phase of development, natural science museums in countries involved in The Belt and Road Initiative have their own highlights.

The first phase of development (1683–1929) was noted for the enrichment and displaying of the collections and the history of technology by natural science museums. As the earliest museum among the In terational symposium on the Development of Natural Science Museums Under The Belt and Road Initiative-participating museums, the Hungarian Natural History Museum (Hungary, 1802) has a long history. Besides, The State Darwin Museum (Russia,1907),

---

[1] MassimianoBucchi, Brian Trench, Routledge Handbook of Public Communication of Science and Technology, Second edition, 2014.

Royal Ontario Museum (Canada, 1914), and State Geological Museum (Uzbekistan,1926) were also inaugurated during this phase.

The second phase of development (1930–1959) was characterized by natural science museums showing contemporary science and enhancing knowledge. The famous Geological Museum of Armenia (Armenia, 1937) was founded in 1937 with almost 900 collections with approximately 14,000 exhibits. The Nikola Tesla Museum in Belgrade (Serbia, 1952) was opened in 1952.

The third phase of development (1960–1975) has a significant effect on the development of natural science museums in that it made science accessible and facilitated knowledge proliferation. In this particular period of time, the Sergei Korolev Space Museum (Ukraine, 1970) was opened to and well received by the public.

The latest phase of development (1976–present) is not without controversy, with some researchers viewing it as a period in which science and technology innovations fully interact with the society. During this ongoing period, many natural science museums have come into being along the routes of The belt and road Initiative and beyond, for instances, the Pakistan Museum of Natural History (Pakistan, 1976), Questacon-the National Science and Technology Centre (Australia, 1988), the National Dinosaur Museum (Australia, 1993), the House-Museum of Viktor Ambartsumian in Byurakan (Armenia, 1998), Museum Herakleidon (Greece, 2004), the Planetarium Science Center Alexandria (Egypt, 2009), the Copernicus Science Centre (Poland, 2010), the Mind Museum, the Philippines (Philippines, 2012), the National Museum of the Republic of Kazakhstan (Kazakhstan, 2014), Science Centre Kenya (Kenya, 2015), the Penang Tech Center (Malaysia, 2016), the Aviation Museum (Nepal, 2017), and Khayyam Planetarium and Science Center (Iran, 2018)... to name just a few.

Natural science museums in countries involved in The Belt and Road Initiative not only have their own unique histories, but also constantly dispense with what is antiquated and evermore begetting what is new and timely in terms of the advancement of science and technology. Meanwhile they also maintain close communication and exchanges with their international counterparts all the time.

In retrospect, there are two key factors that have a strong bearing on the development of natural science museums in China; they may even directly decide the rise or fall, prosperity or decline of the latter. One is the social economic situation, that's the social context of natural science museums in any country; the other is the scale of input into the field by governments at all levels. We are lucky that last year, when we celebrated the centennial of the China Geosciences Museum, which is the first of its kind at state level in the country, President Xi Jinping, sent a congratulatory letter containing extremely warm and encouraging messages. This move of enormous symbolic significance is testimony of the great importance attached by the Chinese government and society to the development of natural science museums; it proved to be a great encouragement not only to the China Geosciences Museum itself but also to the entire field of natural science museums in the country.

In 1949, when the People's Republic of China was founded, there were merely 22 natural science museums in mainland China. Such museums were either government run or privately managed; some of them were owned by Chinese and others by foreigners. With the energetic support from both government and the society, huge progress has been achieved in the hands of the Chinese professionals in the field who are strong to overcome the difficulties and smart enough to find the solutions to various problems. By the year 2015, we have a total of 969 natural science museums in mainland China, which is an exponential increase compared with six and half decades ago.

Besides the general scenario, I would like to share with you 3 cases of museum development which I found interesting when doing the research on Chinese natural science museums.

The first case is the Beijing planetarium. Built in 1957, it's the first planetarium in mainland China. From the picture we can see that it was of small scale at that time. And then in 2004, shortly after our entry into the new century, the Beijing Planetarium undertook an expansion project on its original site, adding an entirely new building to its old structure. The result is the formation of Planetarium A and Planetarium B.

The second case is China Science and Technology Museum. Visitors today are usually amazed by its magnificent building and advanced facilities, but few people can realize that it was very difficult to make it happen in the 1980s. Due to the constraint of financial resources, the exhibition hall was very small when the China Science and Technology Museum inaugurated its first phase construction in 1988. And then in the year 2000 and as the second phase construction of the China Science and Technology Museum, a new exhibition hall was completed and opened to the public together with all the existing facilities, thus greatly enhancing its capacity to serve visitors as a national science and technology museum. In spite of this, it was still a far cry from the ever increasing scientific and cultural demand of the general public. Given this situation, the China Science and Technology Museum started the construction of an entirely new museum in a different location in 2006. Three years later in 2009, the brand new the China Science and Technology Museum was completed for official inauguration inside the public domain of the Beijing Olympic Park, just in time for the celebration of the 60th anniversary of the founding of the People's Republic of China. Even since then, the new China Science and Technology Museum has been feted as a scientific symbol of the Olympic Park with steady increase in visitor numbers—long queues of visitors circling the museum is a normal scene often reported by the media. In 2017 alone, its number of attendance reached almost 4 millions.

The last case is Shanghai Science and Technology Museum. Developed gradually on the basis of the Royal Asiatic Society (1933) and the Shanghai Natural History Museum (1956), Shanghai Science and Technology Museum, first opened in 2001, has become what it is today as one of the largest of its kind in the world. In 2005, the newly opened Shanghai Natural History Museum became the natural history branch of Shanghai Science and Technology Museum. The Shanghai Planetarium currently in construction is scheduled to be opened in 2020 as the museum's planetological arm.

My point in sharing the above-mentioned 3 cases is that in a developing country like China, the development of natural science museums has been a gradual process full of "growing pains"; it started from a rudimentary stage, then steadily grew in strength and finally achieve its thrivingness (from rags to riches, if you like). We have experienced the situation wherein we lacked funding resources, suffered from a want of researches and did not have sufficient professionals. Just like Rome was not built in one day, the fact that today we have almost one thousand natural science museums in China is by no means an easy thing to achieve but a result of decades of unremitting efforts by the Chinese professions in the field. Of the many natural science museums which are represented at the ongoing symposium, some are not big in terms of physical scale; and I do not believe it really matters if a museum is big or small. What fundamentally matters, in my opinion, is to make do with whatever is available and act according to your capability instead of seeking instant successes and craving for the grandiose. To have a natural science museum functioning in your community, in your city and in your country at all is much more important than to suddenly have a big one, which may never could happen at present. As a matter of fact, being small sometimes means more flexibility and more room for improvement; Museum Herakleidon is an example in point. The development of natural science museums, after all, cannot possibly be achieved at one go, it has been and will continue to be a step by step process.

Today we are gathering here in Beijing for discourse on the common development of natural science museums in countries involved in The Belt and Road Initiative; and this would not have been possible were it not with perseverance and dedication of the Chinese Association of Natural Science Museums. Founded in 1980, the Chinese Association of Natural Science Museums is dedicated to academic exchanges and science popularization; and it plays an important role in information communication, theoretical research, capacity building, and innovation facilitation. By the end of 2016, the Chinese Association of Natural Science Museums has 658 institutional members plus 1,264 individual members.

2. New Issues, New Practices and New Challenges Faced by Contemporary Natural Science Museums

Natural science museums are not self-contained ivory towers, but are basic facilities built to provide public science communication services, as well as to serve as pivotal platforms for public engagement with science. The development and applications of science and technology such as information sciences, biomedical engineering, Nano materials and technology, new and renewable energy, marine/geospatial technology etc. have had a broad and profound impact on the society. While we are enjoying the great benefits and conveniences brought by scientific and technological progress and economic development, we are also faced with ethical issues involved in overpopulation, environmental pollution, food safety, genome editing and so on as a result of such progress and development. Now the public begins to reflect on, even to question whether the price we have hereto paid is worthwhile; and this leads to what we call social issue with scientific context and scientific issues of social significance.

Because of the limitation of time and my knowledge, my presentation will focus on the new challenges, new practices, and new problems in regard to 3 issues, namely: sustainability, unbalanced development, science and technology innovation and it's interaction with the society.

2.1 Sustainability

According to Worldwatch Institute Annual Report (2015), human needs are constantly growing, but various natural resources are increasingly becoming too scarce to back up such unconstrained growth in needs, causing the depletion of some of such resources. As such, it is imperative for any healthy society, environment, and economy to take actions and innovative solutions promptly and without delay. Since 2000, the United Nation and other international organizations have committed their groups to launch a lot of initiatives and actions in order to achieve the Millennium Development Goals and adopted the ambitious Sustainable Development Goals in 2015, aka 2030 Agenda. The 2030 Agenda contains 17 Sustainable Development Goals and 169 targets. Each goal has specific targets to be achieved over the next 15 years.

The United Nations Educational, Scientific and Cultural Organization fully recognizes the importance of science for a sustainable future, as well as for creating knowledge; and understanding through science equips us to find solutions to today's acute economic, social and environmental challenges, as well as to achieving sustainable development and greener societies. [1] Based on this understanding, natural science museums strive to build skills and capacities, as well as to send strong messages about the importance of science for sustainable development. Highlighting that these institutions nurture human curiosity, and catalyze research and solutions to help societies meet varied challenges, they also provide platforms for dialogue, understanding and resilience by launching innovative initiatives to promote the learning of science outside the classrooms, as well as mass innovation and entrepreneurship. [2]

Natural science museums need to respond to the growing demand of the state, society and the public, and to constantly innovate the content and format of educational activities in order to promote the practices & concepts related to sustainable development in the entire society. Take the Bibliotheca Alexandrina Planetarium Science Center as an example, every year it dedicates one of the booths in the Science Festivity Village to a PUP-related activity entitled "PSC World Park". The activity, in which animators work every 15 min with 20 students, will be a simulation of the globe with its seven continents. The seven continents will be dealt with as seven different stops. At each stop the scientific content delivered will deal with the most aggravating environmental problems prevailing in the countries of this continent. [3] Great progress has been made in reaching many of these goals.

However, much more needs to be done. Sustainability is a complex and perplexing issue for many nations. It is essential for countries to have a comprehensive approach to deal with it. The same is true of natural science museums, in the face of the sustainability issue, their current

---

[1]　See for details https://en.unesco.org/themes/science-sustainable-future.

[2]　See for details http://www.un.org/sustainabledevelopment/blog/2016/11/unesco-science-museums-vitally-important-for-sustainable-development.

[3]　See for details https://www.iucn.org/content/egypts-planetarium-science-centre-activities-rio20.

practices at the local level are piecemeal and isolated; they are not responsive enough, limited in content and simplistic in mode, hence leaving a lot to be desired. In today's highly interdependent world, individuals and nations can no longer resolve many of their problems by themselves. And that is why we need to work in partnership with others and make concerted efforts in order to make a difference.

How to unite natural science museums globally in order to take coordinated actions? This is indeed a question that merits our in-depth thinking and exploration. Sustainable development impels all countries and their peoples to show concern for the construction of a community of shared destiny for all mankind; and globalization has made cross-cultural dialogues among people of all races possible. All these have combined to bring new opportunities to the field of natural science museums, prolong their reach, break their national and geographical borders, and provide a new way of linkage between them and their visitors all over the world. As such, it is imperative for natural science museums to enhance their global vision and international sense, conduct cross-regional and cross-cultural exchanges with a view to bridging the gap between citizens on one hand and science and technology on the other hand. Natural science museums aim to take up more social responsibilities professionally, and play an important part in addressing global social and environmental concerns. They can give real practical help to these concerns at a global level by increasing proliferation of information and communication technology. With the development of cross-regional and transnational communication, it is a new challenge that requires global cooperation for museums to contribute different skills and capabilities to respond to people's concerns in different parts of the world.

2.2　Unbalanced Regional Development

Over the past decades, the world has witnessed tremendous economic growths. However, the distribution of the created wealth and prosperity is so uneven that people regard the unbalanced economic development as the root cause of the worsening of the social problems and political instability in many regions of the world, as well as of the unbalanced status of educational resources. 1/3 of people worldwide continue to live in low levels of human development. There is a concern in countries of high level of human development, such as China, they are still facing the problem of uncoordinated social and economic development. According to the UN Sustainable Development Goal—Ensure inclusive and equitable quality education and promote lifelong learning opportunities for all—there is a huge gap between

SDG4 goal and reality.

In the matter of unbalanced growth in economy, United Nations Development Programme invests US 1 billion per year. Nearly 24.7 million people (51 percent women) benefited from improved livelihoods initiatives in 119 countries, including economic transformation, natural resource management and early recover. Over million new jobs (36 percent for women) were created in 98 countries. 94 countries implemented measures towards low-emission and climate-resilient development. In regard to unbalanced growth in education, Education 2030 Framework for Action (2015) provides guidance to governments and partners on how to turn commitments into action. The Global Education 2030 Agenda new expanded scope: reaches from early childhood learning to youth and adult education and training; emphasizes the acquisition of skills for work; underlines the importance of citizenship education in a plural and interdependent world; focuses on inclusion, equity and gender equality; aims to ensure quality learning outcomes for all, throughout their lives. [1] In the aspect of unbalanced growth in museum, The International Council of Museum announced "The Strategic Plan (2016–2022)" to enhancing membership value through improved participation, service, communications, and capacity building. [2]

During November 15–17, 2017, the Science Center World Summit, aka SCWS2017, was held in Tokyo, Japan. It brought together science centers/museums representatives and their networks from around the world to a grand global meeting of all of society's stakeholders. Centering on the theme of "Connecting the World for a Sustainable Future", they engaged in wide-ranging, strategic discussions on contemporary global issues and seek strategies for long-term impact. At the conclusion of the summit, the Tokyo Protocol, signed by all the six regional network of science and technology centers around the world, was released. How to realize even distribution of popular science resources and equitable popular science services was one of the hot topics during the summit.

In this regard, the "Science Circus" is particularly exemplary. As an outreach program in science communication operated by Questacon-Australia's National Science and Technology Center—and the Australia National University, it is composed of 25 portable interactive science exhibits loaded in a truck; it targets students, teachers and community citizens. Since its launch

---

[1]　See for details http://unesdoc.unesco.org/images/0024/002456/245656E.pdf.

[2]　See for details http://icom.museum/fileadmin/user_upload/pdf/Strategic_Plan/ICOM_STRATEGIC_PLAN_2016–2022_ENG.pdf.

in 1985, the program has covered 500 communities (including 90 indigenous communities); it has also performed over 15,000 science shows in schools, resulting in 85% of teachers reporting an increase to their students' enthusiasm for science. "Science Caravan" is a similar program conducted by the National Science Museum, Thailand. The program, first developed in 2005, travels 200 days a year, visits more than 40 provinces annually (there are 76 provinces in Thailand, excluding Bangkok).

The achievement is merely the start. For natural science museums in developing regions, the supply of popular science resource is still insufficient. To make the matter worse, their ability to develop resources on their own is very weak; and they lack the channels and mechanisms whereby to access resources that can be shared. In contrast, large or medium-sized natural science museums in economically-developed regions and central cities, boast relatively good quality resources. Under such circumstances, the latter has the responsibility and obligation to provide the former with relevant resources and services by means of cross-regional and museum-to-museum resources-integration, sharing and transfer. This is not a condescending action of charity, but a historical mission and responsibility.

2.3 About Science & Technology Innovation and Its Interaction With Society

Scientific discoveries have consistently changed the world and the way people understand science and the world. There are 3 aspects of this issue. One is where is the frontier of science knowledge? Can the latest scientific findings reflect in the exhibition? The second one is about the advanced technology. Can we use it as a tool to improve the quality of our exhibition and our program? The third one is the interaction of the scientific issue of the social background and the social issue with the scientific meaning.

In their initial development stage, natural science museums were mainly focused on existent and mature scientific discoveries and technological achievements. Now, we must ask ourselves of the question of how to make breakthroughs in our traditional exhibition and display contents; and how to present the latest development in science and technology in a timely manner. For example, the program "Take Your Classroom into Space" conducted by NOESIS - Thessaloniki Science Center and Technology Museum in Greece: The activity is part of the education program of The European Space Agency's (ESA) Directorate of Human Spaceflight and was one of several education activities done during the OasISS Mission (In May 2009, ESA astronaut Frank De Winne flew to the International Space Station for a 6-month stay on board.).

As we know, the progress of science and technology is inevitable. Despite gains, more challenges are ahead of us. First of all, natural science museums need to integrate the state-of-the-art technological advances into their existing exhibitions and displays, as well as their educational programs, so as to avoid lagging behind. Secondly, natural science museums need to take the advantage of such high technologies as IT in their effort to seek innovation in terms of exhibition education, visitor services and user experience, as well as to extend their services to visitors prior to and post of their visit and to the public who cannot make it to the museum. Finally, natural science museums need to set up new platforms for scientific exchanges, bring into full play their unique features and advantages, and promote the public understanding of issues such as science and society and science and ethics.

In conclusion, natural science museums must, by advancing their exhibition content and education programs with the times, respond to the global concerns about sustainable development, and to explain to the general public the social issues with scientific context and the scientific issues of social significance in a timely manner. Correspondingly, the public's new expectations for understanding of and participating in science constitute the new demands that require natural science museums to meet.

3. Solution: Co-operation and Resources-Sharing

In the face of the above-mentioned challenges, my suggestion is co-operation and resource-sharing.

3.1　Co-operation and Resource Sharing-Explorations in China

For the three issues as mentioned above, namely, sustainability, the uneven development, and the interaction between science-technology and the society, the Chinese are not lucky enough to be exempted from any of them. For the sustainability issue, China suffers a lot in terms of air pollution, water shortage and contamination, disorderly urban development... if you browse the media. As for the issue of uneven development, it is still a kind of pain for us not only in terms of social and economic development but also in terms of the distribution of natural science museums, with the majority of them being concentrated in the economically more developed eastern part of the country whereas in the less developed western region they are a kind of luxury.

Over the past years, great efforts have been made by professionals working in the field of natural science museums in their constant search for effective pathways and methods to solve the above-mentioned problem through collaborative sharing and reciprocation. Following are some of their explorations which I would like to share with you:

(1) Case I : The Science Wagon Project. Launched in 2000, the operations of the project were officially transferred to CSTM in 2012. The so-called Science Wagons are multi-functional science education vehicles oriented towards the demands of grassroots science communication. Featuring approximately 20–30 pieces of small scale interactive exhibits (with the exact number depending on the specific type of vehicles) plus multiple exhibition panels, such Science Wagons are mainly intended for the people, especially teenagers living in the remote areas, where natural science museums are not available, as they bring popular science services to township communities, rural schools and remote areas in particular. Thanks to their unique features of mobility and flexibility, the Science Wagons are often referred to as "Light Cavalry" by the local people; and they offer an effective solution to "the last kilometer" problem of science communication, greatly satisfying the demands of the most grassroots public in China. At the same time, they have, to a great extent, also helped mitigate the problem of uneven development of science and technology museums in different regions of the country, as well as between the urban and the rural areas.

(2) Case II. The Mobile Science & Technology Museum Project. Highlighting the overarching theme of "Experiencing Science", the project was first launched in June 2010. A typical Mobile Science and Technology Museum usually has three themed areas and features some 50 interactive exhibits that are divided into 10 sub-theme exhibition zones. These exhibits, combined with science demonstrations, science experiments and popular science film/video shows, require about 800 square meters in terms of exhibition area and could be easily installed within a local gymnasium or convention center to function as a small scale and makeshift science museum for an exhibition period of three months.

(3) Case III. The Rural High School Mini Sci-Tech Museum Project. In order to solve the problem of inadequate public science education services in rural areas, the Foundation for Development of Science and Technology Museums in China has since 2012 raised some-20 million for the implementation of the Nonprofit Rural High School Sc-Tech Museum Project. Directly finding its "home" within a $60m^2$ classroom in a rural high school, such a mini-science museum typically includes 20 pieces (sets) of interactive science exhibits, 4–5 computers with resources loaded from China Digital Science and Technology Museum (for the remote area have no internet), a 3D printers & Multi-Projection, more than 1,000 science publications. Thanks to this project, local teenagers can now have the same beautiful experience of visiting a science museum as their urban peers.

(4) Case IV. Strategic cooperation between the Beijing Planetarium and the Delingha Planetarium & Science Museum. It is a paradigm of interconnectivity cooperation between a big and state-of-the-art science popularization institution in a highly developed city and its much smaller counterpart in an underdeveloped and underserved region. Based in Delingha, a small county in the Qinghai plateau in western China, the Qinghai Delingha Planetarium & Science Museum is the first of its kind in western China specializing in astronomy. According to their recently signed strategic cooperation framework plan, the Beijing Planetarium provides consultation services in terms of exhibition development and staff training for the Delingha Planetarium & Science Museum; in return, the Delingha Planetarium & Science Museum plays host to the stargazing and astronomical observation activities organized by the Beijing Planetarium since the Qinghai Plateau where the former is located boats good air quality and clear skies thanks to less human activities. As a result of this truly reciprocal arrangement, the Beijing Planetarium now has access to good quality astronomical observations. Through such cooperation, both of them gain a series of benefits. It is an opportunity that will help them to enhance their public service capabilities and the quality of science popularization activities.

(5) Case V. Cooperation platform for special effect theaters of natural science museums. Case 5 is about the special effect theatres of natural science museums. Although there are many special effect theaters such as IMAX, the 4D theatre attached to natural science museums in China, we don't have enough capability to produce our own films; therefore, we have to depend on imports from overseas. As a result, a big amount of money is spent on renting movies from big international companies every year, creating a heavy burden for such theaters. Given this unfavorable situation, the Committee of Special Effects Theater (CSET) of CANSM was set up some years ago, which since its establishment, has initiated a kind of collaboration for its members. The committee organizes its members to negotiate collective renting of films in order to lower the rental fees charged by distributors. With coordinated efforts among its members, the committee also succeeds in pressing filmmakers to lower the budget and in helping improve the quality of film production of science museums through joint production. On regular basis, the committee runs training workshops and seminars on mastering the techniques of new equipment for member museums facing the challenge of projector upgrading, transition toward digitalization and computer control. All of these activities are hard to realize by any single museum.

(6) Case VI. The China Digital Science and Technology Museum Project. How to integrate frontier sciences and high technologies into the exhibition content and education programs of a natural science museum? How to develop enriched popular science resources in accordance with the diversified demands of schools, communities and the general public? These are challenging tasks for any natural science museum. In December 2005, China Association for Science and Technology (CAST), in collaboration with the Ministry of Education and the Chinese Academy of Sciences, jointly launched the project now known as China Digital Science and Technology Museum (CDSTM, https://www.cdstm.cn/). This project was designed to organize, develop and integrate the numerous exhibition resources, education programs and popular science films and videos and turn them into digitalized resources for free access and use by institutions of science popularization, nonprofit organizations and the general public. Just now and in his capacity as chief executive secretary of CAST, Dr. Huai Jinpeng solemnly declared that CAST is willing to open these resources to natural science museums in countries involved in The Belt and Road Initiative through the existing cyberspace platforms of China Science Communication as well as China Digital Science and Technology Museum. This will no doubt be a great benefit for all BRISMIS-participating museums and institutions.

What are the outcomes of all these explorations? The following figures and data speak volume for them:

- From 2000 to 2017, as many as 1,445 Science Wagons with 4 types of vehicles have been developed along with 184,000 activities, benefiting their audience of some 206 million.

- From 2011 to 2017, 288 sets of exhibitions have been developed for the Mobile Science and Technology Museum program which have covered 2,261 counties serving 83.82 million visitors.

- From 2012 to 2016, the Rural High School Sci-Tech Museum program had covered 29 provinces, benefiting 1.37 million teenagers.

- From 2006 to 2017, the total resources of China Digital Science and Technology Museum (CDSTM) has reached 9.79 TB with an average of 29.3 million page views per day on its website.

- The Delingha has now become the first city in China in which lessons on astronomy are listed as a compulsory subject for all students in their curriculum.

Figure 1 shows a research done by China Research Institute for Science Popularization (CRISP).The institute issues such survey result of scientific literacy of Chinese citizens every other year. With regard to the visit frequency of natural science museum in the past year, recent data suggest that the population who never visited a natural science museum due to lack of resources locally fell from 57.7% to 22.6%; and the population who visited natural science museums increased from 7.9% to 22.7%. This indicates that citizens now have more chances to obtain scientific knowledge and information by visiting natural science museums; and that there is a relatively high visit frequency of such institutions. Thanks to the construction and development of natural science museums, the problem of rural-urban uneven distribution of popular science resources in the country has been largely resolved and the accessibility to popular science resources has been greatly enhanced, making it possible for public science popularization services to cover all parts of the country and people of all walks of life. This in turn has given impetus to equity, inclusion and efficacy of popular science services in China. Consequently, good social and economic benefits have been obtained.

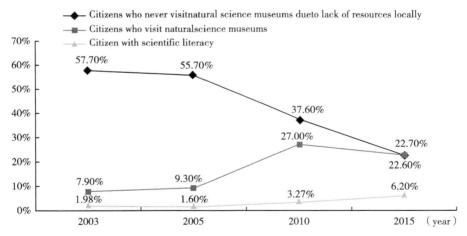

Figure 1　Survey of Scientific Literacy of Chinese Citizens from 2003 to 2015

3.2　The Development Opportunities for Natural Science Museums Offered by The Belt and Road Initiative

And now we have The Belt and Road Initiative. In keeping with the two defining trends of multi-polarization and economic globalization of the world, The Belt and Road Initiative is a development strategy proposed by Chinese president Xi Jinping. The Initiative focuses on connectivity and cooperation among Eurasian countries; and to spur growth in the underdeveloped regions. By deepening policy connectivity, enhancing infrastructure connectivity, increasing trade connectivity, strengthening financial connectivity, and strengthening people-to-people connectivity, BRI has increasingly greater impact on economic, social and cultural processes over national or regional ones

The Belt and Road Initiative encourages all involved countries to forge a partnership of mutual trust, inclusiveness, cooperation, and to persist in learning from each other and expanding humanities exchanges. The rationale of the mission of The Belt and Road Initiative, according to my understanding, can be summarized as below, namely, ① the core idea of peace, development, cooperation, and win-win. ② highlighting connectivity in policy, infrastructure, trade, finance and people-to-people exchanges. ③ as well as building communities with a shared future in terms of interest, destiny and responsibilities. However, my overwhelming interest is in the cultural connotations of the Initiative and the platforms and links the Initiative can help to set up and forge, because they constitute mechanism whereby people as well as natural science museums in countries involved in The Belt and Road Initiative can promote and coordinate with each other in order to achieve win-win cooperation.

Historically, countries along the routes of The Belt and Road Initiative have created different cultures of different patterns and styles. Together, they have formed a colorful Silk Road Culture through communication. In essence, interaction and mutual-learning is an important driver of human progress and the development of world peace.

For natural science museums in countries along The Belt and Road Initiative, we can see similar challenges and opportunities. Frequent and effective exchanges between natural science museums along The Belt and Road Initiative are vital to realize cross-border policy coordination, facilities connectivity, cultural interaction and people-to-people bonds; to enhance mutual understanding of the involved museums' education, exhibition, collection and research in different cultural contexts; to further promote technological development and innovation, which ultimately spurs sustainable social-economic growth and human progress.

3.3 Suggestions and Recommendations for Collaborative Sharing and Reciprocation for Natural Science Museums under The Belt and Road Initiative

Reciprocity and sharing are the solutions we have in mind in response to the new challenges.

Inspired by the principles of constructing an open, inclusive, balanced and universally-beneficial regional co-operation framework, we, together with all participants in the ongoing BRISMIS symposium, are willing to explore the feasibility of building a mutually-beneficial mechanism whereby to facilitate reciprocation and resource-sharing among natural science museums along The Belt and Road. In this connection, I'd like to put forward the following suggestions and recommendations:

(1) Enhancing exchanges and communication

My first recommendation is to enhance exchanges and communication. CANSM is willing to organize a BRISMIS-like symposium every other year, should colleagues from B&I countries feel such a need. We'll do collaboration and fund-raising to host the symposium. Besides, we really encourage mutual exchange programs. That's very important. People need to know each other, and then trust each other, and then work with each other. As this year's symposium has been successfully wrapped up and led to numerous academic outcomes, we suggest we remain committed to being "equal, voluntary and reciprocal" and hold similar symposium every two years on the basis of cooperation, consultation, sharing and being win-win, as a way to build a close partnership among countries involved in The Belt and Road Initiative.

(2) Conducting co-operation projects

My second proposal is to conduct cooperation projects. I think project is the carrier of real cooperation for the real things to happen. I'm very happy that many agreements were signed this morning between various Chinese natural science museums and their counterparts from B&R countries. They are pioneers of cooperation projects who deserve a big applause. We have great expectations for the outcomes of such cooperation projects. There is also a batch of natural science museums in China which have expressed their willingness to provide their quality science communication resources for sharing with their partners in B&R countries. I wish to see more agreements concerning such cooperation projects will be entered into during and after the BRISMIS.

The possibility of and conditions for establishing an alliance for more widespread science communication in countries along The Belt and Road Initiative are all there, we will seize the opportunities and build on the created momentum by developing travelling exhibitions and museum education resources, facilitating regular visits and exchanges among museum partners, conducting in-service training programs for museum staff, working with each other in museum research and collection, jointly holding science festivals and youth science contest, etc.

(3) Building up long-term co-operation framework and mechanisms

As a Chinese saying goes, the last leg of a long journey just marks the halfway point. For the purpose of sustainability, it's not enough to merely have a conference every other year plus some agreements between museums, because it's neither strong enough, nor permanent enough to support this kind of demand of working together for a shared future by this professional communities. In view of this, my third suggestion is to set up a long term cooperation framework. Some of the colleagues, who are in the working panel led by Dr. Wang Xiaoming, Director-general of Shanghai Science and Technology museum, have come up with some really good proposals for long-term and mutually-beneficial cooperation in their months-long process of drafting the Beijing Declaration. I appreciate very much their contributions and proposals, and as a token of your support to their work, I welcome representatives of every participating country to join them this evening for a round table at which the Beijing Declaration will be finalized.

In response to their proposals, CANSM is going to edit and release an e-newsletter of updated information of natural science museums in order to support and promote sustained mutual exchanges among the involved museums. We must make the best use of the major resources of these science-related museums, build a platform for active and effective communication, and develop a cooperative framework and a network of collaboration as a means of achieving more extensive and in-depth cultural and technological exchanges and making contributions to the construction of a community of shared future for all mankind. And I expect tomorrow, when we conclude the symposium, the Beijing Declaration will be approved by everyone in the auditorium.

(4) Capacity building: high-level professionals constitute our core competitiveness

The last but definitely not the least suggestion is on capacity-building. People in my age realize through experiences that high level professionals, especially the young people, constitute our core competitiveness. They are the future of natural science museums. They are the future of the field that we have cared so much all along our professional life. But they need our help, they need the support of those sitting in the front row who are the people in charge of science museums. We need to do capacity building for the young people, provide in-service training corporately for employees in exhibition design, educational activity development, visitor studies, collections and research, etc. We need to do some fund-raising for setting up a special fund to support capacity-building especially for young staff in B&R related museums.

4. Conclusion

The Belt and Road Initiative originates from China, but it belongs to the entire world; it is rooted in history, but it is oriented towards the future; it prioritizes the Eurasian Continent, but it is open to all partners. Countries involved in The Belt and Road Initiative have their own unique historical origins; they also have the tradition of learning from and mingling with each other culturally. Within the context of the New Era wherein science and technology shapes the future, it will be a special landscape if natural science museums in countries involved in The Belt and Road Initiative are united to build a community of shared future; to work to promote the scientific literacy of all citizens of the world, and, above all, to make joint contributions to global peace and development.

*MASSIMIANO BUCCHI, BRIAN TRENCH. Routledge Handbook of Public Communication of Science and Technology [M]. 2nd ed. 2014.

International Symposium on the Development of Natural Science Museums under The Belt and Road Initiative-BRISMIS Forum

Brief Introduction of the Aviation Museum Nepal

Laxmi Basnyat [1]

Firstly, this is a new innovation in Nepal, which gets two separate exhibitions i.e. one is in capital city Kathmandu and the next in Dhangadi, far western Nepal. The Aircraft Museum in Kathmandu opened on November 5, 2017, and Aircraft Museum Dhangadi since September 7, 2014. Both of the museums have been funded by Bed Uprety Trust. The main aim of both museums is to:

- educate and to inspire young Nepalese to join the aviation sector.
- provide information about the history of aviation.
- create awareness about the challenges and opportunities in the aviation sector.
- make a tourist destination.
- assist the cancer patients.

1. Aircraft Museum Dhangadhi

Commercial pilot and former an Infantry Captain of the Nepalese Army, Bed Uprety, created Nepal's Aircrafts Museum to bring tourism and awareness to Nepal's remote far west region, and within its first years of operation, the destination has succeeded in making a mark on all who have made a pilgrimage to the area to visit the charming temple to aviation.

Housed inside the fuselage of a 100−seater aircraft from defunct Cosmic Air, visitors approach the museum's entrance and are greeted by a sign bearing instructions to take off one's shoes. Upon doing so, the door to the aircraft opens and a man dressed in a flight attendant's uniform offers a greeting to all. The tour begins winding its way through the interior of the

---

[1] Laxmi Basnyat: the executive authority of Nepal Aircraft Museum and also working as the Assistant Manager at the Thai Airways International, Kathmandu City offi ce, Nepal. Email: lmibasnyat@gmail.com.

aircraft. During a stop in the plane's cockpit, guides impart the more technical aspects of how one actually goes about navigating and flying aircraft (Fig.1).

Throughout the tour, a profusion of over 200 models of commercial and fighter jets line the plane's walls, all of which were built by local children. These miniature contributions to the museum exist as part of Bed Uprety's community outreach initiative intended to bolster the youth's interest in aviation, all of which is, in turn, fueled by the enthusiasm of the new visitors to the region thanks to the Aircrafts Museum itself (Fig.2).

2. The Aircraft Museum-Kathmandu

The imaginative conversion of this Airbus 300–330 aircraft into Kathmandu's Aviation Museum . As previously mentioned, Capt Bed Uprety is the brainchild behind the conversion of this Airbus 300–330 aircraft too into Kathmandu's Aviation Museum. His inspiration is former US President John F Kennedy, who once said: "Ask not what your country can do for your country but what you can do for your country." New York's main airport, JFK, is now named after him. Capt Uprety has flown various aircraft types in his native Nepal, as well as in India, Indonesia, Singapore, Malaysia, Thailand, Oman, Cyprus, Italy, France and the USA, as well as visited over 80 countries.

Museums can be a major draw for visitors around the world. Some destinations can attract millions of tourists just because there is a famous museum located there. This aircraft had only flown for about eight months and suffered a runway excursion incident at Kathmandu airport in 2015. It was difficult to extricate the huge plane from the runway, and the airport was closed for four days. Capt Uprety used to see the aircraft parked at Kathmandu airport every time he flew in and out, and dreamt one day to acquire the plane to set up a museum in Kathmandu as well to serve not just as a tourist attraction but also educate students about aviation. It took Nepali and international technicians four months to cut the plane into smaller pieces and transport it in trucks at night. It took another nine months to reassemble it here in Sinamangal, very close to the only International airport .

The Bed Uprety Trust is a non-profitable organization that has already set up an aviation museum using an abandoned Fokker 100 in Dhangadi, as early mentioned. The revenue from that museum goes towards assisting cancer patients. This aviation museum at Kathmandu is a sequel to the Dhangadi venture. The trust has partnered with the Civil Aviation Authority of

Nepal (CAAN) to set up this museum. This Aircraft museum –Kathmandu was funded by the savings of Captain Bed Upreti along with an Rs 27[①] million loans from Everest Bank. A total of Rs 70 million has been invested in this project. There are more than 350 miniature models of aircraft inside the museum from the Wright Brother's first aircraft to fighter planes from World War I and II. Grade 9–12 Students from all over Nepal will be allowed free entrance if they have a letter from their schools, and other students can get 50% discount if they come with IDs. Part of the income from the Aviation Museum will go to assist the Bed Uprety Trust which has been funding treatment for cancer patients from the far-west. It will also help the Aircraft Museum in Dhangadi and the statue of Siddhanath Baba (Hindu god) being erected in Mahendranagar, far western Nepal.

Aviation Museum Kathmandu with this huge converted aircraft, cockpit and air traffic control recording, its collection of aircraft models from the Wright Brother's flying machine to the most modern plane, as well as the photo gallery of aerial photographs of Nepal, are all dedicated to the people of Nepal by Capt Bed Upreti.

---

①   RS: Currency name used in Nepal.

The House-Museum of Viktor Ambartsumian in Byurakan

Grigor Broutian [①]

Abstract Viktor Ambartsumian House-Museum represents the life and scientific activity of one of the greatest scientists of 20st century. Viktor Ambartsumian (1908–1996). Although he became famous mostly as an astrophysicist, he was distinguished also as a mathematician and physicist (Fig.3). Ambartsumian was the one who changed the human understanding of the Universe, introduced the idea of the evolution of the Universe. He was also one of the founders of theoretical astrophysics, the creator of statistical physics of stellar systems, stellar dynamics etc. The Museum introduces the visitors the evolution of human understanding of the Universe from antique times up to the end of 20st century, and V. Ambartsumian's biography serves as an example how a man can influence the development of science.

Keywords Viktor Ambartsumian House-Museum Scientific Research Scientific Spirit

1. The History of The Museum

Viktor Ambartsumian Museum was organized in 1998 just two years after Ambartsumian's dead (Fig.4). It is located in his house in Byurakan built by himself in 1950. Formerly it was organized as a department of the Presidium of the Academy of Sciences of Armenia and now (since 2014) it is a department of Byurakan Astrophysical Observatory. Byurakan is a village in Aragatsotn district of the Republic of Armenia 35 km from Yerevan at an altitude of 1,400 m above sea level on the southern slope of Mount Aragats.

---

① Grigor Broutian: The museologist of The Byurakan Astrophysical Observatory, Viktor Ambartsumian House-Museum. Email: gbroutian@gmail.com.

2. About Viktor Ambartsumian

Viktor Ambartsumian was one of the greatest scientists. Despite the fact that he was doing scientific research in 3 different areas of modern science: mathematics, theoretical physics and astrophysics. Ambartsumian became very famous in the field of astrophysics. His works were able to change the radically human understanding of the universe and the processes that occur in it. Thanks to the works of Ambartsumian, modern astrophysics has ceased to be a science describing the heavenly bodies and the physical essence of their radiation and motions, and has evolved into a science investigating the evolution of heavenly bodies and the universe in general. He was a member of almost all the academies of sciences in the world. Also he was the only Soviet scientist who was elected to serve as the President of International Astronomical Union (1961), and International Council of Scientific Unions (in 1968 and in 1970 for the second period as a unique exception). The two most significant discoveries of V. Ambartsumian- the stellar associations and the activity of galactic nuclei (1947 and 1958) showed that the star formation in our Galaxy is not finished yet (it is going on in our days and there are stars with different ages-from very old ones to stars born just less than 100,000 years ago) and the galaxy formation also is not finished yet and new galaxies are being formed due to the activity of the galactic nuclei.

3. The Description of the Museum

The exhibition of the Museum is organized in the rooms of the house (170 m^2), where V. Ambartsumian and his family lived, as there are no additional rooms. There is also 6,400 m^2 garden around the house where the visitors can see the environment that was during the lifetime of Ambartsumian. We have about 5,000–6,000 visitors per year mainly during the spring-autumn period.

On the ground floor of the house, there are four rooms where the exhibition is organized. First is the hall, where a small exhibition is organized concerning the origin of Ambartsumian family. Here the genealogical tree of Viktor Ambartsumian from the end of 17th century until his grandchildren is presented. Besides, there is a brief list and a small presentation of different prizes and awards of V. Ambartsumian. Some diplomas of these prizes are presented here.

From the hall the visitors can see also the lounge (so-called recreation room), where no special exhibition is organized and the original condition of the last years of his life is preserved. The living room (dining room) is the most spacious place in the Museum and the largest number of expositions is organized here. Here the visitors can see small expositions of V. Ambartsumian's handwritings in Armenian, Russian and English (he knew 7 languages). Another small exposition tells about the published works of V. Ambartsumian. Although the complete list of V. Ambartsumian's different publications contains more than 1,000 units, the most essential part of his works are collected in this exposition, where one can see the works of V. Ambartsumian in different branches of modern science: physics, mathematics, astrophysics and philosophy. V. Ambartsumian was also one of the main organizers of the Academy of Sciences of Armenia. He directed the Academy for 50 years just from the beginning: 4 years as the vice-president and then 46 years as the president. And a special exposition tells the visitors about V. Ambartsumian's activity in this field.

A very small exhibition in this room tells the visitors also about Hamazasp Ambartsumian-Viktor Ambartsumian's father, who was one of the prominent figures of the Armenian culture of the early 20th century. He was distinguished as a poet, philologist, philosopher and brilliant translator of works of classic Greek and Latin authors-Homer, Aeschylus, Euripides, Sophocles, and Virgil. The last room on the ground floor is the office of V. Ambartsumian where his scientific library is reserved as well. On the first floor, there are the bedrooms. There are no special exhibitions here. Only the original environment is preserved and one can see the very simple and modest conditions in which the greatest scientist lived.

The Museum is open for visitors 6 days a week (only on Mondays we do not receive visitors). Until now, the museum accepts visitors and provides them with explanations free of charge. (Maybe it will be changed in future).

More than half of our visitors are schoolchildren and students. Besides different schools and various public organizations invite us to give lectures about V. Ambartsumian and his scientific achievements.

We have an annual scientific one-day conference each year on September 18 dedicated to the birthday of Viktor Ambartsumian. The talks at this conference take place just in the garden of our Museum in a temporary outdoor 40–50-seat auditorium organized under the large walnut tree. The topics of the reports are chosen in such a way that they cover the most recent (often-revolutionary) results of modern sciences in non-traditional fields.

The exposition and accompanying explanations of our Museum are organized considering that the most effective way of teaching and upbringing is a personal example that is understandable to all without auxiliary explanations. As such a personal example we use the life and scientific activity of Viktor Ambartsumian.

The presentation of the museum's exposition and accompanying explanations are organized in such a way that our visitors and first of all schoolchildren and students are shown on the example of V. Ambartsumian, how one person engaged only in science, was able to influence the life of a huge country (the Soviet Union), even the world war itself and reach the highest recognition and glory. We try to teach our visitors how much can do a man of science for his country, for making the world better. Thus we try to help our visitor to understand that science in general and fundamental sciences, in particular, are extremely powerful and they help people to make our world better. We represent the life and scientific activity of V. Ambartsumian is a vivid example to show and to prove the young men the importance of scientific studies in different branches of modern science, to encourage them not to be afraid of difficulties, but to love it and try their abilities in science.

Of course, our Museum has also difficulties and problems. The first thing that interferes with the work of the Museum is the limited space. The number of our visitors increases each year (from 0 in the beginning until more than 6,000 for 17 years). We have a large influx of visitors especially in the summer and autumn months and in the frequency at the weekend. The limited possibilities of the Museum make it necessary to limit the duration of visiting visitors to the Museum. And this, of course, complicates the achievement of our goals, which were described above. In addition, a large flow of visitors somehow worsens the storage of museum exhibits, shortening their lives. Thus we have a problem to organize the visiting of the Museum in a different way. As an alternative, we chose to build a new building for the Museum in order to be able to organize there all the explanations and educational part and leave the old house itself as a memorial part, where the visitors can only see the original things and the environments after acquaintance with them in the lecture room. The new building of the Museum will be built near the existing old house. It will be equipped with a large hall for lectures or conferences, several exhibition halls, rooms for the staff of the Museum and other auxiliary rooms. Also, there will be one or two small telescopes on the roof for educational purposes. Now we look for sponsors to start this project. The old building-the original house of V. Ambartsumian also needs major repairs, because it has been built 62 years ago by very poor means and now it has been severely damaged.

Besides we have plans to organize a new museum of astronomy in Armenia, as there is a long history of this science in our country from Neolithic period until the famous Byurakan Astrophysical Observatory. There are too many rocky writings in different mountains of Armenia that obviously have the astronomical and calendric meaning. There are also some archaeological artefacts from various archeological sites in Armenia that have astronomical and calendric explanation. There are also many mediaeval Armenian manuscripts with astronomical texts and texts about calendar problems. We have also copies of some mediaeval Armenian astronomical instruments. Of course, a reach exposition in this museum will be dedicated to the history of Byurakan Astrophysical Observatory, telling about the achievements of V. Ambartsumian and his pupils-B. Margarian, L. Mirzoyian, M. Arakelian and others. So, there is a good basis to organize this new museum. As an additional of our museum can be noted the rich sights surrounding Byurakan. There are too many sights in Byurakan itself and in close vicinity. So this can help to organize touristic tours to the Museum. Thus, we look forward to great projects and expectations.

Science Centre Kenya: Developing Equality of STEM Education for Kenya's Youth

Kenneth Monjero   Charlie Trautmann   Graham Walker[1]

Abstract Young people account for a disproportionately large percentage of the population in Kenya and therefore represent an important and dynamic force in Kenyan society. Of 47 million Kenyans, 50% are under 18 years old. One proven solution for positive development is to invest in education that builds the capacity of youth. The main objective of the current work is to support the development of youth from all backgrounds in Kenya by establishing the country's first science centre. Science Centre Kenya (SCK) is a pilot science, technology, engineering and mathematics (STEM) centre that promotes equal educational opportunity for all youth to become inspired by the STEM (Monjero, et al., 2013). This paper describes the methods that have been used to plan and develop SCK, taking advantage of models that have worked for science centres elsewhere while acknowledging specific needs that distinguish SCK from science centres in most other countries. A SCK planning team conducted a survey to investigate the public's understanding of the science centre concept during the 3rd National Science, Technology, and Innovation Week, held 19–23 May 2014 in Nairobi, Kenya. 70% of the 56 respondents had never heard of science centres, and of the remaining 30% who had heard of science centres, none had actually visited one. After visiting the exhibitions and programs during the event, all but one person (98%) reported having learned at least one new science concept; 60% responded that county and national governments should fund science centres in the future; and all (100%) supported the idea of building a science centre for Kenya. The study found that the majority of Kenyans polled felt they had been left out of science matters and most were

---

①   Kenneth Monjero: Director of Science Centre Kenya. Email: kentrizakari@gmail.com.

unaware of the science that is present around them every day. This study provided the case for starting the first science centre in Kenya as a program of the Kenya Agricultural & Livestock Organization (KALRO) in Nairobi. The vision of Science Centre Kenya is: "A brighter future for Kenya's youth through STEM" with the ongoing mission: "To promote science, technology, engineering and mathematics through interactive science experiences that engage, challenge, and inspire exploration and understanding of the world around us." Experience during the past four years has shown dramatic growth of attendance at the pilot facility and a critical need for additional resources that will enable Science Centre Kenya to provide the exhibitions, programs, space, and visitor amenities necessary to adequately serve the observed public demand.

Keywords　Kenya　science centre development　equal opportunity in education　public participation　capacity building

1. The Benefits of Science Centres

A science centre is an out-of-school (sometimes called "informal") learning institution that, in some ways, resembles a museum. But in contrast with a traditional museum, everything is meant to be touched, handled, and experimented with. It is fully hands-on. It is a space where young learners, their parents, school children, and the general public are scientifically engaged, nurtured and empowered in an interactive and hands-on way through exhibitions, programs, and events on science, technology, engineering and mathematics (STEM). Many of the studies referred to in this paper make reference to research from a range of informal learning environments, such as science centres, art museums, and other types of organizations that make use of interactive exhibitions, programs, events, and other ways of interacting with the public. Results from one informal interactive learning environment in most cases apply to other informal learning environments, and a large international body of research now shows that informal educational experiences, of the type offered by Science Centre Kenya (SCK), promote long-term curiosity and lifelong learning, both in and out of school. Specifically, studies show that interactive science centres increase visitors' knowledge and understanding of science and provide memorable learning experiences with a lasting impact on attitudes and behaviour (Dierking et al., 2002, Leinhardt & Gregg, 2002). Science centres have wide-ranging personal and social impacts and promote inter-generational learning across diverse audiences (Winterbotham, 2005). Science centres promote trust and understanding between the public, the scientific community and provide a positive economic impact for the communities they are located in (Travers & Glaister, 2004).

2. Public Participation in Planning Kenya's First Science Centre

To better understand the public's knowledge of science centres as educational organizations in Kenya, a preliminary study was conducted of visitors during public science exhibits and programs offered during the Third National Science, Technology and Innovation Week, held 19–23 May 2014 in Nairobi, Kenya. The objective of the study was to gain insights into visitor's perceptions, ideas and motivations about the science centre concept. A questionnaire was administered to 56 visitors, who provided feedback after experiencing the exhibits and/or demonstration programs. 71% of respondents were male; 29% were female. 70% of respondents had never heard about a science centre and of the remaining 30% who had heard about science centres, all stated that they had heard of them from schools; none had actually ever visited a science centre. All but one of the respondents (98%) indicated that they had learned at least one new science concept from the exhibits and/or demonstrations they had just experienced. Nearly half (42%) reported that the most attractive activity was extracting DNA from carrots, indicating interest in both the process and content of science as an activity. Over 60% responded that county and national governments should fund science centres in the future and all respondents (100%) stated their support for having a science centre in Kenya. The study concluded that, despite the fact that two-thirds of those polled had never heard of science centres, the concept and demonstrations of science centers are readily appreciated once experienced and are universally attractive to all (Monjero et al., 2013). In summary, the study showed a compelling case for developing a science centre in Nairobi for the youth of Kenya.

3. Pilot Project for Science Centre Kenya

On the basis of the 2014 public survey, Kenya Agricultural and Livestock Research Organization (KALRO) agreed to allocate space for a pilot science centre and resources for initial planning and program development (Monjero, 2017). The next step was to explore the key aspects associated with developing a new science centre, including developing a statement of purpose, mission, structure, and initial funding requirements. These steps were accomplished during 2015–2016 by SCK staff in consultation with a team of internationally recognized science centre professionals. As a result of these efforts, SCK developed its vision and mission statements, created a series of pilot educational programs, acquired a group of science exhibits, and designed a floor plan for a science centre in space allocated by KALRO.

4. Capacity Building

To prepare for its pilot and future planning effort, Science Centre Kenya has built internal capacity by touring science centres and other museums worldwide, and by participating in international conferences and training programs for developing science centres in South Africa, the United States of America, Canada, and Australia (Monjero et al., 2016). The main objective of these capacity-building activities has been to study: 1) how other science centres throughout the world have approached public learning about science, technology, engineering and mathematics, with a focus on youth, and 2) how best to develop a science centre for Kenya that takes advantage of what has been learned in other countries while also taking into account the special needs of Kenya's youth.

Table 1 lists 25 museums that have been visited as part of this capacity-building effort. Most of these have been science centres, while several other types of museums have been visited for additional information on museum buildings, business practices, exhibit development processes, and visitor services.

Table 1: Science centres visited by SCK personnel, as of December 2017

Africa (5)	
Nelson Mandela Bay Science & Technology Centre	Utenag
North West University Science Centre	Cape Town
Sci-Bono Science Centre	Johannesburg
Sci-Enza	Pretoria
Walter Sisulu Botanical	Gardens Mamelodi
Asia (1)	
Nat'l Museum of Emerging Science & Technology	Tokyo
Australia (4)	
Bendigo Discovery Centre	Bedingo

(Continued)

National Dinosaur Museum	Canberra
Questacon Science Centre	Parkes
Wollongong Science Centre and Planetarium	North Wollongong
North America - Canada (6)	
Aga Khan Museum	Toronto
Montreal Museum of Art	Montreal
Montreal Science Centre	Montreal
Montreal Science Museum	Montreal
Ontario Science Centre	Toronto
Royal Ontario Museum	Toronto
North America - United States (9)	
Arizona Children's Museum	Phoenix, Arizona
Arizona-Sonora Desert Museum	Tucson, Arizona
Brooklyn Children's Museum	Brooklyn, New York
Children's Museum of Manhattan	New York, New York
Heard Museum	Phoenix
Liberty Science Center	Jersey City, New Jersey
New York Hall of Science	Queens, New York
Tucson Children Museum	Tucson, Arizona
Wild Center	Tupper Lake, New York

Table 2 lists 11 science centre conferences, museum training courses, and other capacity-building programs that have been attended since 2014 to prepare for the development of SCK. During these conferences and training sessions, SCK personnel has attended sessions to learn about all facets of starting and running a science centre, ranging from how to develop funding and other resources to exhibit development, business practices, diversity initiatives for reaching underserved youth, and maintaining international networks. In addition, these international gatherings have been used to develop a worldwide network of advisers, sources of exhibits, and other resources.

The Walton Sustainability in Science Museums program, for example, helps train science centers in developing countries to work on global challenges that focus on sustainability. The program brings science centre professionals from a variety of countries together to learn about successful educational programming and helps centres develop networks of colleagues worldwide who can offer mutual support, resources, and advisory services. The program provides content-rich kits and other resources. Because science centres are trusted worldwide as sources of information and inspiration for youth education, these resources can be put to efficient use at Science Centre Kenya to promote interest in both science education and the use of science in solving important challenges for Kenya, such as food production, water supply and increasing the supply of STEM-literate citizens.

Table 2: Science centre training programs and conferences attended by SCK personnel, as of December 2017

Africa (6)	
Science Centre World Congress (2011)	Cape Town
Southern African Association of Science & Technology Centres (2013)	Durban
Southern African Association of Science & Technology Centres (2014)	tenhage
Southern African Association of Science & Technology Centres (2015)	Mafikeng
Southern African Association of Science & Technology Centres (2016)	Richard's Bay
Southern African Association of Science & Technology Centres (2017)	Johannesburg
Asia (1)	
Science Centre World Summit (2017)	Tokyo
Australia (1)	
Australian National University Awards Program for Science Centres (2017)	anberra
North America (3)	
Association of Science-Technology Centers (2015)	Montreal
Association of Science-Technology Centers (2017)	San Jose
Walton Sustainability in Science Museums Program (2017)	Phoenix

5. Programs and Exhibits

During its pilot phase, SCK has experimented with a range of programs and exhibits to engage a wide variety youth in the STEM. For example, SCK constructed an outdoor science mini-golf course, with science-related obstacles, to engage teens in science and sustainability. In another project, SCK helped children create, plant, and maintain a vegetable garden. For many of the children involved, this was their first opportunity to plant something and watch it grow over time. SCK is pursuing opportunities like these to attract youth into agriculture, food production, and food science.

Several other programs developed for SCK include:

- Science Circus Africa: Inspiring science curiosity among young Africans through science demonstrations at schools and public venues.

- Young Africans Plan for the Planet: a program carried out in collaboration with Young Australian Plans for the Planet, working with students on sustainable development goals.

- Adventures in Synthetic Biology: a program for youth in partnership with Concordia University (Canada).

- Health Education for Young Learners: a program on health provided in partnership with Junior Medical Academy (United States).

- Home School Programs: STEM support for families who are home-schooling their children.

These and other programs are providing a growing set of opportunities for Kenya's youth to engage with STEM and see themselves having a future in the STEM economy of the future. The public response to these programs has been strong and growing consistently, with attendance at SCK and its programs growing 350% from 2014 to 2017 to a total of over 6,000 visitors annually (Monjero, 2017).

6. Science Centre Space Design

The capacity-building tour has shown that science centers occupy a wide assortment of building types. Some science centres have acquired an existing building and renovated it, while others have built expensive, iconic buildings to serve as tourist attractions. Many of the most successful science centers started small and built in phases over a period of years, rather than creating a large building all at once. In some cases, it was observed that large buildings built all at once had difficult problems, such as more space that could be readily supported, or design flaws that did not support the public use or STEM learning. In a number of cases, large iconic buildings had construction problems that raised costs, leading to reduced exhibit budgets, and less-than-expected use by the public afterward.

Another observation was that the most successful science centres had a number of diverse sources of income (Trautmann, 2017) rather than relying on one or two sources, such as government or a single foundation or donor. In the case of SCK, the support of KALRO for staff support and building use have been critical. Other sources of support, such as foundations and the international community of science centers, have helped provide the programming that has brought SCK to its current level of activity. Other sources of support will be needed to realize the next phase of development.

7. SCK's Vision for Future Development

Now that Science Centre Kenya has successfully demonstrated that there is a high public demand for the types of STEM exhibits and programs it offers, its leaders are developing plans for the next phase of its development (Monjero, 2017). SCK's vision—"A brighter future for Kenya's youth through STEM" – provides the guiding principle for all aspects of the centre's plans, which include the following key elements:

(1) Space: expansion of the current space to provide enough room for the public to view 50 interactive STEM exhibits, participate in educational programs for school groups, and support public use with restrooms, lobby space, staff offices, exhibits shop, etc.

(2) Exhibits: acquire or build 50 additional exhibits on STEM topics to complement those already on display.

(3) Staff: hire 15 additional staff members to provide the exhibits, programs, public contact, promotion, and maintenance of a science centre.

8. Conclusions and Recommendations

A survey of Kenyans indicated low awareness about science centres but a strong interest in the kinds of public science exhibits and programs offered by Science Centre Kenya. In particular, the establishment of a science centre was strongly supported by all Kenyans surveyed, regardless of gender or level of education.

Visits to 25 science centres and other types of museums have provided access to an international network of professionals who are willing to assist SCK with advice, exhibits, and connections to resources. This capacity-building effort also provided an extremely beneficial perspective on how successful science centres are developed and maintained. Attendance at 11 training programs and conferences have similarly provided instruction on how to develop a science centre, along with ready-to-use educational materials and advisors who can assist SCK in the future.

The next step in the development of Science Centre Kenya from the successful pilot stage to opening a fully functional public education centre includes three key elements, including:

• Designing and constructing a permanent science centre building.

- Developing 50 interactive science exhibits.
- Hiring and training 15 additional staff.

Securing the resources necessary for these growth steps is the next priority for Science Centre Kenya.

Acknowledgments

The authors gratefully acknowledge the financial support of Kenya Agricultural and Livestock Research Organization (KALRO), Association of Science-Technology Centers (ASTC), Australian Embassy in Kenya, Australian Department of Foreign Affairs and Trade, Sciencenter of Ithaca, New York USA, National Research Fund of South Africa, National Research Fund of Kenya and the Australian National Centre for the Public Awareness of Science (CPAS-ANU). We also wish to recognize the members of the Science Centre Kenya team, including Roy Gitonga, Peter Wanuthi, Janet Wanjiku and Antony Kinyili for hands-on exhibit development and the delivery of science programs and shows.

References

[1] BONNEY R, BALLARD H, JORDAN R, et al. Public Participation in Scientific Research: Defining the Field and Assessing Its Potential for Informal Science Education. A CAISE Inquiry Group Report. [J]. Online Submission, 2009: 58

[2] DIERKING L D, BURTNYK M S, BÜCHNER K S,et al. Visitor learning in zoos and aquariums: A literature review [D]. Annapolis, MD: Institute for Learning Innovation, 2002.

[3] LEONHARDT G, GREG M. Burning Buses, Burning Crosses: Student teachers seecivil rights. [J]. 2002.

[4] WINTERBOTHAM N. Museums and schools: developing services in three English counties 1988–2004 [J]. University of Nottingham, 2005.

[5] TRAVERS T, GLISTER S. impact and innovation among national museums. [J]. Proceedings of the National Museums Directors'Conference, 2004.

[6] MONJERO, K. Building communities through science-Kenya Agricultural Research Institute. [J]. Proceedings of the 6th Science Centre World Congress, 2011.

[7] MONJERO K. Developing Science Centres in Africa for Equality of Opportunity and Global Sustainability,the case of science centre Kenya [J]. Proceedings of Science Centre World Summit, 2017.

[8] MONJERO K. Do I Belong?Identity and Diversity Issues in Science Centres, serving new audiences, especially those from indigenous and underserved backgrounds inKenya [J]. Proceedings of Science Centre World Summit, 2017.

[9] TRAUTMANN C H. "The Business of Science Centers,"ASTC Dimensions, Association of Science-Technology Centers, Washington, DC: 2017.

[10] MONJERO K, MITEI D, KARIUKI E, et al. Science awareness: the case for a science centre in Kenya [J]. Proceedings of the 2nd National Science, Technology and Innovation Week. Nairobi Kenya, 2013.

[11] MONJERO K, MITEI D, KARIUKI E, et al. Status and Progress of the Science Centre Movement in Kenya and the Region [J]. Proceedings of the 15th Conference Southern African Association of Science and Technology Centres, 2013.

Geological Museum After H. Karapetyan of Institute of Geological Sciences of the National Academy of Sciences of the Republic of Armenia

Gayane Grigoryan [1]

Introduction

Geological Museum of Armenia was founded in June 1937 based on the rich and diverse collections of the prominent geologist, Professor Hovhannes Karapetyan and being adjacent to Institute of Geological Sciences of the Armenian branch of the Academy of Sciences of the USSR. Museum was formed in 1,410 samples, now the museum has more than 14,000 exhibits. Organization of the exhibition was related to the desire of a group of participants of the 17th International Geological Congress in Moscow to get acquainted with the geology of Armenia.

All participants have admired looking exhibits and the Armenian geological excursions. Establishment of the museum was an important event in the geological science of the Republic; one united center for centralization, processing, display and storage of materials was created.

Later the museum collections were daily supplemented with the samples collected by expeditions of the Institute.

In terms of geology, Armenia is one of the most interesting regions in the Caucasus. Here almost the entire stratigraphic sequence of the geological epochs is presented, beginning with metamorphic schists of the Upper Proterozoic Age and ending with thick Quaternary effusive rocks and modern deposits. Due to 80 years museum developed and become richer. The Museum has the following departments: 1.Mineralogy, 2.Paleontology, 3.Petrography, 4.Natural

---

① Gayane Grigoryan: Director of Geological Museum after H. Karapetyan of IGS NAS RA, Armenia. Email: gayane347@gmail.com.

Resources, 5. Volcanology, 6. Mineral Waters in Armenia, 7. Natural Monuments of Armenia.of geological science.

The museum conducts open-air lessons for different age groups, acquainted with the unique geological monuments, Museum Night events, International Museum's Day, commemoration parties of well-known geologists, meetings, seminars visit the villages with movable museum, conduct work in the fields, cooperations with the other museums.

The museum occupies 600 square meters of territory, has 20,120 visitors such as the elderly, students, schoolchildren, pupils from art school, tourists. The number of visitors increases year by year. We conduct educational programs corresponding with the school programs. Organize roundtable with teachers discussing how to support to conduct the lessons in the museums. To organize scientific meetings, discussing, exchanging the best experience about modern ecological problems "World Water's day", "The Earth Planet day", "The Mountains" International day" and so on.

The museum needs sub-constructions: Gift's shop, the microscope for using new technology and technic, radioguide, special supplies keeping the archival documents in corresponding head conditions. The museum can cooperate with the other museums organizing:

There are two natural museums in Armenia.

STATE MUSEUM OF NATURE OF ARMENIA

State Museum of Nature of Armenia was founded in Yerevan in 1952 in the yard of Blue Mosque and was called a Republican Museum of Natural Sciences. In 1960 the museum was renamed to State Museum of Nature of Armenia.

ZOOLOGICAL MUSEUM OF NAA OF ZOOLOGY INSTITUTE OF ARMENIA

The Zoological Museum of the Scientific Center for Zoology and Hydroecology was founded in 1920 by the efforts of the Armenian general Shelkovnikov during the Tsar period.

Geological Museum of Armenia was founded in June 1937 based on the rich and diverse collections of prominent geologist, Professor Hovhannes Karapetyan and being adjacent to Institute of Geological Sciences of the Armenian branch of the Academy of Sciences of the USSR. Museum was formed in 1,410 samples, now the museum has more than 14,000 exhibits.

Geological Museum After H. Karapetyan of Institute of Geological Sciences of the National Academy of Sciences of the Republic of Armenia

Conclusion

(1) Educational practices for pupils, students and the elderly for 7−14 days, Armenia is famous for its geological formation, which attracts both specialists and the tourists.

(2) To organize scientific meetings, discussing, exchanging the best experience about modern ecological problems" World Water's day" "The Earth Planet day" "The Monuments' International day" and so on.

(3) Exchange mineral samples to enrich the museum's fund.

(4) Make a global fund which will support the development and modernization of natural museums. I am sure this meeting will provide great progress and fame for the participating museums.

The Salt Range a Geo-Park of Pakistan

Sakander Ali Baig, Muhammad Waqas, Mian Hassan Ahmed[1]

Abstract Salt Range comprises 6 gorges, well exposed geological, paleontological and easily accessible. These propensities make attraction for the researcher which is unique in the world regarding age diagnostic; Cambrian Trilobites, Permian Brachiopods, Lower Triassic Ammonites, Lower Tertiary Larger Foraminifera and Upper Tertiary and Quaternary Vertebrates generate paleogeographic.

Khewra Gorge (Eastern Salt Range): Along the road between Choa Saiden Shah and Kussuk fort the Pre-Cambrian evaporitic facies and Cambrian clastic sequence with thick carbonate facies in the upper part is exposed.

Nilawahn Gorge (Central and Western Salt Range): The Permian to Early Eocene alternating carbonate and clastic sequence and Miocene to recent molasses are well exposed at Nurpur.

Nammal Gorge (near Mianwali); Chichali Gorge (near Kala Bagh); Lumshiwal Nala at Makarwal and Zaluch Gorge. All these Salt Range gorges are full of such wonderful sites for geologists and are precisely called as a unique field museum of Natural History in the world and a possible Geo-Park of Pakistan.

Keywords Geo-Park   Salt Range   Fossil localities

1. Introduction

The Salt Range is a longitudinal east-west trending trough, bounded on the east by the Jhelum River and River Indus on the west. Beyond the River Indus, it makes a hairpin bend to

---

① Mian Hassan Ahmed: Director of Geological Survey of Pakistan, Email: mian hassan ahmed@gmail.com.

develop a north-south trend. The Salt Range is an active frontal thrust zone of the Himalaya, which rises abruptly from north of the Punjab plains and gently dips in the northern oil and gas bearing Potwar basin. The range derives its name from the occurrence of the thickest seams of rock salt in the world, embedded in the Precambrian bright red marks of the Salt Range Formation.

The salt Range represents an open book of historical geology. It attracts the geologists and paleontologists from all over the world due to its good exposure and easy accessibility. Roadside geology, prominent gorges, Khewra (Fig. 6), Nilawahan, Warchha, Nammal (Fig. 7), and Chichali. The wide variety of geological features and richly fossiliferous stratified rocks. Cambrian stratigraphy, a Permian carbonate succession with brachiopods, the Permian-Triassic boundary, Lower Triassic ammonite bearing beds, Lower Tertiary marine strata composed of age-diagnostic foraminifera are excellently exposed due to lack of vegetation, which provides the fantastic opportunity to study in detail of all the geological aspects. The Salt Range is, therefore, of outstanding universal significance, related to scientific and educational research and essential to be considered as Geopark so that the geological history it contains may be preserved.

2. Previous Work

Gee for the first time published the Geological map of the salt range on a scale 1 : 50,000. Davies and Pinfold recorded Lower Tertiary larger foraminiferas of Salt Range. Waagen worked on the Permian brachiopods; Fatmi studied the Triassic Ceratitids and Grant described Permian trilobites of the Salt Range. Shah compiled the stratigraphy of Pakistan. Rehman published the first stratigraphic code of Pakistan. Kummel and Teichert described the detailed stratigraphy of the Permian and Triassic rocks. Haque described and recorded the distribution of Tertiary smaller foraminiferas of Nammal Gorge, Salt Range. Sameeni in his doctoral thesis established an alveolinid biostratigraphy for the Eocene succession of the Salt Range. Nannofossils biostratigraphy of Upper Paleocene and Lower Tertiary rocks of Salt Range has been discussed by Sameeni and Butt Sameeni and Hottinger.

3. Stratigraphy and Paleontology

3.1 Precambrian

Salt Range Formation is the only rock unit of Precambrian era exposed in the Eastern

Salt Range (Table. 1). It is comprised of thick evaporitic facies consisting of Salt, Gypsum and Marl. A highly weathered igneous body known as "Khewra Trap" is also a part of the formation which consists of highly decomposed radiating needles of a light-colored mineral, probably pyroxene. The formation is devoid of fossils. It is conformably overlain by clastic sequence containing Cambrian fauna, therefore, the age of the Salt Range Formation is considered as Precambrian.

3.2 Paleozoic

The Paleozoic Sedimentary sequence of the Salt Range is of the order of several hundred meters in thickness but with a profound unconformity indicating the absence of Ordovician to Carboniferous systems (Table. 1).

The Cambrian rocks are represented by clastic sequence with thick carbonate facies in the upper part. Cambrian succession contains fossils including; Neobolus warthy, Botsford granulata, Lingulella wanniecki, L. fuchsi, Hyolithes wynnei, Redlichia noetlingi and Pseudothecia cf. subrugosa.

The Permian sequence consists of mixed (cyclic) carbonate and siliciclastic facies and tillitic facies at the base. The fossils consist of abundant bryozoans, brachiopods, bivalves, gastropods, nautiloids, ammonoids, trilobites, fusulinids, conodonts and crinoids. The Permian rocks are famous due to the presence of Products. Fossil contains pollen and spores including Glossopteris and Gangamopteris of Gondwana affinity. The Permo-Triassic paraconformity has been discussed in detail by Kummel and Teichert.

3.3 Mesozoic

Rocks of the Mesozoic era are characterized by important hiatuses that are present at the base of the Triassic, between Triassic and Jurassic, between Jurassic and Cretaceous and on the top the Cretaceous (Table. 2). The sediments are mostly of shallow water marine and continental origin consisting of shale, sandstone, limestone and dolomite. Triassic systems mainly consist of limestone, dolomite, sandstone and shale. The shale unit is highly fossiliferous and contains well-preserved keratitis. Jurassic mainly comprises limestone, shale and sandstone with subordinate dolomitic and ferruginous beds. The Cretaceous rocks are mainly arenaceous and argillaceous. The argillites deposited in reducing environment and contains pyrite nodules and belemnites.

3.4 Cenozoic

The Paleocene to Early Eocene rocks are mostly exposed in the central and western Salt Range (Table. 3). These rocks dip gently northward into the oil-bearing Potwar Basin and overlain by Neogene and quaternary molasses. The unconformity of Middle Miocene to Oligocene magnitude is envisaged as uplift of the area and closure of Tethys. The Paleogene to Early Eocene rocks of Tethyan origin is richly fossiliferous in age diagnostic larger benthic foraminiferas which belong to families Nummulitidea and Alveolinidea. The Neogene and Quaternary molasses record uplift history of Himalaya and are famous for their vertebrate fauna including mammals, birds and reptiles.

Table 1   The Pre-Cambrian and Paleozoic Stratigraphic units and abundant fossil groups of Salt Range (Shah)

Era	Period	Epoch	Lithostratigraphic units		Abundant fossil Groups and their occurrences
Paleozoic	Triassic	Early		Mianwali Formation	Bivalves Trilobites Foraminifera
			Unconformity		
	Permian	Late	Zaluch Group	Chhidru Formation	
				Wargal Formation	
				Amb Formation	
		Early	Nilanwahn Group	Sardhai Formation	
				Warchha Formation	
				Dandot Formation	
				Tobra Formation	
	Cambrian	Early	Jhelum Group	Baghanwala Formation	Trilobites
				Jutana Formation	
				Kussuk Formation	
				Khewra Formation	
Pre-Cambrian			Salt range Formation		

Table 2 Mesozoic stratigraphic units and abundant fossil groups of Salt Range (Shah)

Era	Period	Epoch	Lithostratigraphic Unit		Abundant fossil groups and their occurrences
Cenozoic	paleogene	Paleocene		Hangu Formation	
Unconformity					
Mesozoic	Cretaceous	Early		Lumshiwal Formation	Foraminiferas Belemnites
		Early		Chicahli Formation	
		Unconformity			Mollusks
	Jurassic	Late		Samanasuk Formation	
		Middle		Shinawari formation	
		Early		Datta Formation	
		Unconformity			
	Triassic	Late		Kingriali Formation	
		Middle		Tredian Formation	
		Early		Mianwali Formation	Ammonoids
	Unconformity				
Paleozoic	Permian	Late		Chhidru Formation	

Table 3 Stratigraphic sequence and abundant fossil groups of Cenozoic rocks of Salt Range (Shah)

Era	Period	Epoch	Lithostratigraphic Units		Abundant Fossil Groups and their occurrences
Cenozoic	Quaternary	Recent	Lei Conglomerates		
		Pleistocene			
	Unconformity				
	Neogene	Pliocene	Siwalik Group	Soan Formation	Vertebrates
				Dhok Pathan Formation	
				Nagri Formation	
		Miocene	Rawalpindi Group	Chinji Formation	
				Kamlial Formation	
				Murree Formation	
	Unconformity				
	Paleogene	Eocene	Chharat Group	Chorgali Formation	
				Skaser Formation	
				Nammal Formation	
		Paleocene	Makarwal Group	Patala Formation	Large foraminferas
				Lochart formation	
				Hangu Formation	
	Unconformity				
	Cretaceous	Early		Lumshiwal Formation	

4. Conclusion and Recommendations

Salt Range is rich in several unique geological and paleontological sites of great academic as well as research importance; which are considered for the preservation and conservation. Salt range is a possible geo-park of Pakistan. Significant efforts of paleontologists, Government bodies and awareness program for the locals are required to preserve these sites.

References

[1] PASCOE E H. The early history of the Indus, Brahmaputra, and Ganges[J]. Quarterly Journal of the Geological Society, 1919, 75:138−157.

[2] SAMEENI S J. PaleoParks-The protection and conservation of fossil sites worldwide; The Salt Range: Pakistan's unique field museum of geology and paleontology[J]. Carnets de Géologie/Notebooks on Geology, Brest, 2009.

[3] COLBERT E H. Siwalik mammals in the American Museum of Natural History[J]. Transactions of the American Philosophical Society. New, 1935, 26: i−x, 1−40.

[4] GEE E R. The saline series of north-western India[J]. Current Science, 1935, II: 460−463.

[5] GEE E R. The age of saline series of the Punjab and Kohat[J]. In-Proceedings of the National Academy of Sciences of India, Calcutta, (Section B), 1945, 14:269−312.

[6] SHAH S M I. Stratigraphy of Pakistan[J]. Geological Survey of Pakistan, Memoirs, Quetta, 2009, 21: 381.

[7] WAAGEN W. Salt Range fossils; Productuslimestone fossils[J]. Pal-ontologiaIndica, I(4): 329−770.

[8] GRANT R E. Late Permian trilobites from the Salt Range, West Pakistan[J]. In-Palaeontology, Oxford, 1966, 9(1):64−73.

[9] SAMEENI S J, HOTTINGER L. Elongate and larger alveolinids from Choregali Formation, Bhadrar area, Salt Range, Pakistan[J]. Pakistan Journal of Environmental Science, 2003, 3:16−23.

[10] SHAH S M I. Stratigraphy of Pakistan[J]. Geological Survey of Pakistan, 1977, 12: 138.

[11] REHMAN H. Stratigraphic Code of Pakistan[J]. Geol. Surv. Pakistan, 1962, 4: 8.

[12] KUMMEL B, TEICHERT C. Relations between the Permian and Triassic formations in the Salt Range and Trans-Indus ranges, West Pakistan[J]. NeuesJahrbuchfür Geologie und Pal-ontologie, Abhandlungen, 1966, 125: 297−333.

[13] KUMMEL B, TEICHERT C. Stratigraphy and paleontology of the Permian-Triassic boundary beds, Salt range and trans-Indus ranges, West Pakistan.[J]. University of Kansas, Special Publication, 1970, 4: 1−110.

[14] HAQUE A F M. The smaller foraminifera of the Ranikot and Laki of the Nammal Gorge, Salt Range.

Pakistan[J]. 1956, 1: 300.

[15] SAMEENI S J. Biostratigraphy of the Eocene succession of the Salt Range, Northern Pakistan[D]. Punjab University, Lahore, 1997.

[16] SAMEENI S J, BUTT A A. Alveolinid biostratigraphy of the Salt Range succession, Northern Pakistan[J]. Revue de Paléobiologie, Genève, 2004, 23(2): 505−527.

The History, Facilities and Activities of the Hungarian Natural History Museum

Gábor Csorba [1]

Abstract The Hungarian Natural History Museum (HNHM) is one of the first natural history museums in the world, founded in 1802. With its two affiliated countryside museums it is one of the largest scientific and cultural institutions in Hungary, employing more than 230 staff members. It houses extensive botanical, zoological, anthropological, paleontological and geological collections of more than ten million specimens and over 65,000 primary types, whereas its library contains 400,000 volumes of books and periodicals. The collections uniquely represent the natural history of Hungary and the Carpathians, but highly valuable items originate also from other parts of the Earth. The collections and other scientific facilities of HNHM are used annually by approx. 350 researchers from all over the world. Research activities cover the fields of taxonomy, systematics, ecology, evolutionary biology, epidemiology, conservation biology, geology, paleontology and physical anthropology. The museum's exhibitions attract approximately 300,000 visitors (about 3% of the whole population of the country) annually. The HNHM is successfully participating in several European Union projects, including database management, digitization programs, Research Infrastructure Actions, exchange of scientists, and LIFE+ conservation biology projects. It is a member of the Consortium of European Taxonomic Facilities (CETAF), the International Council of Museums (ICOM), The European Network of Science Centres and Museums (ECSITE) and the Global Genome Biodiversity Network (GGBN).

Keywords collections EU programs Hungary science communication science dissemination

---

[1] Gábor Csorba: Deputy Director of the Hungarian Natural History Museum. Email: csorba.gabor@nhmus.hu.

1. Introduction

The Hungarian Natural History Museum (HNHM) is one of the first natural history museums in the world, founded in 1802. Never forgetting this unique tradition and the defining cultural role the HNHM plays in the society, our mission is to raise and to sustain the awareness towards the recognition and acceptance of the diversity of nature and to ensure the commitment of the public to the preservation of the natural environment. With its two affiliated countryside museums (the Mátra Museum of Gyöngyös town and the Bakony Museum of Zirc town) the HNHM is one of the largest scientific and cultural institutions in Hungary, employing more than 230 staff members.

The HNHM houses extensive botanical, zoological, anthropological, paleontological and geological collections of more than ten million specimens and over 65,000 primary types, whereas its library contains 400,000 volumes of books and periodicals. The collections uniquely represent the natural history of Hungary and the Carpathians, but highly valuable items originate also from other parts of the Earth. These irreplaceable assets render the museum the ultimate hallmark of bio- and geodiversity in Hungary and in Central Europe. The collections and other scientific facilities of HNHM are used annually by approx. 350 researchers from all over the world. Research activities cover the fields of taxonomy, systematics, ecology, evolutionary biology, epidemiology, conservation biology, geology, paleontology and physical anthropology. The museum's exhibitions attract approximately 300,000 visitors (about 3% of the whole population of the country) annually.

2. Highlights of the Collections

The Department of Zoology comprises one of the largest zoological collection in Europe with almost 8 million objects. It includes significant material from all over the world with collections from the Balkan Peninsula, Inner and Southeast Asia, the Far East, whereas historical material from New Guinea and East Africa is truly unique worldwide.

The Department of Botany includes 2 million specimens, mostly herbarium sheets. Its collections contain remarkable historical material dating back to the mid-18th century including Linnaeus specimens, Central European herbaria and fossil plant remains.

More than hundred thousand individually registered fossils from the Carpathian Basin are housed in the Department of Paleontology. These include the very first dinosaur remains from the Upper Cretaceous of Hungary.

The Department of Mineralogy and Petrology gives home to about 80,000 objects. The majority comes from Hungary; however, a significant amount is accommodated from all around the world. Its meteorite collection and lunar rock samples are recognized for their outstanding universal value.

The Department of Anthropology contains the remains of around 50 thousands individuals representing the whole Post-Pleistocene human population of the Carpathian Basin. The Hungarian Neanderthal Man remains and the naturally preserved mummies from the town of Vác, Hungary are among the most notable objects.

3. Brief History

The museum dates its origins back to 1,802 with the establishment of the Hungarian National Museum by Count Ferenc Széchenyi, who donated his 1,700 volume library, manuscript and coin collection to the nation. In the same year, his wife donated the very first natural history collection that contained selected and valuable minerals and dried plants from Hungary. In 1,811 the first paleontological and zoological collection found their way to the museum. Later on, in the age of the rising Hungarian aristocrats, and in line with rising patriotic feelings of the Reform Period of the 19th century, thanks to donations and purchases, the collections started growing fast.

1869 was the year when the first data regarding the number of visitors were accounted; nearly 65,000 visitors looked up the entire museum (with history and art collections included). Owing to the diversity of the ever-expanding material, the natural history collections had to be made independent of the National Museum. As a consequence, independent zoological, mineralogical and paleontological, and botanical departments were founded in 1870. Thanks to the greater independence, more and more specialists joined the departments and the growth became even more intense. At the turn of the 19th-20st century, the number of natural history objects exceeded 1 million specimens.

During the Hungarian Revolution in 1956 the museum suffered its greatest losses in the history. Its famous Africa exhibition and the majority of the mineralogy and paleontology collections were lost due to fire caused by Soviet artillery-shots. The Department of Zoology, accommodated in a separate building, was also hit by a firebomb, and hundreds of thousands of specimens perished. In the following decades the collection strategy focused on acquisition in order to replace its lost collections and exhibitions. Several collecting trips were organized to Africa and "exotic socialist countries" like North Korea, Vietnam and Mongolia. As a result, materials from these regions are the richest of their kind worldwide. Beginning in the 1980s more countries in Asia, Africa, South America, Australia and Oceania became available destinations for collecting trips.

At the end of the 20st century, the Hungarian Natural History Museum was still lacking an individual and permanent building which could be identified as such by the public. It was only in 1994 when the Hungarian government decided to provide a permanent location for the museum, and designated the building of its current location for its permanent exhibition and certain collections. According to plans of that time all the collections would have been accommodated at one location by the year 2002. In 1996 our permanent exhibition of history and ecology "Man and Nature in Hungary" was opened in the newly reconstructed building. In 1999, a brand-new loft of the building was opened for the public, and the Department of Anthropology together with two sections of the Department of Zoology, i.e. the collections of mammals and birds could occupy some of the most up-to-date museum premises of Europe. In 2010 our current permanent exhibition, "The Variety of Life" was opened.

In 2013, two countryside natural history museums became affiliated as member institutes to the HNHM in Budapest: one in Gyöngyös town, Northeast Hungary, the Mátra Museum, and the other one in Zirc town, West Hungary, the Bakony Museum.

Unfortunately as a result of changes in political decisions concerning the museum other collections were not able to be moved to this location but rather remained in their separate premises. To find a permanent place for all the collections and exhibitions is now the biggest challenge the museum has to face.

4. Education Activity

Our Department of Exhibitions organizes almost 500 educational events annually, reaching 44,000 persons, while the Department of Education holds almost 800 one-hour classroom lessons to children between 4~18 years old with topics defined in the formal education programs and related to the scientific scope of our collections. In the Mátra Museum an EU-funded project was launched in January 2018 to reach and engage disadvantaged groups by introducing educational methods elaborated for them, and by utilizing specific ICT tools developed for educational purposes.

The HNHM has a long history of taking part in higher education either through teaching activities of its staff members or by housing and supervising MSc and Ph.D. projects of student at different universities and colleges. Currently, HNHM serves as an auxiliary university department to the University of Veterinary Medicine, Budapest and is responsible for zootaxonomy, zoosystematics and zoology field courses. Beside these activities in every academic year almost 20 seminars take place in the scientific collections of HNHM.

The HNHM is successfully participating in several European Union projects, including database management, digitization programs, Research Infrastructure Actions, exchange of scientists, and LIFE+ conservation biology projects. It is a member of the Consortium of European Taxonomic Facilities (CETAF) and the Global Genome Biodiversity Network (GGBN).

5. Conclusions

Among several socio-economic factors, the rise of new cultural expectations significantly influences cultural consumption. Not only do museums compete for consumers with each other but also with other cultural suppliers—and with the digital world. According to the 2013 Eurobarometer report on cultural access and participation, an overall decline in cultural engagement is seen between 2007 and 2013. In Hungary the ratio of those, never participating in cultural activities rose to 54% within this period. The boom in information technology of the 1980s and 1990s brought about changes in the use of spare time and created the so-called Y and later Z generations, who are not able to live without computers and internet, and whose spare time rather meant playing computer games. In 2017 the Hungarian Natural History Museum has recognized the role of gamification, a technique employed successfully in so-called escape rooms in Hungary. In order to approach the youth, the museum, therefore, created two escape rooms within the museum public area based on collection material. By encompassing new ways of implementing exhibition, disseminating knowledge and providing access to collection material, this brand new service accomplishes the mission of transferring cultural values while satisfying the needs of youth at the same time.

During the scientific, curatorial and other work on natural history collections, a vast quantity of digital data is produced and stored. Digital data are both essential tools and products of a modern research and collection based institute. They complement the physical objects and are parts of the collection. Their societal advantages are beyond doubt. They raise awareness for natural history and biodiversity topics, provide information validation for education (updates), provide data to the public, increase public interest in natural sciences and the museum, and last but not least they promote citizen science. However, our main task is to find both financial and human resources to complete tasks with regard to digitization, digital storage and the provision of access to them. The museum is going to possess diverse kinds of databases, and, therefore, it will have to put measures on issues such as access (limited or free), privacy, ethics (regarding software licenses), reproduction of data, ownership and rights, licensing, attribution and credits.

Nowadays the importance of taxonomy (identification and classification of natural history objects) is rather underestimated. However, global challenges like the loss of biodiversity and the anthropogenic impacts on ecosystems can be tackled only if expertise to understand biodiversity is available. As the only institution in Hungary with such an expertise and collection material, it is our mission to convince the public and policymakers of the importance of taxonomy and systematics with its connections to economy, health and everyday life. As a natural history museum we must tackle global issues like climate change, biodiversity crisis, emerging infectious diseases, or biological invasion. As a renown institution with considerable collections and specialized taxonomists HNHM should take a leading role in identifying special and generalized problems and finding the way to co-operate with the appropriate governmental offices and with other scientific organizations and institutions.

Biodiversity and its protection are tightly linked to a societal change which can only be achieved by a strong investment in environmental education. The right training and capacity building need to be provided today to improve competencies of teachers in the sector of biodiversity education, effective at raising the level of biodiversity literacy for teachers and students, motivating both to learn about biodiversity, to engage in conserving Europe's biodiversity and to ignite their passion for science.

6. Acknowledgements

We are very grateful for the organizers of the First BRISMIS and to the leaders of the Chinese Association of Natural Science Museums for making possible the participation of the HNHM's representative in the event. Our special thanks are due to Emese Jókuthy for her expert help during the preparation of the manuscript.

Science, Art and Mathematics

Paul Firos [1]

Abstract Herakleidon is a small Museum from a small country, trying to make its way in a comparative environment where cultural production is not always welcome. Over the last ten years, we managed to correlate art, science and mathematics in a unique educational program designed to make math learning easier, to cultivate the love for the arts and to make science easy to understand. Thus, we are working together with education specialists, mathematicians, art teachers and we are adopting our work to the student's needs.

Keywords Herakleidon Museum Art Mathematics Education Educational Program Science Escher

It is a great honor for me to represent Greece in this Forum. I looked at the maps depicting the many paths of the ancient Silk Road and to my surprise, the westernmost destination was indeed Greece. And here we are again twentyfive hundred years later reenacting what our ancestors initiated. This time, however, we are gathered here for a different but very important trade, not to exchange goods but culture and ideas.

1. The Museum

Greece has a population of about ten million citizens, probably less than the combined annual attendance of the top five museums represented here today. Yet we attract more than thirty million visitors annually, who come to Athens to visit the Acropolis, its museum, the

---

① Paul Firos, Apostolos Papanikolaou: Director of the Museum Herakleidon Athens, Greece. Email: hyperpol@gmail.com.

National Archaeological Museum and all the sites that constituted the citystates of Ancient Greece.

What may surprise you is that there is no tradition of privately funded museums in Greece. When my wife and I decided to found the Museum Herakleidon, we were surprised to discover that only a handful of other such institutions existed and, most importantly, no legislation was in place to categorize us as a Bona Fide museum. So we opened our doors as a Not-For-Profit organization dedicated to Culture, which is a good way of describing any museum in the world (Fig. 8). Indeed this lack of infrastructure helped us organize in the most efficient ways our various activities. I wish at the time we had an equivalent of BRISMIS to help us overcome many of the obstacles we encountered on the way.

What we do today with the New Silk Road initiative at the time we founded our museum we were forced to discover by ourselves how efficient and productive collaboration between museums is. We quickly learned to reach out to our peers. BRISMIS offers this to us on a plate.

2. The Education

I once invited the Director of the Louvre in Paris, one of the most attended museums in the world, to visit our museum in Athens. I was surprised by the two comments he made. The first was that the office of our museum director was larger than his, the second comment, and more important, was that he wished the number of visitors to the Louvre was less so that he could devote more resources towards the museum's goals and less towards crowd control.

In other words, big or small, with great attendance or not, museums must focus on their goal, which is Education. We are educating the adults who come and visit our museums, but, most importantly, we are educating the children who come either as individual visitors or, preferably, as part of a classroom.

Education is the primary objective of Museum Herakleidon and we are very proud of maintaining a 1 to 1 ratio of students vs. regular visitors. We have four student classes daily, going on simultaneously. Our teachers are all University graduates and the educational material is always derived from the current exhibition.

We have worked firmly upon the bounds of Mathematics and Art for more than a decade.

The bounds of Mathematics and Art were defined 2,500 years ago by Plato in the tenth book of his dialogue Republic, where, having previously (in the fourth book ibid.) divided the psyche into two parts, a superior (logic) and an inferior (willful/ wishful) one, he went on to regard Mathematics as connected to the reasoning (logical) part while Art as connected to the inferior (emotional) part.

Having studied Plato's treatment of Art, I stand assured that if Plato was to come to life and lay eyes on Escher's works, he would recognize in the latter's face one of the artists that he so passionately sought for his ideal State: *Seeking those creators that are intelligently able to search out the nature of the good and the decent*, while he rejected the imitators who drew "merely holding a mirror up to nature", thus ascribing it superficially, external characteristics. These are the words that Plato used some 2,500 years ago to place artistic representation up on a high pedestal, refusing to consider Art a realistic-slavish depiction of reality. In another of his dialogues, "Philebus", his description of the absolutely beautiful is reserved for the kind of drawing that uses geometrical shapes as its basic components. Many took this as a warning sign for the appearance of 20st century Modern Art, which, in the words of Paul Klee, "does not reproduce the visible; rather it makes visible".

For ten years now, we are teaching courses in mathematics and art, pairing a scary thing, math, and a fun thing, art and we are doing it successfully. Our teachers are discussing with their students the golden ratio, fractal, perspective, music, rhythm and harmony, symmetry etc. using the works of artists such as Escher, Vasarely, Carol Wax or referring to monuments that can be found nearby in Athens, Parthenon for example. Escher and the others can be found in our collections and Parthenon is located a few hundred meters from our museum but art is full of hidden or obvious mathematics and the golden ratio can be found anywhere in the architecture. That's why we always find a way to corelate the ongoing exhibition with science or mathematics making learning easier and maths adorable.

3. Educational Program

Through video projection and by means of a parallel tour through the history of Art on the one hand and that of Mathematics on the other, students are guided to search out the points where these two aspects of human thinking and acting meet and interact. Emphasis is placed on Greek geometric art, classical art (the Parthenon-proportions - golden ratio), the analysis of linear perspective (renaissance), the geometry of modern art (cubism, constructivism, Bauhaus) and, last but not least, the contemporary, "mathematical art" of fractals. At the same time, through selected works of M.C. ESCHER and V. VASARELY, students are introduced to nature and indeed the very philosophy of significant mathematical concepts. Targeted tours through the running exhibition are connecting mathematical ideas or scientific concepts with specific exhibits.

The whole program is designed accordingly to the states school program of mathematics and is adopted every time in the current exhibition. For example, at the moment we are teaching Chinese paper folding, the geometry of a dragon, Chinese calculating tools such as abacus etc (Fig. 9). Generally speaking, the structure of our educational program is as follows:

Part 1, in the classroom for 60 minutes

1st Grade - Secondary School: Seeking the geometrical background of specific, appropriately selected art (Escher, Vasarely, Carol Wax etc.) Students are invited to explore the structure of the selected prints and, if possible, reproduce them solely by means of a pencil, ruler, and compass. In doing so, they will clearly have to mathematically deconstruct the images, thus entering the realm of mathematical concepts.

2nd Grade - Secondary School: An introduction to the fundamental geometric transformations (reflection, central symmetry, transport, rotation).

3rd Grade - Secondary School: Associating the distinction between "being and appearing to be" of Euripides' tragedy "Helen", which students are taught at school, with the philosophical as well as mathematical "being and appearing to be". The association is intended to lead to an awareness of the necessity for the use of reasoning and proof through appropriate paintings in which "intuition" can be misleading.

4th Grade - Secondary School: a) The use of appropriate paintings that are in fact optical illusions leads to an empirical and, thereafter, theoretical extraction of the criteria governing parallelograms. b) Escher's print "Verbum" (=speech, logos) is the visual stimulus for a brainstorming session over the definition of the mathematical concept of the ratio (proportion) and its philosophical overtones. c) Introduction to non-Euclidian geometries.

5th Grade - Secondary School: a) Study of the geometrical variables of the basic curves in the alterations of the grading grid. b) Through Escher's tessellations, we study the regular and semi-regular division of the plane (through regular and semi-regular polygons respectively). c) Zeno's paradoxes constituting the vaulting horse for an introduction to the concepts of limit and self-similarity.

6th Grade - Secondary School: a) Study of the geometrical variables of the basic curves in the alterations of the grading grid. b) Appropriate paintings are being used with a view to negotiating the concepts of limit, infinite and infinitesimal, discreet, continuum and they're philosophical overtones. c) The "countability" of the natural numbers set, the density of the rational numbers set and the "overcountability" of the real numbers set. d) The concepts of function and inverse function. e) The concept of mathematical structure.

Part 2, in the exhibition hall, 50 minutes

A guided tour through the exhibition allows not only interaction with the works of the artists or the artifacts that have been mentioned but also discussions spurred by the students' own remarks.

Part 3, in the classroom, 10 minutes

Upon returning to the classroom, students complete the evaluation-feedback form, where they anonymously record their comments on the program just attended, including brief assessments of the overall activity and of what new things they have learnt, as well as feelings or thoughts that puzzled them.

4. Our Future/conclusion

The BRI will undoubtedly bring us together and foster greater cooperation between Science Museums. There is much to be gained from sharing information and exchanging marketing knowledge and know-how. But in my personal experience this may not be enough. Two museums do not always have something in common to exchange. Sometimes it takes three or more museums in order to "trade" collections. We solved this problem by creating a separate entity whose goal is to facilitate such exchanges. As we speak, there are four collections brokered by us that are traveling in Europe and the USA. The principle is quite simple: a museum identifies a traveling collection and sets the terms and conditions under which it is willing to lend it to other museums. We then use our international museum connections to find institutions willing to participate in this exchange. We would welcome the addition of your collection to our list under your terms and conditions and we will do the rest. This kind of cooperation between museums is what makes us dynamic. Collaboration among our museums is not only vital, it is our moral obligation towards the communities we serve.

References

[1] Allentuck M E. Herbert Read, The Philosophy of Modern Art[J]. Art Journal, 2015(3): 302–303.

[2] M C ESCHER. Regular Division of the Plane, 1958.

[3] DORIS SCHATTSCHNEIDER. The Polya-Escher Connection[J]. Mathematics Magazine, 1987, 60(5): 293-298.

[4] SCHATTSCHNEIDER D. The Plane Symmetry Groups: Their Recognition and Notation[J]. American Mathematical Monthly, 1978, 85(6): 439-450.

[5] H S M COXETER. The Non-Euclidean Symmetry of Escher's Picture'Circle Limit III'[J]. Leonardo, 1979, 12(1): 19-25.

National Dinosaur Museum, Canberra, Australia

Thomas Kapitany [①]

Abstract The National Dinosaur Museum, Canberra, is a fun, educational, interactive Museum, not a research museum. It is a place for adult, children and students to have fun, interact with geology and, by using social media such as Facebook or WeChat, to promote the business whilst making it enjoyable and interesting. Despite its relatively small size, the Museum has been able to achieve high visitor numbers through maximizing interaction, engagement and education opportunities. Opportunity for engagement and interaction lead to free publicity and an increased social media presence as the public share their experiences.

A focus on effective retail sales has allowed the National Dinosaur Museum to turn over a healthy profit which is then used to increase our collections, displays and stock. To ensure continuity of the Museum space, many display specimens are positioned throughout the shop. The Museum has achieved a very high turnover of stock and excellent profit margins through the purchase of large quantities of bulk geological and non-geological items direct from the suppliers in China and by having a relatively large retail section in relation to the overall size of the Museum.

Keywords Social Media Facebook WeChat Dinosaurs Interaction Education

1. Introduction

The National Dinosaur Museum is situated in Canberra, Australia's capital city (Fig. 10). One of the premier tourist attractions in the area, the museum was established in 1993 and since

---

① Thomas Kapitany: director of the National Dinosaur Museum. Email: tomk@crystal-world.com.

then has grown into one of the largest permanent displays of dinosaur and prehistoric fossil material in Australia. With a key focus on education, interaction and entertainment, the museum offers a wide variety of activities to schools and the general public. A number of exhibits include life-like animatronic dinosaurs as well as a large number of realistic replicas both within the Museum and around the grounds.

2. National Dinosaur Museum-Overview

The museum has about 10,000 m² of the public display area and very large retail store within the premises, which provides funds to allow sufficient acquisition of new specimens, displays and stock, as well as significant regular improvements of the museum. Our shop makes it about $1 million a year in retail sales and occupies about 2,000 m².

In addition, there is a large outdoor garden filled with realistic dinosaur replicas which is open to the public without any payment (Fig. 11).

We are open seven days a week to public and evenings by appointment for students and private functions.

2.1   Interaction

A key to the success of the National Dinosaur Museum is the focus on providing a high level of public interaction with many of our specimens in both the garden and within the premises.

Engagement of the public and students is achieved by allowing children and adults to interact with the dinosaur models and by encouraging them to touch the dinosaurs, rocks, meteorites, minerals and fossils. Displays, and even simple items like bins, can be used to create photographic opportunities for the public. Apart from providing excellent entertainment for the participants, it also leads to effectively free marketing and promotion of the museum on social media such as Facebook and WeChat.

To engage children we regularly offer free activities during their school holidays for further participation and interaction, such as the opportunity to dig for their own fossils.

We are often used by media and TV programs when they are doing a news story about dinosaurs, fossils and ancient geology. We encourage this as again, it provides free advertising and promotion for the Museum, not unlike that of the social media by our visitors.

2.2 Education

Education is extremely important for us and we regularly give public educational tours and school tours for up to 1,000 children per day. These can include after hour and night options. We take also take our dinosaurs on tour around the state of New South Wales and the territory of Canberra to educate the public about Dinosaurs and our Museum.

Many of our staff are students who study palaeontology or geology at university. By employing passionate, knowledgeable and engaging presenters we hope to inspire the scientists of the future.

2.3 Retail

Unlike many other small museums, who do not have a large retail presence or focus, the National Dinosaur Museum has dedicated one fifth of our floor space to a retail area. We have found that this large and successful enterprise allows us to expand and increase the quality of our collections and also to employ more staff.

Our shop includes display specimens as well as retail sales specimens so it appears to be part of the museum rather than the separate outlet (Fig. 12).

The National Dinosaur Museum purchases many of our non-geological specimens from Yiwu Zhejiang Province, China which allows us to do mark-ups of between 500 to 1000%. By purchasing directly from the suppliers rather than local wholesalers (which would only allow us a 100% markup), we are able to able to have this higher profit margin yet and still retain attractive pricing for our visitors. With this very large profit we are able to buy more specimens for the museum.

3. Conclusion

Touch, play, interact, fun, share - five words are at the heart of the National Dinosaur Museum.

Despite its small size, the National Dinosaur Museum is extremely successful and profitable due to engaging the public's hearts and minds with our fun, hands-on, educational interactive approach and displays. We have a large emphasis and impact on Social Media which gives us free advertising and promotion (including a large proportion directly from our visitors sharing their experiences on social media).

Free social media advertisement in turn leads to greater visitor numbers and increased door sales. Interestingly, for every $1 AUD from entry fees, we also take $1.10 AUD in retail sales. This excellent retail revenue is due to sourcing much of our extensive range of toys, games and collectable items directly from the suppliers in China. Our large retail space is an excellent source of revenue for expanding our collections, displays and staffing.

In conclusion, a small Museum CAN be profitable by making it fun! If it is fun, the public wants to come and share their experiences on social media. Hands-on, interactive and engaging displays and photo opportunities make the museum very popular. When people have fun they tend to buy and spend more which, in turn, leads to the success and viability of the Museum.

4. About the Author

Thomas (Tom) Kapitany grew up with a love of collecting all things related to Natural History. After completing a Bachelor of Science degree at the University of Melbourne, majoring in Botany and Geology, Tom went to work with his brothers and set up a retail nursery called Collectors Corner (http://www.collectorscorner.com.au).

Collectors Corner, a 5,000m^2 premises, specializes in rare plants from around the world such Cacti, Orchids, Bonsai, Carnivorous Plants, Bromeliads as well as Crystals, Fossils and Meteorites.

It was in this shop that Tom started to sell Fossils, Minerals and Meteorites from his collection, eventually opening Crystal World & Prehistoric Journeys in 1999.

Tom is the Executive Director of Crystal World, the largest importer, exporter and manufacturer of gems, crystals, minerals, fossils, meteorites and unique specimens in Australia. It is located on his 2 hectare property in Devon Meadows, Victoria and is devoted to all things geology. Tom has over 20 staff as well as a facility devoted to cutting, polishing and processing rare stones, crystals, Australian rough stone and fossils (including dinosaurs and stromatolites), exporting rock to many countries around the world as well as the local market.

Much of the rock and fossils processed, sold and exported by Crystal World comes from Tom's 40 mining tenements and exploration licenses run by his company, Australian Mineral Mines.

From these tenements he is able to mine crystals, Quartz, Molybdenite, Mookaite, Variscite, Stromatolites, Amethyst, Fossils, Hematite, Marra Mamba Tiger-eye, Tiger iron, Prehnite and other commercial and high-grade mineral and fossil specimens.

Thomas is one of 4 Directors managing the National Dinosaur Museum in Canberra, as well as consulting for international Natural History Museums in the UK, China, USA and New Zealand.

In his role as an International Museum Consultant he has recently been working closely with the design and development of a $55 million dinosaur park in the Hubei Province.

As well as an interest in Plant Molecular Biology, primarily genetic engineering, Tom specialises in Deep Time Geology traveling the globe extensively collected rare and ancient rocks for museums and research scientists.

In the past 20 years he has traveled extensively throughout Asia, Europe, Pakistan, India, North and South America collecting rare items of Natural History.

From these travels, Tom picked up important learned basic language skills including German, Mandarin, Vietnamese, Spanish, Japanese, Bahasa, Portuguese and French.

Many of his travels have been "off the beaten track", including a trip into the Karakoram Mountains of Pakistan (Western Himalayas) with armed police escorts to the Khyber Pass, Mexico on numerous trips and Tom was even fortunate to visit the Giant Crystal Cave in Naica, Mexico. Tom also went into the jungles of North Vietnam, as well as underground in Antimony mines in China, where even the Chinese guide had trouble with the language!

Tom travels extensively in Australia to the outback visiting his mining projects in all Australian states and many of the commercial mines. Tom also owns Australia's largest Bonsai Nursery with his brother Jeno.

National Museum of the Republic of Kazakhstan

Alipova Saya ①

Abstract The National Museum of the Republic of Kazakhstan is the youngest and largest museum in Central Asia. The museum has been created in the framework of the "Cultural Heritage" State Program on behalf of the President of the Republic of Kazakhstan Nursultan Nazarbayev.

July 2, 2013, the Decree of the Government of the Republic of Kazakhstan No. 675 was issued on the establishment of the Republican State Institution "National Museum of the Republic of Kazakhstan" of the Ministry of Culture of the Republic of Kazakhstan.

Keywords Astana Cultural Heritage Silk Road Central Asia Kazakh Golden Man

1. Introduction

The museum is located on the main square of the country - the Independence Square, which harmoniously blends into the single architectural ensemble with the "Kazakh Eli" monument, the Independence Palace, the Palace of Peace and Harmony, the "Hazrat Sultan" cathedral mosque and the National University of Arts. Many values identified during the "Cultural Heritage" State Program constitute the invaluable fund of the National Museum of Kazakhstan.

The museum building is an eye catcher with its unusual external form. The largest unique museum complex has an area of 74,000 sq.m. and consists of seven blocks with a variable number of stories to the ninth floor. Exhibit space occupies 14 rooms with a total area of over 14,000 sq.m (Fig. 13).

The museum is fitted out with equipment of international standard, one uses modern

---

① Alipova Saya: head of the scientific and methodical service Republic of Kazakhstan. Email: muzei_nauka@mail.ru.

exhibition technology for expositions: a unique curved screen with a special content, working for two halls, a media floor, a dynamic layout of the central part of modern Astana, numerous media outlets, holograms, LED-technology, touch-sensitive kiosks, and a multimedia guide providing information in three languages.

The National Museum of Kazakhstan is composed of the following halls: Hall of Astana, Hall of Independent Kazakhstan, Hall of Gold, Hall of Ancient and Medieval History, the Hall of History, Ethnography Hall, Halls of Modern Art. The structure of the museum on studying the national heritage is represented by the Research Institute. There are also facilities for a children's museum, children's art centre, two showrooms, restoration workshops, laboratories, professional depositories, a scientific library with a reading room, a conference hall and souvenir stalls.

2. Session Heading - The Future of Natural History Museums: Collection, Exhibition, Education and Technological Advance

2.1　Hall of Ancient and Medieval History

Kazakhstan is a country with rich historical and cultural past which is located in the heart of Eurasia. Long-term surveys conducted by Kazakh scientists across the country helped to gather rich and unique material perfectly illustrating the ancient and medieval history of Kazakhstan, which can be seen in the exhibition hall on the second floor of the National Museum of the Republic of Kazakhstan (Fig. 14). There are more than 500 exhibits out of 5,000 on view, which remain deposited in the holdings and are gradually replenished by the researchers.

The museum's archaeological collection gathered by scientists introduces visitors to the history, economy, religion, crafts, the architecture of tribes inhabiting the territory of Kazakhstan from ancient times to the Middle Ages.

The exhibition consists of four sections: the Stone Age, Bronze Age, Early Iron Age and the Middle Ages. The artifacts have been arranged in thematic and chronological order.

A horse was first domesticated on the territory of Kazakhstan. In Botaisk settlement in the north of Kazakhstan a large number of artifacts have been discovered which are associated with the domestication of a horse. That fact played a vital role for people living in Kazakhstan for interrelationship and further development of cultures of the entire Eurasian continent.

In the Bronze Age Kazakhstan was one of the major centres for extraction of nonferrous

092

metallurgy functioning in 2nd - beg. of 1st millennium BC.

Original economic-cultural type of nomadism started to take shape In the early Iron Age, and it replaced the agricultural and pastoral settled life of the Bronze Age.

In the Middle Ages, the territory of Kazakhstan enters the Western Turkic Khanate. Medieval nomads being the Turks left stone sculptures for their descendants, which are silently looking at us keeping the mysteries of the past.

From the 6th century A.D. the territory of Kazakhstan was crossed by the Great Silk Road. The exhibition hall displays objects from other countries that have been found in the medieval towns of Kazakhstan.

In the middle of the 14th century, a powerful state was established in Central Asia by Emir Timur, and it included the South of Kazakhstan. Construction of towns and mausoleums began in the unprecedented scale for that time.

The exposition ends with the restored medieval streets and iconic architecture of Kazakhstan, where the wall shows a magnificent architectural complex of Khoja Ahmed Yasawi. The mausoleum was built on Timur's order. It followed the goal of spreading Islam, strengthening his position among the nomads and glorifying his power. Entering this part of the exhibition a visitor can get to the so-called medieval market and plunge into the atmosphere of the time.

Touch-sensitive stands to introduce Kazakhstan's ancient and medieval architecture and the towns along the Silk Road.

2.2   History Hall

History Hall displays the centuries-old history of the Kazakh people; it is full of important historical and sometimes crucial moments. The exposition begins with the presentation of the Kazakh state emergence in East Dasht-i-Kipchak, Semirechye and Turkestan. Next topic to be highlighted is the topic of the Hundred Years War of the Kazakh people with Dzhungars, which left an indelible mark in the minds of the nation. A special place in the history of Kazakhstan takes the Great Silk Road, which contributed to strengthening economic, cultural, social and political ties between nomadic and settled population. The Institute of biys is of tremendous value for shaping statehood and legal culture of the Kazakh society. Great Kazakh Biys left a rich spiritual heritage having not become obsolete to the present day.

Severe and tragic events fill the history of Kazakhstan in the 20th century. They are the years of Soviet power and the subsequent famine years of 1932–1933, political repression in 1937–1938, the participation of Kazakhstan's population in the Second World War and the Soviet-Afghan military conflict (1979–1989), nuclear tests at Semipalatinsk test site, the December events of 1986 in Alma-Ata city. Despite all the hardships falling to a lot of the national science and culture were under development. There was development of virgin lands and industrial growth together with exploration of new mineral deposits which contributed to the emergence and growth of new areas and cities.

2.3 Hall of Ethnography

Ethnography Hall located on the third floor of the National Museum is dedicated to the traditional culture of the Kazakh people and occupies the central place in the exhibition space.

Exposition hall is made as a complete familiarity with the traditional material and spiritual culture of the Kazakhs, their economic order, in which the overwhelming importance belonged to nomadic herding combined with settled agriculture, where demonstration of fishing and hunting objects reveals features of multifaceted economic and cultural lifestyle of the Kazakhs.

Traditionally felt dwelling of the portable type used by nomadic Kazakh people – *kyiz ui* (Fig. 15), displayed in the exposition, is one of the cultural dominants having absorbed all aspects of social, cultural, legal and regulatory forms of the society's life. Spiritual culture of the Kazakh people is shown by objects of musical art, household and ritual culture, attributes of religious beliefs and practices.

Ethnography hall is equipped with an innovative technique that allows showing material and spiritual culture of the Kazakh people. In the centre of the Hall above the *kyiz ui* there is a projected effect of the sky, a circumferentially mounted circular screen (Circle Vision), which shows a video with images of a traditional village. These audio-visual techniques create the effect of reality and presence of a viewer in a Kazakh village under traditional natural landscape environment and allow you to experience a sense of belonging to a priceless centuries-old culture of the Kazakh people.

In the hall a wall projector shows visuals of the culture and traditions of the Kazakh people, based on documentary photos and video footage. Visitor also can get complete information about the design and build order and installation of the traditional felted home from the video projected on the glass (Glass Vision). A special place in the exhibition hall is occupied by a hologram presenting the visitor a unique work of art of Kazakh masters.

2.4 Hall of Gold

Hall of Gold is a treasury and collection of gold and unique highly-valued artifacts found in the territory of Kazakhstan.

The exhibits of the Hall are annually refilled with unique items made of gold and precious metals discovered within the state program called "Cultural Heritage".

Gold jewelry of Scythian-Saki culture is renowned throughout the world. Imagination is staggered by rich burials of Saki chiefs in royal burial mounds. Everywhere researchers recognize originality of the Scythian-Saki culture, with its hallmark called "animal style". The purpose of the Hall of Gold exhibition is fostering and showing Kazakhstan's gold masterpieces to the world community, which were made in the Scythian-Saki animal style.

On display, there are gold products covering the time period from the Bronze Age to the Golden Horde. The main part consists of gold ware from the Saki archaeological finds.

Hall of Gold is divisible into two sections. The most striking manifestation of the culture of early nomads in Saki time, aesthetic approach to the world around them, is the "Scythian triad" - standard models of weapons, horse harnesses and art objects. Each of these cultural components is impressively represented in the objects found in the famous Saki cemeteries.

A special place in the exposition is given to reconstruction model of genuine human grave in burial structures. Good examples are the models of the king's barrow Baigetobe and the famous Issyk burial.

Following the principle of historicism and respecting the chronological order the Hall of Gold exposes such rarities as a gold earring from the Mayer burial (East Kazakhstan), more than 200 apparel items of the "Golden Man" from the grave called Taldi-2 , 7th – 6th centuries BC (Karaganda oblast), and among them there are more than 1,800 gold ornaments in the form of flakes of fish and more than 20,000 small gold ornaments.

Golden ornaments of Kargali and Zhalauly treasures are displayed as splendid patterns of metal and ancient Usun and Kangly people. The main exhibit at the exhibition the "Golden Man", which was found during excavations of Issyk barrow by a group of archaeologists with the leader Kemal Akishev, is a striking sample of Saki art's "animal style" (Fig. 16). That find dates back to 5th – 4th centuries BC. and includes more than four thousand gold products made in various techniques like forging, stamping, engraving and granulating.

By metal treatment, the scientists equate gold jewelry of ancient nomads to the world's masterpieces. Today the Issyk "Golden Man" is known throughout the world has become a sensation and a symbol of independent Kazakhstan.

The latest "Golden Man" – a Sarmatian woman – was found in West Kazakhstan in Taksai-1 mound complex and it is one of the most surprising findings of archaeologists, which manifests the wealth and power of the ancient Scythians.

3. Conclusion

The museum has been developing various kinds of excursions - survey and thematic, philosophical, special programs in the form of interactive sessions and gaming excursions.

The National Museum aims to become a modern intellectual cultural institution, a place for the analysis, comparison, reflection, discussion, and evaluation of statements on the historical and cultural heritage of Kazakhstan. Contemporary museum is always an open dialogue with the visitor. This museum has done everything possible to make its visitors active participants in the conversation with the history.

References

[1] Minbay D. National Museum of the Republic of Kazakhstan: first steps[R]. Astana:Foliant. 2015.
[2] Minbay D. Annual report on the results of the National Museum of the Republic of Kazakhstan[R]. Astana, 2016.
[3] Alipova S. Actual problems of museum activity.[J]. Cultural Heritage, 2017, 2 (71): 78–80.

The Role of NGOs in Publicizing Astronomical Science

Mohamad Hadi Tabatabaee Yazdi   Shahram Abbassi
Iman Ahadi Akhlaghi [①]

Abstract   Astronomy has been of major interest among ancient Iranian scholars during the history. A quick review of major contributions made by Iranian scientists to science in general and astronomy, in particular, is presented. Contemporary situation is then explained focusing on relative motivation happened in the holy city of Mashhad (N-E of Iran) and around. Current activities and plans under consideration by national NGO's and local private sectors for reinforcing the positioning the astronomical activities and programs are then highlighted.

1. Introduction

In many ancient civilizations, the sky and what is visible in it, has had a key role in motivating people to think about the unknowns. As one of the oldest natural sciences or even the mother of other sciences, astronomy taught people how to use their observations in converting a simple thought to a brilliant discovery.

The Iranian civilization is home to the many advancements in the history of astronomy. Dating back to the Parthian period (about 100 BC), there are some evidence that Babylonian mathematical astronomy and astral omens are studied mainly in the east of Iran. Other evidence is also found in the next centuries. These ancient astronomical outcomes were utilized until the

① Mohamad Hadi Tabatabaee: Director of Engare Physics Co., Mashhad, Iran. E-mail: Haditaba@yahoo.com.
Yazdi Shahram Abbassi: Department of the Physics, Ferdowsi University of Mashhad, Iran. E-mail: abbassi@um.ac.ir.
Iman Ahadi Akhlaghi: Department of the electric engineering Sadjad university of Technology. E-mail: I_A_akhlaghi@sadjad.ac.ir.

middle of the ninth century. Most important and outstanding outcomes in that era were belongs to astronomers such as Ahmad ibn Mohammad ibn Kathir al-Farghani and Mohammad ibn Musa al-Khwarizmi. In the next tenth to thirteenth centuries, the Almagest was taught all over the Middle East. Although many improvements were introduced to this book, however, the basics were not challenged. We can name Abd al-Rahman al-Sufi as the person who identified the Large Magellanic Cloud and also made the earliest recorded of Andromeda galaxy in 964 AD, (tenth century), Al-Biruni who estimated the radius of the Earth and pointed out its movement. Another famous poet, mathematician and astronomer in this period were Omar Khayyam whose works on algebra is well known. He also was head of an observatory that was built in his time. It was at that era that the Jalali calendar which is the most accurate calendar ever devised was reformed. The Jalali calendar is the base of official Iranian calendar.

In the 21st century and with the growing the youth publication in Iran, inspiring the beauty of science and spreading scientific ideas among the public may affect the quality of people's lives and also help them to understand and respect the universe in a better way.

2. Contemporary Works

In the Place of birth of Omar Khayyam, the City of Neyshabur and in the first year of the 21st century (more than 1,000 years after Khayyam) Khayam Cultural and Scientific Society that is an NGO began to build a very special and huge Project named Khayyam Planetarium and Science Center. A 20 meters dome screen powered by Zeiss projection Technology and a 4k dome theater including a Hybrid optomechanical projection, in a showroom with more than 300 seats, surrounded with 5,000 squaremeters of museum area was a dream that began to come true for the scientific society in Iran. Everybody believed that this project will start a revolutionary movement especially in natural science education (Fig. 17).

Neyshabur with the population of two hundred thousand people is located near Mashhad (the second Major city of Iran). Mashhad is very strategic location because of the Shrine of Imam Reza, the eighth Shia Imam where it hosts millions of people visiting the holy shrine every year. This gives Mashhad city and surrounding (like Neyshabur) a great opportunity for promotion of science activities and also investing in relative activities such as planetarium and science center projects.

Unfortunately, things haven't happened as properly as it was expected. The Khayyam mega project Planetarium that was supposed to be opened in 2005, was influenced by the economic sanctions against Iran and the Zeiss Company that was the technology vendor was not able to ship the goods and facilities to the building site. So the project went into a long period of suspension.

However, the scientific movement was alive in the meanwhile and still growing but mostly concentrated in the hand of universities. In the year 2003, the Mashhad Astronomical Society was launched by the University Professor, Mohammad Taghi Edalati and his students. The aim was to have an active NGO for science promotion in Mashhad. Prof. Edalati was also a board member of Khayyam Planetarium where he hoped to make appropriate reforming in order to showcase his pilot projects. He also planned to engage many graduate students for proving valuable material and content as well as preparing them for conducting and operation of the planetarium program before the opening ceremony of the great Khayyam planetarium project. As he passed away in 2005, his followers continued his way for promotion of the astronomical science by working for the NGO that he started. This newly reconstructed Astronomical Society of Mashhad (ASM) conducted many clubs and other monthly activities with thousands of participants attending each year. It was only due to prof. Edalati and his successor's activities that in the year 2007, Mashhad was recognized as a city having the most active amateur astronomical societies in Iran and this title was announced in the occasion of the same years' of the International Astronomy day (Fig. 18).

These activities soon found lots of audiences and attracted the attention of investors as well as government educational authorities. Activities made by ASM encouraged some of the governmental and semi-governmental organizations to cooperate with them. Some of the private companies started to invest in ideas which could be of tourist interests such as science museum or science centers with the help of ASM and its partner's organization such as Physics Teacher Association. The Mashhad municipality opened several subjective cultural centers in the city. Examples are Mashhad's Astronomy house and Science Center of Koohsangi Park etc.

Gradually, the idea that first started in the field of astronomy was extended to other subjects such as physics and medicine and students started to demonstrate complicated concepts with powerful multimedia language. They could convince their audiences having different background and knowledge, that science is not too difficult to understand by the normal public and it could also work as fun and recreation for fulfilling their spare time food place to spend money. In one of the ASM interactive activities performed in the largest parks in Mashhad, more than hundred thousand people visited the performances during two weeks time in the spring of the year 2007.

As a result of ASM activities, many similar projects were implemented with full support received from the government, NGOs and private sectors by different. Examples are Professor Bazima's Science Center, Permanent Science fair center in the central park, The Razavi Museum, the Great Museum of Khorasan Razavi, etc.

The vast growth of such projects had an unpredictable adverse effect that worried the scientific society. Unlike many other countries, soon after implementing some of the so called first generation science centers, newly established projects were started independently and without receiving the necessary guidance and supervisions from university and experienced professors. This could have some misleading consequences in the way of transferring fake knowledge to the public made by unauthorized and invalid oral or documented resources.

In order to prevent divergence that could take place from the original idea, in 2015, it was decided to start building a joint venture intermediate level planetarium project with collaboration ASM and a knowledge base private corporation (Engareh Physics co.) within the Mashhad university campus. The main idea was to act as a leading group for investors and interested visitors by showcasing the concepts and standardizing relevant activities making use of direct supervision received from appropriate department and professors who are easily accessible at the one of the highest ranked Iranian universities. This goal is about to happen shortly whereas all prerequisites including very sophisticated and illustrative contents and high tech instrumentations are prepared and installed at the site and whole work is supposed to be opened to the public in near future (Fig. 19). The fact that cannot be neglected is to mention that because of many unfavorable beliefs, the project could not be successful for absorbing enough financial support from the government as well as did not take advantage of receiving the proper amount of donation by individuals and private parties. Most people and volunteers prefer to spend their money by direct help to poor people or construction mosque and schools in remote villages.

The project developers welcome all kind of contributions from individuals and institutions being interested in the project objectives and hope that this could somehow lead us to compensate the lake of useful work for the growth of knowledge for our children and mankind new generation in the country and worldwide.

3. Questions (Suggestions)

(1) Could it be so that people of all cultures and nations have access to knowledge and

have the opportunity to understand the elements of science?

(2) Why should politics affect peaceful projects like Planetarium Projects?

(3) How can science museums in the world come together and cooperate in creating new content?

(4) How to create an organization with the goal to support non-governmental projects in developing countries independent of all international and political situations.

(5) How to develop some multimedia devices so that joint events and activities could take place in relevant science museums.

Legacy of Nikola Tesla

Branimir Jovanovic [1]

Abstract Nikola Tesla was a Serbian-American inventor and engineer, who is best known for his contributions to the design of the modern alternating current electricity supply system. After the discovery of the resonance frequency of the planet Earth in 1899, the basis of his global wireless problem of the power project, Tesla saw a revolution in all spheres of human life which would be yielded by his and similar innovations of global character. After the rejection of the leading American financiers to further invest in his project, which seemed to be too ambitious and unrealizable, Tesla started to think more critically about the trends of contemporary development in the world. This was the time of the 1,920s and 1,930s when he spoke about the need to preserve energy resources, search for new, renewable and alternative energy sources, strive for a more modest development with less consumption, aspire a culture oriented to universal mankind values and true education.The value of Tesla as an authentic and early thinker of globalism is unknown to the broader public and even undiscovered and unrecognized among experts.

The Nikola Museum, situated in Belgrade, Serbia, preserves complete Tesla heritage including a larger number of personal items used by Tesla during his life. These items are today divided among several of the Museum's collections. The Museum's greatest treasure, however, is Tesla's personal archive whose documents, manuscripts, scientific notes, calculations, diagrams, drawings and letters make up a total of 163,911 call numbers. In 2003, in recognition of the universal significance of Nikola Tesla and his inventions, UNESCO added Tesla's archive, as part of the moveable human documentary heritage, to the Memory of the World Register, the highest form of protection of cultural assets. The entire collection of the Museum has been digitalized

---

[1]   Branimir Jovanovic: Director of Nikola Tesla Museum, Serbia. Email: branimir.jovanovic@tesla-museum.org.

and also microfilmed. This procedure secures the permanence of the material for future generations.

Keywords Nikola Tesla Museum Legacy Museum True sustainability culture Belgrade Serbia

1. History of the Nikola Tesla Museum

After the death of Nikola Tesla in New York city in January 1943, American court awarded custody of his property to Sava Kosanović, the son of Tesla's youngest sister Marica. Sava Kosanović was a Serbian politician, publicist and diplomat who, at that time, was living in New York as a member of the Royal Yugoslav Government-in-exile.

Following his death, Tesla's entire property was packed, sealed and handed over to the Office of Alien Property Custodian. His belongings were transferred from the New Yorker Hotel to the Manhattan Warehouse and Storage Co. where some of Tesla's property was already stored.

On the initiative of Sava Kosanović, all Nikola Tesla's personal property and writings were shipped to Belgrade, where Kosanović subsequently presented them to the state.

Packed in sixty packages, suitcases, metal trunks and barrels, the legacy of Nikola Tesla arrived on the ship Serbia in the port of Rijeka in September 1951. The material was then transferred by train to Belgrade, where it was stored in the Belgrade University Faculty of Electrical Engineering. In June 1952, it was moved from the Faculty to the Genčić Villa at 51 Proleterskih Brigada, as the street was then known. That address is now the Museum.

The responsibilities and goals of the Nikola Tesla Museum are defined by its 1953 statute: to preserve the scientific and personal legacy of Nikola Tesla; to continue collecting and conserving documentation and personal items connected with the life and work of Nikola Tesla; to maintain a permanent exhibition of material from its collections and to organize and facilitate research into that material; to publish the works and writing of Nikola Tesla and, as copyright holder, to authorize publication, reprinting and translation of these; alone or in collaboration with local and international scientific and educational institutions and individuals; to encourage and support scientific endeavor and research in the technical sciences in which Nikola Tesla worked, and to publish such work.

Following the founding of the Museum and the organization of the archive material, work began on preparing the permanent exhibition. This was carried out during the period from 1953 to 1955. The design for the reworked ground floor of the Genčić Villa was prepared in 1955 by architect. The Nikola Tesla Museum in Belgrade was opened to the public on October 20, 1955. It was the first technical museum in Yugoslavia. The opening presented the permanent exhibition, which gave visitors the opportunity to see models built accurately according to Tesla's drawings. Perhaps the most celebrated of these demonstrates the effects of a rotating magnetic field. The Egg of Columbus, which had amazed visitors to the 1893 World's Fair in Chicago, was also displayed for the Belgrade public. Also on display were Tesla's first induction motor, a model hydroelectric power plant which illustrated Tesla's polyphase transmission system, various generators and transformers, and a remote-controlled model boat. The most popular exhibit for visitors today is the Tesla coil with antenna, which was the basis for the fluorescent light.

The opening was attended by many famous figures from the world of science and culture. The guests included the then US ambassador to Belgrade, James Riddleberger.

The next important ceremony in the Museum was in the second half of 1957 when Milica Trbojević, the daughter of Tesla's sister Angelina and heir to Sava Kosanović, presented Tesla's ashes to the Museum for permanent preservation. The Yugoslav Embassy in Washington had handed over the urn containing the ashes to Charlotte Muzar, who arrived with it in Rijeka on the merchant ship Triglav on July 13, 1957. Four days later it arrived in Belgrade. A new space on the ground floor of the Museum was prepared for the ceremonial handover of the urn.

The museological work of the Nikola Tesla Museum may be seen as beginning in 1957 when the urn received by the Museum went on display as a permanent exhibit.

Since October 9, 1969, the Museum has been the property of the City of Belgrade, having been transferred from the federal government by an agreement promulgated in the Official Gazette of the Yugoslav Government.

2. Tesla's Heritage

The Nikola Tesla Museum is today, by any criteria, a scientific and cultural institution which is unique in Serbia and the world. It is the only museum preserving the original and personal legacy of Nikola Tesla. Its holdings include the following exceptionally valuable collections:

• more than 160,000 original documents

- more than 2,000 books and periodicals
- more than 1,200 historical and technical exhibits
- more than 1,500 photographs and glass photographic plates of original technical items, instruments and devices
- more than 1,000 plans and drawings.

2.1 Archive

The Museum preserves a larger number of personal items used by Tesla during his life. These items are today divided among several of the Museum's collections. The Museum's greatest treasure, however, is Tesla's personal archive whose documents, manuscripts, scientific notes, calculations, diagrams, drawings and letters make up a total of 163,911 call numbers. This material, created over a long time (1856–1943), includes documents which are extremely diverse, both in content and in form. Among them are handwritten pencil notes, printed business cards with and without annotations, cancelled postage stamps, accounts on the cheapest paper, checks and printed forms with handwritten contents, and typewritten texts with duplicates, as well as diplomas written in colored ink on parchment, a charter with an intaglio stamp, India ink drawings on tracing paper, blueprint copies of plans and many other items. Given the age and variety of this material, it may be said that most of it is relatively well preserved. The entire archive is kept in 548 boxes (which are the basic units for processing and storage of the material) and catalogued in seven units according to the subject matter of the documents.

2.2 Library

A special collection—the Personal Collection of Nikola Tesla-contains monographs and periodicals from Tesla's legacy and many loosely bound press clippings. Because of their exceptional importance, they are treated as a special section of this collection (Tesla's Press Clippings).

Nikola Tesla's personal library consists of 975 titles (in 1,172 copies), monographs (books) and 2,435 individual issues of 347 periodical titles (magazines and newspapers).

2.3 Press Clippings

As part of Tesla's legacy, there also arrived from America perhaps the most valuable part of the library collection - press clippings. While the books and magazines owned by Nikola Tesla allow us to assess his interests in the field of science and his literary taste, the clippings give us insight into the social context and the kind of response he received, both from the public and in professional circles.

The collection of press clippings and newspapers from the personal legacy of Nikola Tesla consists of bound and unbound clippings, pages from newspapers with articles marked for clipping, unannotated pages and entire issues of newspapers. The bound clippings are in 57 individual albums and it is estimated that there are more than 20,000 of them. An original record of these is kept in a card index according to the subject. Tesla instructed his associates and secretaries to clip articles from periodicals, most of which dealt with his work and activity in Europe and the USA. The articles are primarily classified by subject (energy, electrochemistry, teleautomatics, X-rays etc.) and, within that, chronologically. Every article is annotated by hand with the source of the clipping. Detailed inspection of this material has established that Nikola Tesla also engaged a specialized press clipping service.

As well as articles relating to Nikola Tesla, here may also be found research on other scientists and inventors who were working in the same fields of science and technology.

The methodical way in which these articles are classified bears witness to Tesla's great devotion to, and skill at, collecting and utilizing information. If we also include all other sources of collecting and exchanging information, it becomes clear that Nikola Tesla created an analog version of what today we call the Internet. That is to say, he had his own organized network of information which was indexed, described and arranged in an appropriate place within the entire system of his sources.

As well as containing unique information about Tesla's life and work, these articles also reflect the social and scientific climate in which Tesla conducted research. For any historian of science, they are an essential source of information about events in science and technology at the end of the nineteenth and the beginning of the twentieth century, all available in one place.

2.4 Collections

The museum corpus consists of nine collections:

Mechanical Engineering.

Electrical Engineering.

Fine & Applied Arts.

Small Technical Items.

Chemical Technology.

Medals.

Textile & Leather.

Memorial Items.

Tesla's Personal Items.

Within this corpus, more than 1,200 items were officially registered. Among these collections, original technical objects in the field of mechanical and electrical engineering are preserved, as well as personal items and Nikola Tesla's clothing, his medals and decorations and the artifacts of fine and applied art from his legacy.

3. Conclusion

As well as containing unique information about Tesla's life and work, Tesla's legacy also reflects the social and scientific climate in which Tesla conducted research. For the historian of science, they are an essential source of information about early development especially of electrical engineering at the end of the nineteenth and the beginning of the twentieth century, all available in one place. Tesla is today more and more recognized among the broad public, especially among young people.

Institutions all over the world are yet to recognize Tesla contribution to electrical engineering.

Because of the great popularity of Tesla among young people educational potential is great. The mission of the Nikola Tesla Museum to help research and education!

Promoting International Multicultural Exchanges with the Automobiles

Yang Rui [1] , Dou Limin, Liu Yujie Xiong Wei

Abstract Beijing Auto Museum opened in 2011, which is a national non-profit museum themed on the automobile, combining the functions of museum, exhibition gallery, and science & technology museum. The architectural style of the museum is like a bright eye, implying opening our eyes to the world and future. Beijing Auto Museum sets up three galleries and one area according to theme selection mode of "science-technology-society". It has broken the boundaries among countries and brands, offers a glimpse into the history of automobile development over last one century and displays the start stage of Chinese automobile industry and how it develops and becomes stronger. It also reveals the profound effect of the automobile industry on human civilization and society. With automobile as the media, the Chinese stories will be told to the world, from the point of view of museum hosting multicultural integration, and ideological concept and management methods of Chinese museums are to be shared. With the Chinese-French cultural exchange as the demonstration, China and France are working together to make progress toward the beautiful and far-sighted goal of integrating people, cars, and society via The Belt and Road Initiative.

Keywords Beijing Auto Museum multi-culture international communication

1. Introduction

Over thousands of years, we can explore China ancient chariot culture, learn more about automobile culture with different cultural backgrounds in various countries in the world, and

---

[1] Yang Rui: Director of the Beijing Auto Museum, China. Email: 56185826@qq.com.

even can travel back in time to think about how automobiles will change our lives in the future. Beijing Auto Museum strives to prompt the multicultural exchange between countries with the automobile as the carrier, which is discussed by taking Sino-French Cultural Exchanges as an example. Beijing Auto Museum narrates the creativity and friendship of China and France with a case of the automobile's return to the Silk Road and China chariot culture regarded as the Chinese contribution and Chinese traditional cultural spirit, and tells the integration of automobile culture between China and France and also tells how Beijing Auto Museum can prompt the dialogues and exchanges among countries, cities and people based on the multicultural activities of museums under The Belt and Road Initiative in future.

2. Beijing Auto Museum Under the Comprehensive Proposition of "People, Automobile, Society"

2.1 Integration Development of Culture, Technology, Education and Tourism

Beijing Auto Museum fills the blank of the national special thematic museum, combining the functions of the museum, exhibition gallery, and science & technology museum (Fig. 20). Located in Fengtai District, it has a total construction area of 50,000m². Since its opening in 2011, the museum has been open all year, has provided services to five million visitors. Moreover, Beijing Auto Museum practiced the integration development of culture, technology, education and tourism, and gained the series of honors such as AAAA National Tourism Attraction, National Automobile Culture Promotion Base, the First National Service Standardization Model Institution in National Museum Industry, the honor of National Popular Science Education Base for four years in a row, Beijing Patriotism Education Base, the Advanced Grass-roots Party Organization in Beijing. Beijing Auto Museum advocates the harmony development of man-automobile-society for the purpose of "running the museum in an open doorway, Integration into society".

2.2 Seeing Things With a Global Vision in a Topic Selected Mode "Science-Technology-Society"

As per the topic selection mode "science-technology-society", to dismantle the boundaries between countries and brand names, the exhibition is set up with three galleries and one area to display the history of world automobile development in the past century and the development process of China automobile industry, which reveals the tremendous impact on human civilization and society brought about by the automobile industry. Breaking the traditional ways of display, the exhibition applies high-tech methods such as new media and mechanical-electrical integration to lead visitors to learn automobile culture and knowledge through their own experience, uses diversity exhibition and scientific methods including sound, light, image and interactive machinery to meet the need of visitors, then achieves such an ideal of "completely integrating technology museum with museum", and transforms historical value, scientific and technological value and cultural value into social value. The project "exhibition and display system technology development and application" have won the third prize of China Automotive Industry Science and Technology, which represents the highest level of development of China's auto industry. Taking basic exhibition as the source, Beijing auto Museum plans temporary exhibitions in combination with important current affairs nodes and social hotspots to promote Chinese independent brand and new technology, spreads Chinese traditional culture and core values, and plays the role of international cultural exchange via automobile as the intermediary, in Sino-France, Germany, Russia, U.S. cultural exchanges with (Fig. 21).

2.3 Collections Fill up the Blank

There are more than 10,000 pieces of collections in the museum including those collected and repaired vehicles, which are national automobiles from China, Germany, US and so on (Fig. 22). It extends to 6 categories, 21 sub-categories, including vehicle composition, literature and model (Fig. 23~Fig. 27). It includes classic vehicles in the development process of the automobile and some brands with important influence in the development process of the automobile society. The first accreditation and grading organization for vehicle collection are set up in Beijing Auto Museum, and then the idea of "repairing the old as the old" is put forward. Based on this, we completed the reparation of 1976 Shanghai SH760A as the representative of the carrying body, and the 1968 Hongqi CA773 as the representative of the nonload body and the 1987 Beijing BJ212L as the representative of off-road vehicle repair, then achieved such a goal "let the cultural relics live". Beijing Auto Museum makes up the first domestic Standard for Repair Technology, Technology and Acceptance of Vehicle Collection to fill the blank of domestic vehicle collection restoration.

2.4   Automobile Multicultural is Placed in a Wide Range of Social Connection

The cultural exchanges and educational activities of the museum have been constantly being carried out throughout the year, giving full play to the advantages of the industry. Through the integration with the society, it is fully placed in a wide range of social connections. In 2017, we continued innovating ideas, curating thematic exhibitions, jointly developing science projects with educational institutions, planning cultural activities jointly with the media, and jointly promoting public services with the government. Beijing Auto Museum attended the 2nd Sino-French Cultural Forum-The Belt & Road: Cultural Exchange and sharing between China and France and planned the branch forum "Regional Attraction for Museums", taking this opportunity to spread Chinese story and prompt the memorandum signing of sister cities between Fengtai and Lyon. Through planning "the story of Lei Feng - a car soldier" and propagating social positive energy, the project "Lei Feng Preach Volunteer Service" won the gold medal of the third China Youth Volunteer Service contest. We built Beijing New Energy Automobile Display and Experience Base to advocate the harmony and beauty of "man-car-life-society", also pushed out a series of activities about "Green Beijing, Protect the blue sky", and planned a series of thematic exhibitions of "car-city-people", from the angle of car, seeing history, humanity, technology, city changes and social progress, and letting people love the city. In the night of museum, we listened to the sound beauty of the museum and used the sound to resonate and arouse best wishes, inviting the old generation of artists such as Chen Duo to bring a sound feast in the way of the reading. The sound of history echoes in sound waves to call the sound of the future. The sound of the museum is spread farther by the way of cross - boundary dialogue.

2.5   Breakthrough the Traditional Educational Idea and Emphasize "Hands-on" and "Experience"

More than 50 visible, audible, touchable and participable interactive exhibitions have been implemented in the galleries of Beijing Auto Museum to achieve an organic combination of form and content, technology and art, encourage audience participation, and make visitors feel that science is funny and fantastic. Beijing Auto Museum strives to plan a complete set of educational systems to train future automotive social personnel at a different level, in which eight levels should be considered: focus and target, available resources, main audiences, content, time scope, marketing promotion, activity evaluation and training needs. It focuses on the combination of exhibitions and education and carries out the "secondary development" of multi-dimensional exhibits, guided by learning objectives, combined with the realization of cognition and audience cognition, combined with daily life and social concerns. It supplies quality education to the primary and middle school education level of service-oriented, based on the school education, carries out research and practice of the course development and resource utilization of Museum, and explores the establishment of covering "theme visit" and "Museum teaching", "school-based curriculum development" and "the cultivation of innovative talents" four levels of the Museum - school education cooperation mode. Facing the development of service industry in the automotive professional education level, based on the popularization of popular science activities, we set up the platform for innovation and practice, train theoretical and practical comprehensive automotive talents, and provide services and support for the long-term training and construction of Chinese automobile talents. Facing the mass education level and serving the automobile society, China has entered the automobile society. Meanwhile, the automobile civilization needs to be improved. The contradiction between automobile and society, automobile and people is more and more obvious. Traffic congestion, environmental pollution and dangerous driving are becoming more and more serious. As an automobile thematic museum, it is duty bound to participate in the era of automobile civilization construction. The patriotic education level inherits the Chinese spirit. The automobile is not only a technological giant but also a collection of humanistic spirit. It contains a plot that condenses the heart and condenses the nation. In 2015 when the whole nation commemorates the victory of the war of 70th anniversary, in view of our military vehicles, we planned a special exhibition "Cavalry Military Vehicle", in which, with the national spirit of the Anti-Japanese War as a source, the spirit of patriotism is spread, and through the development of own-brand vehicles reflects China's comprehensive strength, the sense of national pride is enhanced.

3. Automobiles: An Intermediary for International Multicultural Exchange

3.1 The Vehicle has Been the Response to Humankind's Dream of Transportation Since Ancient Times, and Beijing Auto Museum is the Carrier of History and Heritage

China is one of the countries to have first invented and made use of the wheel. It is said that 5,000 years ago, the Yellow Emperor invented wheeled vehicles in China, so the Emperor was also called Xuan Yuan Yellow Emperor (Fig. 28). Xuan means an ancient curtained carriage with a high top in the front; Yuan, the longitudinal component of the vehicle, refers to the two front poles that harness the animal. Legend has it that while the Yellow Emperor was in battle against his adversary Chi You, the latter caused the rise of a great fog so as to confuse the Yellow Emperor's forces. However, since the Yellow Emperor had invented the south-pointing chariot, which served as the compass, his forces did not lose their sense of direction and were able to claim the victory. Later developments to the south-pointing chariot led to the invention of the drum carriage: with two wooden figures inside the carriage, a drum would be beaten once for each li (about 0.5 km) of distance that had been covered. These two types of vehicles were always part of the Emperor's processions, and they were also used for drawing maps and measuring the area of land.

Over 2,500 years ago, the great ideologist and educator Confucius made a tour of numerous kingdoms in a carriage, spreading Chinese culture and ideas (Fig. 29). He once stated, "A kingdom with a thousand chariots must rigorously and sincerely carry out major affairs of state while faithfully honoring its commitments; it must be economic in terms of spending, take proper care of its officials, and must be mindful of the agricultural season when seeking work from the peasants." These were Confucius' ideas on how the kingdoms of dukes and princes should be governed, and such ideas are still seen as important in modern governance. As can be seen, carts and horses in ancient times were of utmost importance in daily life socially, economically, and militarily.

2,200 years ago, China's first emperor, the First Emperor of Qin, united all of China under one rule for the first time (Fig. 30). He standardized certain aspects of the tracks and vehicles throughout the empire, including the distance between carriage wheels, the size of wheels, the width of roads, the clear division between parts of the road dedicated to vehicles or pedestrians, and the linking of all the roads in the cities, thus creating an expansive and well-ordered transportation network. Such a measure made travel much more convenient, in turn spurring economic development. The standardization of vehicles, written language, and measurements during this time initiated China's development in the aspects of organization and order, which benefitted mass production, and it served as a vivid and pragmatic measure in bringing about standardization in a broader sense to ancient China. In short, the car, in all its forms throughout history, has been an important witness of China's contribution to the world and serves as a record of the evolution of mobility in human civilization.

3.2 The Automobile has Been a Bridge Between East and West, and Beijing Auto Museum is the Carrier of Creativity and Friendship

Beijing Auto Museum preserves records not only of the admirable creativity of the Chinese civilization but also great contributions to humankind around the world. The museum includes a model of the world's first motor vehicle, which was built by the French artillery engineer Nicolas Cugnot in 1769. His steam car recalls a time when humankind was embarking upon a new era of vehicular transportation, that is, one without the need for horses. Excitement is often seen in the face of French visitors upon seeing an invention from France in our museum. They sometimes release a sigh of satisfaction upon seeing the respect in Beijing for French innovation and creativity, and it serves as a source of surprise and pride for them.

Beijing Auto Museum also documents the story of the world's first motorcar race that spanned Asia and Europe in 1907. Beginning in Beijing (known as "Peking" at the time) and finishing in Paris, it was a legendary feat for these two continents brought about by a joint Sino-French effort. The model of the route at the museum has touched the hearts of thousands of visitors, and according to statistics, nearly 90% of those who have seen it originally had known very little about this bit of history. For my part, life has become very closely linked to France. In 2007, it was the centennial anniversary of the Peking to Paris motorcar race. In commemoration of this first-ever motorcar race connecting East and West, as well as to publicize awareness for the fight against Alzheimer's disease, a Citroën 5HP built in 1925 was driven by Elisabeth Pete and Fabien Hamm of Des Route et Des Hommes along that same 1907 route. It took one and a half months for them to travel the route of 16,000 kilometers that passed through nine countries. And thus, through the mutual cooperation between people, the car, and society, this magnificent feat was once again accomplished. In the end, that same car was bought back by Groupe PSA and donated to Beijing Auto Museum to be part of its permanent collection, making it the first foreign automobile donated to the museum. The car is now on permanent exhibition at the museum, and it stands as a witness to the friendship between the Chinese and French people.

3.3 The Automobile has Opened a Door Between China and France, and Beijing Auto Museum is the Carrier of the Cultures and Pursuits of China and France

Using the automobile as a tool to promote international cultural exchanges, Beijing Auto

Museum serves as a bridge for car culture exchanges between China and the whole world. In response to this calling, the Sino-French car culture exchange is working toward creating an indissoluble bond between Beijing, Paris and France. 2014 marked the 50th anniversary of the establishment of diplomatic relations between China and France, and on that day worthy of commemoration by the people of both nations, cooperation between the two nations was again initiated in the area of car culture exchange. The Chinese people welcomed the Mulhouse-based MuséeNational de l'Automobile as well as the cultural exchange series of events planned out by the MuséeNational de l' Automobile. The exhibition "French Automobile Body Styles of 1891–1968: Imagery of Artistic and Professional Achievements" furthered academic exchange between the two museums in the area of antique car restoration techniques. The MuséeNational de l' Automobile, the most well-reputed of European car museums, allowed the people of China to witness the French people's boundless passion for art, the pursuit of perfection in technology, and sense of responsibility and mission toward the guardianship of cultural achievements.

Both museums are cooperating in this series of exchange events based on the theme of "Art, Technology, and Professional Achievements". In 2014, the exhibition "Have a Ride in Hongqi: The Chinese Automobile History Since 1949" was held in France, bringing the Chinese car culture to another country for the first time ever. Moreover, the admittance of a Chinese Hongqi car to the MuséeNational de l' Automobile collection is the first instance of an Asian car being accepted into the museum, and it is also the first Chinese-produced vehicle to be placed in a collection in Europe, the land of auto industry roots. In telling the story of the origins of the Hongqi brand, Chinese cultural elements integral to the brand were related to telling visitors the story of hard work and innovation that has gone into the brand.

On the heels of the 50th anniversary of the establishment of diplomatic relations between China and France, the Sino-French car culture exhibition "Indissoluble Bond between Beijing and Paris" held in 2016 allowed visitors to look back upon car culture exchanges between the two countries and gain an understanding of the indissoluble bond between France and China, their cities, their cars, and their people. The persistence of the connection between the Chinese and French car cultures is also due in large part to the efforts of former Chinese Ambassador to France Wu Jianmin. Ambassador Wu stated, "Exchange must be established through a common language in order to touch people, and here, the car has indeed become a vehicle for our dialogue." This exchange was a key to finding solutions to certain questions and became resolute that cars can indeed be used to promote exchange and communication between people, cities, and even nations.

3.4 The Automobile has Connected our Nations, and Beijing Auto Museum is the Carrier of the Cultures and Integration of China and France

Museums have been honorably described as "a book with the power to open up the world of the human spirit." History has been the object of examination for the museum, but even more important has been the inspiration for the mind. Since its opening in 2011, Beijing Auto Museum has been open all year, provided services to five million visitors, and held over 500 cultural and educational events. We believe the value of a museum lies not in how grand or modern it is or how many exhibition pieces are in its collection but in the concepts it holds to, its vision, its values, the inspiration visitors obtain, and the spirit and culture they experience.

The diversity of the car culture has allowed for variance in the way the museum presents culture. Cultural exchange is the foundation for the Sino-French cooperation. By connecting the Chinese and French people's hearts through cultural exchange, their relationship has been nourished and injected with vitality. Besides the use of cars as a way of bringing these two nations together, flowers have also played a similar role. Before the French flower festival Fête du Muguet in April of this year, Beijing Auto Museum held a lecture entitled Art and Fragrance, in which the culture and history of Guerlain were introduced. Coincidentally, that was also the time of China's traditional Flower Festival. In this way, through the use of cars, we promoted the use of flowers as another way to forge a link between our countries. Thus, at the time of the French Fête du Muguet, the Chinese Flower Festival was held in Huaxiang (which translates as "land of the flower") of Beijing's Fengtai District. Huaxiang, known for 800 years for its flowers and which happens to be the location of Beijing Auto Museum, became the site of integration between the local flower culture and car culture, enhancing the magnetism between the unique and rich cultures of China and France and allowing for the sincere appreciation of certain quintessential components of each other's cultures. Those at the event made flower arrangements with the guidance of a flower art expert, heard Mr. Boillot, president and CEO of the French cosmetics company Guerlain, speak about Western concepts of fragrance, enjoyed the delicate taste and fragrance of Tang dynasty-style pan-fried tea, were treated to flower cakes from the time-honored Gongyifu brand, and experienced a number of Flower Festival traditions. Linking East and West and creating the dialogue between China and France in such way allowed all present to experience the charm of the flower cultures of both countries. Upon seeing this, museums must pursue diversity and the understanding and friendship between two nations at the national, municipal, and personal levels can be furthered through such efforts.

The cultural exchanges between China and France are founded upon a mutual trust. By sharing a common dream and pursuit, our two nations have created cars and related products of unprecedented quality that are full of the poetic creativity of the imagination, such creativity being a trait shared by the Chinese and French people. As a witness to the friendship between the Chinese and French and as a vehicle to carry on our cultures, in turn promoting cultural event cooperation and exchanges, Beijing Auto Museum has made use of the car in actively promoting Sino-French car culture exchanges to bring about dialogue and strengthen the bridge of communication between Beijing, Strasbourg, and Lyon. During the 2nd Sino-French cultural forum in 2017, Fengtai and Lyon signed a memorandum of sister city. By doing so, Beijing Auto Museum will be able to promote more exchanges and cooperation between our countries and cities in the realms of culture, education, tourism, and the young generation (Fig. 31).

3.5　China and France are Working Together in Moving Toward the Beautiful and Far-Sighted Goal of Integrating People, Cars, and Society

Christine Carol, the founder of Yishu 8, an art philosopher, and a council member at the MuséeRodin, for long devoting herself to Sino-French cultural and art exchanges. Yishu 8 is located in the former Sino-French University in Beijing. Ms. Cayol has said, "This is the story of a dream. A unique place that has, in the end, turned my dream into reality. From my youth, I had dreamed of having a house, friendly and open, where artists and all kinds of people may feel at home. This place exists for the sake of creativity, innovation, passing on legacies, and facilitating exchanges…. If a bridge is to be built between China and France, then we must personally walk this bridge…. Everything I have done in China has met my expectations. I have achieved my Chinese dream. Yishu 8 is just that." With China's The Belt and Road Initiative, we will jointly build a common destiny for humankind. The dream we all hold to is the realization of the utmost of both the Chinese Dream and the World Dream. We look forward to continuing working together toward the Sino-French Dream and the World Dream! We also look forward to the continuation of cars carrying people and people carrying ideas. All of these things will work together in propelling our progress toward the beautiful and far-sighted goal of integrating people, cars, and society (Fig. 32).

People are constrained by nationality while cars are not, which provides a broad space for the international exchange via the automobile culture. In addition to developing cultural exchanges between China and France in 2016, during the International Museum Day and China Tourism Day, by the China tourism year opportunity, Beijing Auto Museum also planned a "life on wheel – Sino-American Auto Culture Photographic exhibition", via 150 paintings collection of photographic works from the 5 international photography artists, let audience observe the USA, a country on wheels and review of car culture in each period of the development trajectory of the United States and also interpret car culture of the global largest auto market via life on the wheel. Besides, in 2015, the Night at the Auto Museum, a public service project held in the museum plaza outside the regular business hours, brought together the social resources and introduced Urban Science Festival for the first time. Surprisingly, all the science programs drew roomfuls of the audience, especially those live experiment demos presented by the guest scientists from the United States and the UK. Cars and science are inseparable in that automotive invention employ many scientific principles; science is also an international language. These provide a strong theoretical foundation and a broad horizon for an auto museum to pursue international cultural exchange through automobiles. In the future, by employing cars as a medium and a communication tool, we can organize more cultural exchange events on similar commemorative occasions between China and Germany, Italy, UK and Korea, and let China and the world use automobile culture as an exchange language to develop a dialogue.

4. Mutual Sharing of Resources in Science and Technology Should be Carried out to Solve the Contradiction Between Unbalanced and Inadequate Development and the People's Ever-growing Needs for abetter life

To further enhance the capacity of exhibition and education, collection research, leisure tourism and international exchanges of Beijing Auto Museum, and expand the coverage and influence of cultural dissemination and scientific popularization, Beijing Auto Museum compiles the "Beijing Auto Museum 13th Five-Year Development Planning", in order to fully implement the fundamental task of rich spiritual and cultural life of the people, construct "human-vehicle-society" culture system, promote the construction of contemporary museum education system, improve the management level of the museum, provide high quality service for the audience, provide new cultural forces for sustainable development, and run a good cultural cause with the people's satisfaction. With the aim of "running museum in an open doorway, Integration into society" to determine the development concept and goals, Beijing Auto Museum strives to become a demonstration platform of standardization for industry services, a public service platform to highlight the quality of the city, a platform for integration of science, education, culture and tourism, a popular science platform for the dissemination of automobile science and civilization, and a platform for the communication and dissemination of the world's automobile culture. Beijing Auto Museum will undertake the mission and responsibility in displaying the automobile culture and representing China to exchange with the world and becoming the leader of automobile culture and science popularization, and then promote the harmony and development of "human-vehicle-life-society". With a clearly-defined vision to provide "first-class service, top-scale exhibition and top-notch efficiency", Beijing Auto Museum devotes to become a China Auto Museum of the world's advanced level with a value system centering on "legacy, innovation, cooperation, excellence, responsibility". Combining the work of the museum, and based on solving the contradiction between the growing needs of the people and the inadequately balanced development, the strategy direction can be tried in three aspects: ① Based on the carrier of exhibition, exhibits for the use of "diversified development", select the carrier of shared resources, fully apply "Internet plus" thinking, and use new media means to break the boundaries. ② "Secondary development" based on contents is widely placed in the social needs. It studies the orientation of school curriculum, studies the core value orientation of the society, and breaks through the boundaries of the single discipline. ③ Based on personal training, carrying out the planning of "horizontal and vertical development": the commentator to the researcher training, the science popularization staff to the science instructor training, breaking through the limit of the single line.

5. Conclusion

Every paper must have a conclusion section to restate the major findings and suggest further study.

Generally, according to interaction with the society and connection between countries, the museum is to disseminate technological and cultural ideas, to tell the stories of world evolution and Chinese civilization by means of scientific and technological presentations, and to narrate the story of the cars, of the human race. As an important hall for protecting and inheriting human civilization, museums are bridges connecting the past, the present and the future. They play a special role in promoting the world's multi-cultural exchanges and mutual learning. Based on the model of the Sino-French car culture exchange, Beijing Auto Museum will work to promote more international car culture exchanges, relay the legacy of the spirit of the Silk Road, and give new significance to this new era of cooperation, thus providing momentum for contributions toward achieving the purpose of the Belt and Road.

References

[1] Zheng Yi, Education Activity Research of Museums[M]. Shanghai: Fudan University Press, 2016: 138.

[2] Ge Yuchun, Strengthening the effectiveness of the educational function of the exhibits through the "secondary development"[J], research from Museum of Natural Sciences, 2017(2): 42–43.

Science Popularization Method of Beijing Nanhaizi Elk Park Museum

Bai Jiade[1]

Abstract With the ecological ethics education and environmental education as the main line, through creating the popular science creativity carriers, expanding the route of popular science communication and enriching the popular science education activities, Beijing Nanhaizi Elk Park Museum benefit the building of "Humanities Beijing" and has received wide acclaim from visitors.

Keywords Popular Science Creativity   Route of Communication   Mode of Activity

Beijing Nanhaizi Elk Park Museum (Elk Park) is the first outdoor ecological museum in Beijing, undertaking the popular science publicity and educations in ecological knowledge, ecological awareness and ecological culture. On the basis of the construction of ecological civilization and ecological moral education, by integrating the science popularization resources, through innovation of popular science carriers, expanding the route of popular science education and perfecting the model of popular science education activities, the Elk Park has formed a series of unique and thought-provoking popular science education projects, which have achieved good results in popular science education.

1. Research on Interactive Popular Science Facilities, Innovate Popular Science Carriers

The popular science facilities not only spread scientific knowledge and scientific ideas to the public but also promote the improvement of visitors' scientific literacy through experiencing

① Bai Jiade: Director of the Beijing Nanhaizi Elk Park Museum. Email: baijiade234@aliyun.com.

the scientific and technological achievements. Elk Park focuses on the unique ecological museum building in the world. By making use of the natural and ecological advantages in the park, it has continuously developed the interactive popular science facilities, which have formed 5 themed exhibition areas including "Elk Return Memorial Garden" "Wildlife Homeland", "Ecological Civilization Garden" "Customs and Culture Garden" and "Low Carbon Science Garden".

There are several popular science facilities in the Elk Return Memorial Garden, including the science popularization plank road, science popularization walls, elk cultural walls, seats with Tang poetry, elk scientific discovery memorial stele, Qianlong hunting figure relief, Duck of Bedford sculpture, stone-made elk antlers, elk legend exhibition and elk memorabilia science column. Those facilities in the Elk Return Memorial Garden not only supports safe and comfortable environments to visitors but also affords the scientific knowledge to the public. The biodiversity billboards are arranged on both sides of the plank road permanently; the ingenious drawings describing elks are painted on the scientific walls, and the verses recording elks are written on the culture walls and seats. In this way, visitors can feel the elk's cultural glamour through these facilities. By taking time as the main line, the elk scientific discovery memorial stele, Qianlong hunting figure relief, Duck of Bedford sculpture, stone-made elk antlers, and elk legend exhibition have recorded the pictures that the elk was scientifically discovered, extinct in China, was transferred into Europe, and returned back to China, which tell an elk's countless frustrations to people and also confirm that only a prosperous country can bring prosperous elk population.

The facilities in the wildlife homeland combine the cute animal modeling to highlight the "childlike". The sculpture of shark fin and bird's nest is warning people to avoid consumption of them as the food, which will hurt both of nature and human health. The bird migration globe, honeycomb, gecko climbing wall, wolf hole and birdcage provide a playing area for children and let them understand the close relationship between animals and humans. "Three wise monkeys" is telling children the Chinese traditional culture: "See no evil, hear no evil, and speak no evil".

Through the construction of the world elk small pavilion, wetland bird watching platform, Olympic mascot animal sculpture square, elk silhouette style Chinese traditional nature protection poems and paintings and extinct animal cemeteries, the elk park has formed the ecological civilization garden. The elk silhouette sculptures imitate elk's movement of running, jumping and eating. There are nature protection poems and paintings on the side of elk silhouette sculptures such as "A younger crow feeding the old one" and "A lamb kneels to suck its mother's milk", which try to show the animals' moving stories and advise people to protect animals and nature. The extinct animals' tombstone, arranged in the form of domino, records the extinct animals' name and extinction time. Some tombstones are falling down, trying to impress the endangered species' plight to visitors.

The Elk Park has formed a customs and culture garden by means of construction of the Nanyou autumnal scenery stone, "Welcome" stone, a pair of stone couplet on watching elk platform, Chinese zodiac bronze statues and cultural bridges. Chinese zodiac bronze statues are the classic interactive popular science facilities, made of bronze, which imitate the bronze heads depicting the 12 animals of the Chinese zodiac in the Old Summer Palace. These animal sculptures take on vivid shapes, and there are 12 different musical instruments (used in a Buddhist or Taoist mass) in their hands. The cement base to support bronze statues was written these animals' routines that can make people learn the Chinese traditional culture when they visit.

The low carbon science garden is an inimitable themed exhibition area, where environmental protection motto stone seats, carbon and ecological footprint pathways, and low carbon lifestyle labyrinth can be found.

The low carbon lifestyle labyrinth uses cypress as the fence and establishes little stone (with a different lifestyle on it) on the pathway. For instance, you can find such the sentence: "If you use the throwaway chopsticks, it would destroy a forest. So, you can't pass". Such edutainment attracts kids to grope the correct way and deepen the concept of low-carbon life. Visitors can calculate the carbon emission by different trip mode and lifestyle, and can also figure out the forest resources for consumption of the carbon through turning the sundial's stone dish.

The popular science facilities are very popular with the visitors, and the bird migration globe and the elk scientific discovery memorial stele were granted the patent by the state Patent Office. These well designed, distinct theme and creative popular science facilities impressed visitors and laid a strong foundation for popular science education.

2. Focus on the Popular Science Writing, Expand the Popular Science Mode

Popular science writing is a bridge to popularize science knowledge and scientific achievements, which must be based on the scientific research. Elk Park focuses on the popular science work creation and makes efforts for the spread. The staff of Elk Park is writing popular science books, designing the exhibitions, telling the elk's story to visitors freely, spreading the popular science on the Internet, and building the brand of the popularization of the science of Elk Park. More than 20 popular science books written by the staff of Elk Park have been published in recent years, including *Elk and Elk Park, Elk Park, Earth Ethics*, etc. More than 100 papers written by the staff have been published in the newspapers, and are favored by readers. The leaders of the Park were also engaged in the popular science work and went to many places around the country to give various series high-level speeches.

The Park has realized that the Internet would be an important platform for the popular science. Thus, we integrate kinds of popular science resources, to set up the special column of Elk Park's stories, *Impact of Elk Park, Stunning Photographs of Elk Park and Stereograph of the Park*. Citizens can read lyric prose that describes the landscape of the Park, watch the beautiful pictures that show the biodiversity of the Park, and watch the videos that introduce elks through the special column. The Park puts up the themed exhibitions about the extinct animals, biodiversity and legend of elks, during the month of technology, the week of technology, and the week of peace. These exhibitions show the relationship between human and nature, the meaning of biodiversity, the history and biologic characteristics of elks. People intensify the biodiversity & nature conservation ideas through learning the exhibitions. These exhibitions have obtained the "Popular Science Day Award" granted by China Association for Science and Technology. Visitors can understand the popular science facilities well through the free explanations by the staff. The announcer's emotional explanations moved visitors and made people realize the importance of harmony and unity for people and nature and experience the profound influence of the environmental protection for future generations.

3. Carry out Colorful Popular Science Activities; Improve the Way of Science Popularization

For providing a better service to the citizens, Elk Park enriches popular science activities, widens the popular science education idea, and forms unique scientific resources, which attract the public and improve their scientific literacy. Elk Park works with communities and schools nearby to carry out the popular science activities. For example, staff went into Yinghai First Primary School to show the "Animal Extinction" exhibition; set up an environment protection interest group for Sihezhuang Primary School; carried out the "International Biodiversity Day" knowledge contest for Tiangongyuan Community.

Elk Park cooperated with the media actively, and through the media, publicized the popular science: with China Weather Network, jointly organized the "Protect the Elk's Wetland" activity; with Sina Network, jointly organized the parent-child activity; with CCTV, shot the film *Find Elk*; with BTV, shot the film *Nanhaizi Elk* Park. People are interested to learn scientific knowledge through these films, and the Elk Park improves the popularity at the same time.

Elk Park also takes the World Earth Day and the World Environment Day as an opportunity to carry out the educational activities of the juvenile ecological ethics. With "Friend of Nature", Elk Park co-organized the "Memory of the Extinct Animals and Watching Birds" activity. The popular science drama "United Nations of Animals" and "Summer of Elk Park" took part in the "National Popular Science Day - Beijing Home Campaign 2012". Elk Park also used the social big classroom and other platforms to carry out large-scale public welfare activities and launched "Biodiversity Digital Photo Exhibition" "Elk's Vicissitudes of Life" "Primates' Popular Science Exhibition" and other minor special exhibitions.

Working with international environmental groups, Elk Park expands its influence through the development of popular science activities. In recent years, Elk Park, in cooperation with "Earth Education Institute (USA)".

In general, Elk Park aims at building the unique ecological museum of the world, through improving the popular science facilities, expanding the popular science mode, and carrying out colorful popular science activities, receives more than 40,000 person-times on average every year, got Beijing "Social Classroom Advanced Award" and "International Week of Science and Peace Contribution Award", and was named the "Beijing Outstanding Popular Science Education Base".

Research on Exhibition and Popularization of Science of Tianjin Natural History Museum

Huang Keli [1]

Abstract This paper focused on the development of exhibition and popularization of the science of Tianjin Natural History Museum. Based on collection resources, Tianjin Natural History Museum associated the national policy with the exhibition and expanded the connotation of the exhibition. Tianjin Natural History Museum also concentrated on the popularization of science and constructed a system to develop the popularization of science. Tianjin Natural History Museum explored a mode that cooperated with the primary school to promote the effect of science popularization.

Keywords Natural history museum exhibition the popularization of science education

The predecessor of Tianjin Natural History Museum is Musée Hoangho Paiho (HHPH), which was founded by Émile Licent, a French Catholic priest and a natural historian, in 1914. From 1957, the name "Tianjin Natural History Museum" was used. From 1997 to 1998, the expansion of Tianjin Natural History Museum completed at the Machang Road site. It was moved once again in 2013, and at the beginning of 2014, the new Tianjin Natural History Museum was officially opened to the public.

The new Tianjin Natural History Museum is located in the Cultural Center of Tianjin, covering an area of 57 thousand square meters, with a total building area of 35 thousand square meters. By taking "HOME" as the main theme, the permanent exhibition consists of "HOME-exploration", "HOME-life" and "HOME-ecology" three topics and twelve plates, including popular science theater, natural science classroom, science popularization

---

① Huang Keli: Director of the Tianjin Natural History Museum. Email: 764715304@qq.com.

activity area, living butterfly garden, dinosaur dig valley, 4D cinema and other functional areas. A comprehensive natural history museum has been formed, integrating collection and research, display and experience, cultural exchanges and science education, cultural tourism and leisure.

Tianjin Natural History Museum was recognized as the first batch of the national first-class museum in 2008 by the State Bureau of Cultural Relics. In 2013, it was awarded the first batch of the social practice base of science popularization and education for elementary and middle schools in China, with average audience number of nearly 2 million per year. Tianjin Natural History Museum has always adhered to the concept of public cultural service based on audience and culture benefiting people, creating fine display and excellent science education activities, providing high-quality Museum cultural experience for audiences.

1. Exhibition Features and Development Status of Tianjin Natural History Museum

1.1 Permanent Exhibition: "HOME"

Tianjin Natural History Museum is the only natural history museum in the world with the theme of "HOME".

"HOME-life" exhibition takes the origin and evolution of life as the main line, and the events in evolution as the clue, the evolution time as the order, the combination of ancient and modern as the performance characteristics, showing the earth "home" from the different perspectives of major events in the history of hundreds of millions of years of life. Moreover, it tells the "home" story by using the narrative techniques and shows the ups and downs of magnificent lives on earth "home" by the exploration and discovery from the perspective of the new concept of the international natural history museums.

"HOME-ecology" exhibition is based on more than 200 pieces of rare wild animal specimens donated by the famous American philanthropist Mr. Kenneth Behring, using large landscape, generous form, artificial landscaping and background painting with various modern display methods, to show the real scene of vivid animal life, and reproduce the representative world wild animals and their ecological environments.

1.2   Temporary Exhibition: Magnificent Natural Scenery of Silk Road

At the end of 2016, in response to the national "Belt and Road"(B&R) Initiative, Tianjin Natural History Museum launched a temporary theme exhibition "Magnificent Natural Scenery of Silk Road". The exhibition area is about 600 square meters. With the presentation order of thematic units and the order of geographical location, we take the important cities on the ancient Silk Road as nodes and show the natural scenery and natural heritage of the Silk Road from the natural and historical perspective. Each node of the ancient Silk Road is connected by the camel elements. In the end, it shows the great future image of the B&R Initiative in the new period by using high-speed rail, aircraft, network and other modern traffic and communication tools in the modern silk road.

The exhibition theme is to show the ecological countries along the Silk Road and the city's natural and natural heritage, with many popular science activities as the theme of "the nature on Silk Road ", aimed at promoting social public concern and showing the significance of the B&R Initiative by providing the experience for the public.

1.3   Musée Hoangho Paiho (HHPH) Reopening

HHPH, located in Tianjin Foreign Studies University, is the predecessor of Tianjin Natural History Museum, which, famous all over the world in the 1,920s to 1,930s, had been in a long dormancy since the beginning of 1,940s. In 2014, HHPH launched a comprehensive renovation and restoration work of the north building, which was completed and reopened to the public in January 2016.

The new exhibition of HHPH is based on the reparation and restoration, by repairing the old as before, inheritance, development, and innovation as the core concept in design, and consists of three parts: restoration display, cultural history, and open warehouse, displaying nearly 20 thousand pieces of natural specimens and cultural relics, having won the title of Annual National Top Ten Exhibitions 2016.

"Restore" is the core of the restoration display of HHPH, and beyond it. In the "restoration", kit keeps improvement in design and exhibition arrangement, with a plenty of comparisons of the image and text data of the times E. Licent, faithfully reproducing the building pattern, style, and exhibits details of HPHH a hundred years ago. Every excellent collection is showed out in this reopened exhibition so that the public can enjoy the fun of scientific discovery and the spirit of scientific exploration from the rich and layered exhibition.

Furthermore, a special exhibition was added to tell the history of HHPH, with more than

20 thousand pieces of cultural collections showing out, and the organic integration between relics and literature resources, causing people to think the opportunities and enlightenment for the scientific discovery and the development of museums in modern China by the western rational scientific spirit spread to the East, giving full play to value of cultural relics and historical literature in the interpretation of history and scientific research.

In addition, reopened HHPH opens the specimen warehouse and research area to the public in an all-around way and strives to explore the intrinsic value of specimens and cultural relics, and fully guides the ideology of "Let the cultural relics live" throughout the practice of science popularization. While making the public close to the exhibits, it also presents the story of the museum's work and the story behind the collection, so as to achieve the purpose of "visiting, experiencing and thinking".

2. Development Status of Science Popularization Activity in Tianjin Natural History Museum

In the opening ceremony of the Nineteenth National Congress of the Communist Party of China, the General Secretary Xi Jinping pointed out that we should strengthen cultural confidence, promote the prosperity of socialist culture, promote the development of cultural undertakings, improve the public cultural service system, and put forward new tasks and new requirements for cultural workers. As the national first-class museum and the national science education base, Tianjin Natural History Museum always adhere to expand the depth and breadth of the education service function, carry out the relevant requirements about museum education service in the "Museum Regulations", take the science education as a priority among priorities, and realize the normalization, specialization and diversification of educational activities. Since the opening of the new museum in early 2014, Tianjin Natural History Museum has received nearly 2 million people including more than 600 thousand young audiences per year. There are more than 1,000 public tour guide services per year and more than 300 kinds of popular science education activities with various forms.

2.1 Creating the Unique Experience in the Theme Exhibition

The collection is the foundation of all the work of the museum. The exhibition is the important the link between collection resources and educational activities. Therefore, every year, our museum develops a variety of theme exhibitions based on the collection resources, such as "Golden Monkey Spring" "Rich South China Sea" "Magnificent Natural Scenery of Silk Road", and on this basis, a series of thematic science popularization activities are carried out, such as "Monkey Spring Festival special events" "Sea Flower - the South China Sea exhibition special activities" and "Nature on the Silk Road".

At the same time, we actively organize outreach exhibitions, such as "Flying Flowers - Exhibition of world rare butterflies" with the Hebei Museum and Tianshui Museum, innovatively linked live butterflies flying and science drama into the exhibition, broaden the cooperation way, form and rich connotation of popular science education activities, and harvest good social repercussions.

2.2 Building a Perfect System of Exploring Activities

Our museum has more than 10 excellent popular science activities, such as "Yutian" lessons, living show, winter and summer camps of popular science, popular science drama, a night at the museum, etc. The annual average number of science popularization activities is over 300. While focusing on the dissemination of knowledge, it also incorporates interest, novelty and artistry, especially the development of interaction, and makes full use of museum resources to open up a scientific camp for the public, especially teenagers.

The Natural Science Lecture is scheduled every week, according to the natural science, hot topics, and school curriculum. The well-known experts are invited to carry out the rich and colorful lecture in the museum. Popular science drama, in a more vivid and intuitive way, spreading scientific knowledge, environmental education and other ideas, has performed more than 200 times every year, especially popular among young audiences. "Yutian" lessons, living show, and other activities provide a more intuitive participation experience for the audience. The various forms of science education activities complement each other and jointly build a distinctive quality event system.

Particularly worth mentioning is that in 2016, our museum launched the "Night at the Museum: sleeping with Dinosaurs" series of science activities, creating a precedent for the overnight museum, combining natural science, field exploration, parent-child activities and experience and exploration. By breaking the traditional promotion mode, using Internet+ communication function, it innovatively added online live links, to achieve online and offline interaction. Although only 40 participants in the museum, there are tens of thousands of interactive personnel online, which greatly broadens the audience and opens up new channels for museum science education activities.

2.3 Establishing a Long-term Mechanism of Museum-School Collaboration

By combining the advantages of collection and expert team of Tianjin Natural History Museum, selecting some primary and secondary schools as a pilot project, giving full play to the social function of museum as the "second classroom" of school education, we plan to build our museum as a practice base for primary and secondary school students. A natural science classroom was set up, including the experience area, interaction area, book area, exploration area and specimen exhibition area, to form a unique multiple experience mode. Furthermore, our science education team combines the characteristics of the new museum exhibition, adopts the model of the union between school teachers and the science popularization personnel of the museum, and aims at the study, development and design of school-based curriculum for the primary and secondary school needs. Since 2016, by the carrying out of Museum-School Collaboration, over 12 schools moved classes into the museum, such as Tianjin Nankai Science and Technology Experimental Primary School, Endeli Primary School and so on. The students regularly experience and explore in the museum, learning in practice, and feel the charm of nature and science.

Out of the museum, the science popularization team of Tianjin Natural History Museum has entered more than 40 schools, according to the characteristics of students' age, making the learning manuals of dinosaurs, birds, butterflies, and ancient human, and science course, science exhibition, model making, science drama, handmade, interactive games and other forms of activities into the schools. Through the "Museum-School Collaboration", a positive guidance for young students has been given. A series of courses have been designed and developed to supplement and perfect the knowledge system of young people under the premise of ensuring the original content, fully exerting the advantages of both sides and achieving win-win results.

Specimens are the foundation of all the work of the natural history museum, and education is the soul. Tianjin Natural History Museum always puts popular science education in the first place, further implements the development concept of "innovation, harmony, green, opening, and sharing", and devotes to improving the public scientific quality. We will take B&R Initiative as an opportunity to strengthen exchanges and cooperation with the museums of the countries and regions along the line, and constantly improve the science education and the social-cultural services.

Cozumel Planetarium: A Tourist Experience

Abstract Cozumel Planetarium belongs to one out of 4 nodes of the Planetarium Network of the state of Quintana Roo. It is located on an island in the Mexican Caribbean, whose main economy is based on tourism. It receives hundreds of people per year who want to enjoy its turquoise blue beaches and the great biodiversity that make Cozumel a special destination to enjoy. Within this context, the Cozumel Planetarium has among its objectives not only to be a place of interest for the local community, but also for foreigners. For this purpose, it has carried out a series of strategies that make the Cozumel Planetarium not only an educational landmark within the island, but also a tourist point to enjoy.

Keywords Planetarium Fulldome Science Workshops Tourism

Introduction

Out of the almost 40 existing planetariums in Mexico, four of them are in the Quintana Roo state, thus being the state with the biggest number of planetariums in México and having more than some countries, such as Chile, Belgium, Denmark among others. The Science and Technology Council of Quintana Roo has envisioned the planetariums as a space for divulgation of science, technology and the culture of the region, making a special emphasis on astronomy, biodiversity of the zone and the Maya legacy.

Cozumel is an island in the Caribbean Sea off the eastern coast of Mexico's Yucatán Peninsula, opposite Playa del Carmen. The economy of Cozumel is based on tourism, being the

---

① Milagros Varguez: Staff of the Cozumel Planetarium. Email: mvarguez@frutosdigitales.com.

main cruise destination in Latin America. Beside white sand and turquoise blue waters, one of the big attractions of Cozumel Island is its planetarium. Cozumel Planetarium, Cha'an Ka'an, is a space that favors the gathering of people, idea exchange and the development of several activities for the enjoyment of both, national and foreign visitors in the environment that favors knowledge.

1. Background

1.1   History and Origin of Cozumel Planetarium

Cozumel is the third biggest island and the second most populated of Mexico measures approximately 48 km measured from north to south and 14.8 km as measured from east to west and is located to 20 km of the Quintana Roo coastline in front of Playa del Carmen. Cozumel Island does not only count with blue crystalline water and white sand covering its beaches but also has the second largest reef barrier in the world, only below Australia. For this reason, it is acknowledged in the whole world as a first-class destination for scuba dive and snorkeling. It receives annually around 3 million visitors only from the cruise ships, making the island one hot spot of tourism in Mexico. But it is not all about tourism in Cozumel, we also spread science.

Nowadays, the work of dissemination of scientific knowledge has become in a focal point of attention, reason for us to find a permanent communication of the great public of the advances in science, technology and innovation, as well as to attract the youth to the scientific work in a recreational way, with the goal of establishing a permanent link to these activities and strengthen their skills and vocations. On August 20th, 2015, Cozumel Planetarium, Cha'an Ka'an, opened its doors to the public for the first time, with the goal of being a bridge between science and the Cozumel island people (Fig. 33).

The Planetarium of Cozumel is an effort that goes beyond its own conception and role played in the creation of the Planetarium Network of Quintana Roo. The Planetarium of Cozumel, Cha'an Ka'an, is the third node, preceded by *York' ol Kaab* in Chetumal, built in 2011, and the *Ka' Yok'* in Cancun in 2013. The fourth and last node, located in the city of Playa del Carmen, opened its doors in December 2015. Generally speaking, these planetariums share many similitudes in their construction and the areas of work; nevertheless, each has a different theme that gives an identity to each complex. The name of the Cozumel Planetarium, Cha'an Ka'an, was chosen by the community and means in Maya, "to observe the sky".

1.2 Cha'an Ka'an Planetarium Description Amount Spent

The planetarium of Cozumel was made with an investment of almost 3 million dollars of the National Science and Technology Council of México (CONACYT). The land of 10,000 m^2 located close to the Puerta Maya dock, the one that receives the biggest cruise ships, was donated by the State Government of Quintana Roo.

The planetarium was built and equipped as a playful, friendly place for Cozumel habitats and as another attractive feature of the island. This space is meant to endorse wholesome activities, exchanges of ideas and development of several activities for the enjoyment of visitors, in an environment that endorses scientific knowledge and innovation, as well as the preservation of the Maya culture legacy and biodiversity of Quintana Roo state.

Cha'an Ka'an has some features that differentiate it from the rest of planetariums not only in the state but also in all of Latin America. Stated not only by its theme rooted in Mayan Archaeoastronomy, biodiversity of the island and ancient navigation, that are within its museum, but it also has the only immersion full dome on 3D. Having a total of 95 seats, the planetarium has six projectors working entangled to show the presentations on 2 and 3 dimensions. The public has several available movies to choose from in topics as astronomy, care of the environment, culture, cutting-edge technology and discoveries, etc. Beside this great attraction, it has an observatory; two didactic rooms equipped for workshops, an auditory as a dedicated space for researchers to share some of their scientific work with the community, and a virtual reality room. It also has a didactic room of water and a nature interpretation center, both focused on addressing issues related to the conservation of our natural resources.

2. Social Impact

In the two years that we have been working, we have given our services to almost 25,000 people that have made at least one activity within the planetarium, like assisting to a movie projection in our dome, astronomical observations, conferences, workshops, special astronomical events, etc. Although the acceptation of the Planetarium of Cozumel has been well received by the people of Quintana Roo and its visitors, there are challenges to meet at short and long-term to achieve a higher number of visitors to the planetarium. Amongst those challenges we have that the public is more demanding and exigent; for this reason, we have the task of generating now and more attractive activities for the community and exploring which way we must follow and which one we should avoid. In the planetarium Cha'an Ka'an, we are permanently looking to make a better offering of activities not just for kids, but also for youth in general, adults, housewives, entrepreneurs, students, professors, researchers as well as the other sectors of the population as well as the visitors of the island.

Another important goal to meet is to create content specially addressed for each scholar grade. In other words, to offer schools content according to the current subjects being treated in class and thus, to help to make more meaningful the topics learned in the formal education system. With respect to the last point, it not exclusive for the workshops, but also the use of the dome as a tool for explanation in the matter of geography, history, physics, math, among others. With this, it is expected to create an institutional bond with educational institutes and to have them to visit us regularly throughout the whole year.

It is important to create and to reinforce programs that allow a better approach from the planetarium to the community of Cozumel, such is the case of the "Planetario en to Parque" (planetarium on your park) program, same that has had good acceptance and it is the confirmation that it is necessary to create a more formal program in which we're not just to visit the most popular parks, but all parks in Cozumel or at least try to make it happen. In order to create interest in the people to visit us, it is also necessary to "take out" the planetarium outside the complex and be able to demonstrate ex situ some of the activities that we regularly do within the building of Cha'an Ka'an. Besides, this will allow us to reach the population whose economic conditions may be an impediment to visit us.

Given that Cozumel is a point of major importance for cruise ship tourism, one of the most important partnerships we have are those with *tour operators* so that the activities of the planetarium can be offered amongst the available services for cruise ships on the Cozumel island. This would allow is in a way, to turn the planetarium not just in a cultural meeting point, but also in a touristic one that draws the attention of tens of visitors looking for something more than sand, sun and the regular activities carried on the island.

Another point of importance that has been carried by the planetarium team, is the bonding with other institutions, which has derived in the establishment of activities in partnership with institutions such as the National Bureau for Protected Natural Areas, the Foundation of Parks and Museums of Cozumel, the Cozumel Birding Club, among others. The collaboration with other institutions has allowed to the Cozumel Planetarium to reach other sectors of the population and to adventure in different environments beside astronomy, but just as important for those in the island. In addition, this cooperation with other institutions has not only been local but has also given way to other institutions at the international level, such as the case of China.

At the end of the month of November 2017, a collaboration memorandum was signed between the Beijing Planetarium and the Cozumel Planetarium. This agreement brings important benefits for the BRI in terms of exchange of opinions on promoting public awareness and understanding of astronomy; development of personnel exchange program for staffs, science communicators, or researchers involved in directing and/or managing respective planetarium activities; promotion of joint planning, research projects, symposia, or special exhibitions, etc.

3. Cha'an Ka'an: Tourist Reference

Although the Cozumel Planetarium has managed to position itself in the collective imagination of the local community of the island, there are still many people who have not visited the planetarium or even do not know that it exists. This forces us to create and renew our strategies, however, this is complicated when it is necessary to take into account that there are different public goals: the general public, school groups and tourists. In previous lines some of the activities directed to the first two groups have already been commented, however, the offer addressed to foreigners is different.

Tourists who visit Cozumel come in search of sand, sun and turquoise blue water, so treating astronomical issues does not become attractive. Therefore, it is necessary to create an attractive offer in English that motivates the visit of tourists to our planetarium. The main strategy that has been followed is the dissemination of the scientific knowledge of our ancestors, the Mayans. For this, museum tours have been created in order for visitors to be able to learn more about Mayan cosmogony, Archaeoastronomy and navigation, among others. We also have workshops such as "Mayan Numbers", "Mayan Calendar" and "Mayan Mathematics". The astronomical observations for tourists are oriented towards stories about the Mayan skies, the Mayan zodiac and Mayan cosmogony. The problem with this last activity is that it depends on the weather, therefore sometimes clouds or rain forbid the activity from taking place.

The workshops and the museum tours are not enough attraction for tourists, so we would have to create a complete product that would be unique for our foreign visitors. For this reason, we thought about the production of a fulldome show that could gather part of the scientific and cultural legacy of the Mayans and this is how the first fully animated fulldome show made in Mexico was born, *Mayan Archaeoastronomy: Observers of the Universe* (Fig. 34).

This program was produced by the consultancy in science communication *Frutos Digitales*, with the support of the European Southern Observatory (ESO) and it is narrated in 4 languages: Spanish, Chinese, English and Portuguese. It has spread freely in over 150 planetariums in over 30 countries in the world. *Mayan Archaeoastronomy: Observers of the Universe* shows, by means of a tour through six different Mayan archeological sites, the importance of the relationship between architectonical orientation and the movement of certain heavenly bodies. With this fulldome show, it has been possible to create an attractive scientific-cultural product both for tourists interested in knowing more about Mayan culture, as well as for hotels and tour operators that seek alternative tours for their clients.

Although there has been a good response with these initiatives, there is still a need to position the product more and ensure that these activities can be sold to cruise passengers while still being on board, which is the biggest challenge.

4. Conclusion

Cha'an Ka'an, the Cozumel Planetarium is a space that makes possible the science communication, as well as the integration of the community with respect to the learning process, the debate, opinion, making of choices in key topics regarding science, technology and innovation as well as the strengthening of the scientific culture in Cozumel.

Although the Cozumel Planetarium enjoys a good popularity among the local community, it is still necessary to create strategies that allow it to reach more people. Likewise, it is necessary to create and renew continuously the activities to provoke a greater attendance by the general public. In the case of tourists, since there is a product, it is necessary to position it and create links with tour operators, hotels, travel agencies, taxi drivers, etc. Also, for the Cozumel Planetarium is important to take advantage of the arrival of visitors coming from the cruise ships. To do this, it is necessary to create an alliance with the tour operators of the cruise ships so that the tour can be sold from on board.

The challenges are big for a planetarium and even more when science seems to have no presence or importance, however, we must take advantage of all the contexts, however far they may seem. It is necessary to create an emotional link between our target audience and science since it is precisely through this bridge that science presents a more attractive face. While people find some direct link with scientific knowledge, their interest and acceptance will be more favorable.

References

[1] PEMBROKE. PINESFL E L.Méride, Mexico Targets $7 Million Boost by Hosting the FCCA Cruise Conference & Trade Show[EB/OL]. (2017-10-17)[2017-12-1]. http://www.f-cca.com/press/17-FCCA-Cruise-Conference-Boost-Business. html.

Science Circus Diplomacy from the Australian National University and Questacon

Graham Durant   Will J Grant[①]

1. Introduction

The Australian National University and Questacon have an over 30 year history connecting Australians from all walks of life - and many others around the globe - with science. Famed for their shared outreach program the Master of Science Communication Outreach/Shell Questacon Science Circus, both have a wealth of experience connecting people with science and using science outreach to build a better world. Delivering hands-on science experiences across regional and remote Australia for over 30 years has led to the development of tried and trusted methods — using easily portable pop-up science exhibitions and simple and easily available materials — to deliver science shows and experiences suitable for many different venues and audiences. The experience of delivering the Science Circus in Australia is relevant to many countries and has led to its use in "Science Circus diplomacy".

2. The Australian National University

ANU is Australia's national university. A world-leading university in Australia's capital city, Canberra, it has a unique mission among Australian universities - providing research and

① Graham Durant: Professor of Questacon, Australia's National Science and Technology Centre. Email: Graham.Durant@questacon.edu.au.

Will J Grant: Professor of Australian National Centre for the Public Awareness of Science, The Australian National University. Email: will.grant@anu.edu.au.

education to both government and the nation at large, and connecting Australia with the world.

3. The Australian National Centre for the Public Awareness of Science (CPAS)

The Australian National Centre for the Public Awareness of Science (CPAS) is the world's most diverse - and Australia's oldest - academic science communication centre. Established as an ANU centre in 1996, its mission is to encourage a confident democratic ownership of modern science nationally and internationally by increasing science awareness in the community, fostering public dialogue about science, and improving the communication skills of scientists. Through research led to education, our students become skilled communicators who can engage people with the science, technology, or medical information that is most relevant to them. The Centre's research investigates the ways science is being communicated in the public arena, new ways to excite the public imagination about science and methods to encourage informed decisions about scientific issues that concern us in the 21st century.

4. Questacon – Australia's National Science and Technology Centre

As Australia's National Science and Technology Centre, Questacon's purpose is to excite and motivate through inspirational learning experiences. Questacon's assets comprise the Questacon Science Centre and the Ian Potter Foundation Technology Learning Centre in Canberra, a suite of traveling exhibitions, world-class informal learning engagement programs, and above all a creative, loyal, passionate and diverse workforce working in partnership with some of Australia's leading organizations. Questacon's core product is hands-on exhibits, science shows and people to people interaction offering high-quality engagement. Questacon's activities are based in Canberra with outreach programs extending across Australia working with partners to reach millions of Australians each year. Questacon also plays an important role in international engagement in the science centre sector. Questacon operates with a mix of government funding and income earned through ticket and shop sales, exhibition hire and sponsorship. A new Questacon Foundation, once established, will help extend Questacon's reach into geographically and socially isolated communities through philanthropic support. Our vision is for a better future for all Australians and the world through engagement with science and technology.

5. The ANU Questacon Shell Partnership

The Australian National Centre for the Public Awareness of Science and Questacon - the National Science and Technology Centre were projects born of a single vision. Questacon started life as a collection of hands-on demonstration exhibits to help to teach physics. It became a public facility in an unused primary school hall and then in 1988 Questacon became Australia's National Science and Technology Centre in a purpose-built building supported by Australian and Japanese investment. The Science Circus history can be traced to the 1985 innovation of then Questacon director Dr. Michael Gore to take the exhibits of the early Questacon on a tour of regional Australia. The Science Circus has been operating ever since and over 30 years has visited every part of Australia. The Science Circus is now the longest and furthest traveled science centre outreach running in the world. Renowned Science Communication academic Brian Trench has described the program as one of the first programs in science communication in the world. In 2014 the Shell Questacon Science Circus was recognized by the Global Telefonica Fundacion Education Challenge as one of the top 20 programs for the promotion of scientific and technological careers.

Since those early beginnings, the program has recruited graduate students from scientific disciplines from around Australia to be educated in key science communication skills via postgraduate programs at the ANU (currently the Master of Science Communication Outreach), and then travel around Australia in the Shell Questacon Science Circus, delivering science shows for school children and public exhibitions of hands-on science activities using ultraportable exhibits.

From the very beginning, the national scope of the program has been crucial - recruiting new students for the program from around the country and then making sure that the program travels to as much of the country as possible. It is very much a highlight for the students in the program to travel to regional, rural and remote areas of Australia never normally seen by most Australians.

When they arrive in Canberra at the beginning of each year, the new Master of Science Communication Outreach / Science Circus students undertakes a comprehensive year of education combining theoretical and practical experiences.

6. Impact of the Science Circus

The outcomes of the Science Circus have been well documented. While many key outcomes of the program are quantitative - such as number of graduate science communicators produced (over 400), number of shows delivered (over 15,000), number of kilometres travelled, number of towns visited (over 500, including more than 90 indigenous communities) number of school children inspired (millions) - the bigger qualitative impacts are also crucial.

The Science Circus was essential in the development of Science Communication as an academic discipline in Australia, and influential globally. Here McKinnon and Bryant recount the story of an early Science Circus student:

Politely but forcefully they criticized the program for sending them out underprepared— "Have you any idea of what you're doing?" said the spokeswoman. They had a point; the only training they had received was in the science of the shows and exhibits and that first trip had relied on their enthusiasm and attractive personalities. Bryant and Gore immediately responded by adding courses in public speaking, writing simple English, creating shows, and even more training in presentation and academic background. It was the first attempt in Australia to provide a disciplinary underpinning for science communication—there was no "best practice" to emulate.

This development led directly to the foundation of the Australian National Centre for the Public Awareness of Science in 1996, and more broadly to the development of Science Communication as an industry in Australia over the last two decades.

The development of the Science Circus in Australia over 30 years required the ongoing development of a number of simple hands-on science exhibits that were easily portable, as well as simple but effective science demonstrations and workshops not requiring complex apparatus that could be delivered in regional and remote settings. It is now possible to set up a pop-up hands-on science centre with 50 exhibits in less than two hours in any location accessible with Questacon's truck. In more remote offshore island settings, the Science Circus scholars have to take all of their props in a small rucksack on a light plane to deliver science shows and workshops.

The lessons learned in Australia are applicable to regional areas around the world and the well-trained enthusiastic presenters, the ultra-portable exhibits, exciting science shows and capacity-building workshops in any combination are at the heart of Science Circus tours to different countries and the basis for Science Circus diplomacy. On occasion the exhibits tour on their own, sometimes it is just a small number of presenters. There is a Science Circus model that suits any situation and any budget.

7. Science Circus Diplomacy

7.1 Science on the Move 1996–8

Copies of the Science Circus exhibits were produced for a special touring exhibition to several South Pacific Islands funded by UNESCO, AusAID and the Australian Government Department of Foreign Affairs and Trade and presented as "Science on the Move". Former Science Circus scholars working at Questacon delivered a multi-faceted program of science shows and workshops to over 42,000 participants in the Solomon Islands, Vanuatu, Fiji, Kiribati, Tuvalu, the Marshall Islands and the Cook Islands.

7.2 Science on the Streets, Dili, East Timor 2002

Following the successful tours into the 9 South Pacific countries, UNESCO asked Questacon to develop and deliver a pilot science education program for Timor Leste. Science Circus exhibits were set up on the streets of Dili and a series of science demonstrations delivered by former Science Circus scholars.

7.3 India 2005

Questacon Science Circus coordinator Lish Hogge was invited to present simple science demonstrations at the International Workshop on Demonstrations in Physics organized by the National Council of Science Museums, India to celebrate the World Year of Physics 2005.

7.4 Thailand 2006

Questacon was invited by Australian Education International to contribute Science Circus exhibit to a booth at the Thailand National Science and Technology Fair in August 2006 and had the pleasure of a visit by Her Royal Highness Princess Maha Chakri Sirindhorn.

7.5 South Korea 2006

Questacon Science Circus exhibits were enjoyed by some 50,000 South Koreans at the

Korean Science Festival. Questacon trained local Korean staff in the science exhibits and continued discussions with key stakeholders in Korea.

7.6 Science Circus Presenters in China 2010

Two Questacon Science Circus presenters toured Beijing, Shanghai and Guangzhou to contribute to the Australian Pavilion's Science Week program at the World Expo in Shanghai, sharing approaches to communicating science with staff at the China Science and Technology Museum, the Shanghai Science and Technology Museum and the Guangdong Science Centre.

7.7 Abu Dhabi Science Festival 2012

Former Science Circus scholars working at Questacon were able to support the Science Festival in Abu Dhabi with a series of science shows and practical workshops.

7.8 Science Circus Vietnam 2013

Questacon was invited by DFAT to tour an exhibition to Vietnam as part of the celebration of the 40th Anniversary of the start of diplomatic relations with Vietnam. During April 2013, Science Circus Vietnam consisted of a three-city tour of Hanoi, Da Nang and Ho Chi Minh City (HCMC).

7.9 Science Circus Japan 2014

2014 was the 25th anniversary of Questacon as Australia's National Science and Technology Centre. Questacon was established in 1988 by a generous bicentennial gift to the people of Australia from the Government of Japan and Keidanren (Japan Business Federation). The 2014 Science Circus Tour Japan recognized Japan's investment in Australia and Questacon. The 2014 Science Circus Tour Japan performed interactively, engaging science shows to over 14,000 people in Tokyo, Minamisanriku, Morioka, Kuji and Misawa.

7.10 Science Circus Africa 2015

The Science Circus Africa project builds on more than a decade of intermittent activity in Southern Africa. Led by the Australian National Centre for the Public Awareness of Science with support from Questacon, Science Circus Africa is an ongoing project to help develop science centre activities in African countries. Science Circus Africa is an Australian Government funded program taking engaging, fun science to schools, teachers and communities in five countries in southern Africa, while also training and building capacity in African staff and organizations. The key program components are science shows, teacher professional development (through workshops and distribution of resources/books), a DIY science exhibition and staff training. In total, the program reached 41,367 people in five countries.

7.11 Science Nomads

Amongst Australia's various overseas aid projects, the International Volunteers for International Development program has created opportunities for former Science Circus scholars to work in a number of countries including Namibia, Indonesia and the Philippines, assisting in organizational capacity building and presenting science demonstrations. Other former Science Circus scholars, Questacon and ANU staff have undertaken projects in various countries including Myanmar, Thailand, South Korea, Mongolia and Brunei.

8. Conclusion - Impact of the Science Circus

The success of the Shell-Questacon Science Circus has meant that there has been a constant stream of international visitors who have visited Australia to experience the Science Circus in action. It is a wonderful example of a three-way partnership between a university, a science centre and a commercial business. It has led to similar science centre outreach and science communication training models in other countries. It has led to the development of other outreach projects within Australia. The impact of the Science Circus has also been the subject of a number of Ph.D. and other studies.

Delivering hands-on science experiences across regional and remote Australia for over 30 years has led to the development of tried and trusted methods – using easily portable pop-up science exhibitions and simple and easily available materials – to deliver science shows and experiences suitable for many different venues and audiences. The experience of delivering the Science Circus in Australia is relevant to many countries and has led to its use in "Science Circus diplomacy". Our vision at ANU and Questacon is to explore how the wonderful history of Science Circus science exhibitions, shows and diplomacy can be expanded to change the lives of children, students and adults throughout the world.

References

[1] RIAN. All staff welcome[EB/OL]. (2016-2-6)[2017-12-1]. http://www.anu.edu.au/news/all-news/all-staff-welcome.

[2] TRENCH B. Vital and vulnerable:Science communication as a university subject[M]. Netherlands:Springer, 2012:241-258.

[3] BRYANT C, Gore M M. The evolution of a masters course in scientific communication:Some reflections on experience at the Australian National University[M]. Quebec City, Quebec:Multimondes, 1999:141-158.

[4] MCKINNON M, BRYANT, C. Thirty Years of a Science Communication Course in Australia:Genesis and Evolution of a Degree[J]. Science Communication, 2017, 39: 169-194.

Art Culture Nature: Royal Ontario Museum

Burton K. Lim [①]

Abstract Royal Ontario Museum is Canada's largest museum that incorporates both natural history and world cultures. The ROM has over 12.5 million specimens and artifacts in its collections with a strong emphasis on Canadian material, but also a broad international scope of research. The Ontario Provincial Government's Royal Ontario Museum Act of 1912 established in Toronto 5 different museums with separate directors for Archaeology, Geology, Mineralogy, Palaeontology, and Zoology. Each of these museums had distinctive origins, but all had early connections with the University of Toronto. The original museum building opened in 1914 along the University's Philosopher's Walk near the Provincial Legislative Assembly. In 1955, the ROM was unified under one director and now has 23 major collections and research foci in natural sciences, archaeology, and arts. Last year, the museum had its highest attendance record of more than 1.3 million visitors, which was the most for any cultural or scientific institution in the country. Collection and research strengths are traditionally centered on Ontario and Canada, but antiquities were also originally prominent and in the late 1950's there was a further emphasis to broaden the scope to throughout the world. The public space at the ROM includes 27 permanent galleries of world cultures and natural history, 2 hands-on discovery galleries for children, and 4 major display areas for temporary exhibitions. There is an initiative to develop in-house exhibits to tour at other venues in Canada and abroad. A recent example is "Out of the Depths: The Blue Whale Story", based on two whales that washed ashore in New foundland. The ROM salvaged both skeletons and plastinated one heart for display in a major exhibit to celebrate Canada's 150th anniversary since confederation, which will now travel across

① Burton K. Lim: Curator of the Royal Ontario Museum. Email: burtonl@rom.on.ca.

the country and overseas in the coming years.

Keywords Blue whale exhibition   Natural history museum   world cultures museum

1. Introduction

The Royal Ontario Museum (ROM) is one of the world's largest museums with the dual mandate of conveying science and culture to the public through research, collections, and galleries. It is located in Toronto, the most populous Canadian city, and is an agency of the Provincial Government of Ontario under the Ministry of Tourism, Culture, and Sport. Historically, the ROM has averaged almost one million visitors per year, but last year was the highest attendance ever at over 1,300,000 people, which was a 23% increase over the previous year. Most (58%) of the admissions are people from the Greater Toronto Area and the U.S. represents the highest foreign visitors at 13%.

2. History

Officially established in 1912 by the Royal Ontario Museum Act of the Provincial Legislative Assembly, the ROM opened to the public two years later with the completion of a new building on the northeast corner of the University of Toronto's Philosopher's Walk (Table 1). The main entrance was on Bloor Street West, which is a major thoroughfare of the city. The Province and University shared equally in the costs of operating the museum, but it was under the governance of the University of Toronto with an original staff of 20 employees. The ROM began as 5 separate museums concentrating on Archaeology, Geology, Mineralogy, Palaeontology, and Zoology – each with its own director.

It was the bold vision of primarily two individuals, Edmund Walker and Charles Currently, who were the acknowledged driving forces behind the genesis of a provincial museum. Walker was a prominent banking executive and philanthropist with the passionate idea of founding a cultural and scientific institution in Toronto. Currently was a collector of art and archaeology who was responsible for acquiring many of the earlier artifacts of antiquity, mainly from Egypt, that eventually formed the collections of the ROM. He was also appointed the first director of the Royal Ontario Museum of Archaeology and held it until his retirement in 1946. Other collections such as Zoology had close associations with the University of Toronto's Biological Museum established in the mid-1850's primarily for teaching purposes. In addition, the provincial Education Department's Normal School for teachers' collection was eventually transferred to the ROM. The official governance under an association with the University of Toronto ended in 1968 when the ROM became an agency of the Ontario Government.

Table 1　Chronology of major events at the Royal Ontario Museum in Toronto, Canada

Year	Milestone	Description
1912	Royal Ontario Museum Act	Establishment by the Provincial Legislature
1914	Museum opens	Building located beside the University of Toronto's Philosopher's Walk
1933	First expansion	Queen's Park wing opens creating an H-shaped structure
1951	Canadiana building	Dr. Sigmund Samuel endows a new home for Canadian decorative arts
1968	Planetarium built	$2 million donations from Samuel McLaughlin to build the planetarium
1978	Second expansion	Start of renovation for the new curatorial centre and Terrace Gallery
1987	Gardiner Museum	Ceramic art museum across the road comes under ROM management
1994	Heritage galleries	New Canadiana Gallery opens in the main museum building
2001	Renaissance ROM	Revitalization project begins culminating in the Crystal opening in 2005
2017	Queen's Park entrance	Historic Rotunda entrance re-opens as the alternative to Bloor St. access

Buildings

The museum has experienced several periods of expansion beginning in 1933 with the addition of a new building and entrance on Queen's Park to the east of and connected to the original Philosopher's Walk structure to form an H-shaped configuration. In 1951, the Sigmund Samuel Building was endowed by its steel-industrialist namesake and built south of the ROM across from the Provincial Legislature to house the Canadiana collections, which he was instrumental in acquiring. The first planetarium in Canada was built beside and managed by the ROM in 1968 with a donation from Samuel McLaughlin, who was an automotive magnate. A major expansion occurred on the main property in 1982 with the opening of a new building for collections and research in the south (now named the Louise Hawley Stone Curatorial Centre) and the new Terrace Gallery in the north of the older H-shaped buildings to form an interconnected rectangular structure. The height of the expansion phase occurred in 1987 when the 3-year old Gardiner Museum of Ceramic Arts located across the road came under the administration of the ROM. But by 1995, a contraction in size began when the Planetarium was closed and the property sold to the University of Toronto. In 1997, the Gardiner Museum became independent again with an endowment from its namesake benefactor. The Sigmund Samuel Canadiana building was sold in 2000 to the University of Toronto and the collections moved to the main museum. This trend in reduction was reversed in 2007 with a transformative project that replaced the 1982 Terrace Gallery with the opening of the Michael Lee-Chin Crystal with 6 new galleries and moving the entrance back to Bloor Street West (Fig. 35).

3. Contemporary

3.1 Leadership

The Royal Ontario Museum is governed by a 21-member Board of Trustees responsible for policies, procedures, and the assets of the museum, which are held in trust for the people of Ontario. The daily operations of the ROM are managed by a senior administration led by the Director and CEO with support from Deputy Directors and Vice Presidents. Josh Basseches is the current Director who is rebranding the ROM as North America's interdisciplinary museum of art, culture, and nature. There is over 300 full-time staff working at the museum, in addition to part-time staff and more than 1300 volunteers.

3.2 Vision and Challenges

The stated vision of the ROM is to be recognized globally as an essential destination for making sense of the changing natural and cultural worlds. Digital and social media are new areas that the museum is using to increase exposure further afield online and also to help boost admissions through the front doors. The museum website (www.rom.on.ca) had over 3.7 million visits last year, which represented an increase of 20% compared to the previous year. Social media engagements increased by 70% with almost 150,000 impressions across all platforms, such as Facebook, Instagram, and Twitter.

Maintaining the aging infrastructure is an ongoing challenge. The curatorial building housing the collections and research facilities were completed in 1982 so it has become not only dated but also expansion space has now been used in most areas making future growth an ongoing challenge. Similarly, information technology and infrastructure require updating and significant financial investment, including in the area of digital assets of museum specimens and online availability.

3.3 Collections and Research

There are over 30 curators at the ROM collecting specimens and doing research in art, culture, and nature. In the past year, fieldwork was conducted in about 30 countries throughout the world and more than 100 publications were written in scholarly journals and books.

Some collection highlights in World Cultures include Canadiana with holdings of both images about Canada, and decorative arts and material culture produced in Canada. Chinese antiquities are an important and comprehensive world-class collection that is ranked among the top 10 collections held in museums outside of China. In Natural History, the Burgess Shale collection constitutes the world's largest and most complete sampling of Canada's premier fossil locality, which is most significant for illustrating the early evolution of animals during the famous Cambrian explosion half a billion years ago. Frozen tissues are the fastest growing collection across several life science disciplines and one of the top 10 in the world. It comprises over 110,000 samples that are used for DNA and genetic analysis to study rare and endangered species, assesses biodiversity, and help build the Tree of Life.

3.4 Communities

The museum has substantial and broad outreach into the surrounding communities in the Greater Toronto Area. The Royal Ontario Museum Community Access Network (ROMCAN) helps to ensure inclusion of as many people as possible despite any social, financial, or cultural barriers. Partnerships with 62 non-profit organizations enabled the distributing of 100,000 free admission tickets to the museum. One of the largest citizen science projects in Canada is the annual ROM-led Ontario Bioblitz initiative with the collaboration involving other local institutions to document as many species in the Toronto-area watersheds. About 250 professional scientists were assisted by 400 amateur naturalists to document 1,400 species last year in the Credit River region. Creative public programming is another innovative way of attracting different audiences to the museum. Popular events include Friday Night Live, which is aimed at a younger crowd and introducing them to the museum.

3.5 Education

In the previous year, the ROM welcomed more than 110,000 students and teachers from schools across Ontario with guided tours, hands-on labs, maker activities, and online resources that are linked to the Ontario education curriculum. And we have an outreach program that extends beyond our physical building, including an inflatable and portable dome that is a popular traveling planetarium that projects a simulated night view of the constellations and was loaned to schools, libraries, community centres, as well as other museums.

3.6 Membership and Philanthropy

There are almost 150,000 memberships to the museum ranging from individuals to families, including 2 types of patron members who receive the highest level of exclusivity such as behind the scenes tours. The ROM Governors is the fundraising arm of the Royal Ontario Museum Foundation with its own independent Board and management team to support collections, research, and exhibits at the museum. For example, of the >30 curators, 8 are endowed chairs.

3.7 Galleries

The permanent galleries at the ROM have approximately 30,000 objects on display. There are >30 galleries showcasing art, archaeology and natural science, including 17 of world culture, 10 of natural history, 2 hands-on discovery galleries, and 4 temporary exhibition spaces. The children-focused Hands-on Biodiversity Gallery and Discovery Gallery were the 2 most popular at the museum last year. These discovery galleries include programming that features a mix of play-based activities and multisensory experiences with primarily touchable specimens, but also some live animals.

3.8 Exhibitions

There were 7 major exhibitions at the ROM last year spanning art, culture, and nature. Three exhibitions that exemplified this diversity include the visually stunning artwork of glass in Chihuly: From Sand, From Fire, Comes Beauty; the 5,000 year old cultural ritual of Tattoos: Ritual, Identity, Obsession, Art; and the grandeur of nature with Out of the Depths: The Blue Whale Story, which was a new and original ROM exhibition. In addition to the Blue Whale exhibit, Anishinaabeg: Art & Power and The Family Camera were 3 in-house exhibitions developed for the Canada 150 anniversary celebrations in 2017.

3.9 Traveling Exhibits

Producing ROM-original traveling exhibitions is an area that the museum is actively pursuing. For example, this year's A Third Gender: Beautiful Youths in Japanese Prints had a critically acclaimed second installation at the Japan Society in New York City. And Pharaohs and Kings: Treasures of Ancient Egypt and China's Han Dynasty was a collaboration, based on ROM objects, with the Nanjing and Jinsha Site Museums in China that attracted 1 million visitors.

Out of the Depths: The Blue Whale Story is the next traveling exhibition the ROM is promoting. It was developed as a modular exhibition to fit full-size block-buster spaces, but also mid to smaller-size venues depending on availability at receiving institutions and the exhibit components desired. The full exhibition includes the real 26-metre-long skeleton of the Trout River, Newfoundland, blue whale that was salvaged in 2014. In addition, the plastinated heart of the Rocky Harbour the blue whale is an impressive 1.5 metres in height, the largest preserved heart of any animal known – sure to stir the imagination and curiosity of visitors (Fig. 36).

There are 9 components of the Blue Whale exhibit, including the introduction with an immersive multimedia 3-screen video presentation detailing the tragic event of 9 blue whales found dead in the ice-pack of the Gulf of St. Lawrence in Canada and the ROM recovering 2 of the skeletons for science and educational purposes. The next section is the big reveal of the skeleton and interactive discussion of size – the heaviest animal that was ever lived. In the life and biology section is a sound chamber detailing the infrasonic calls that blue whales use to communicate over hundreds of kilometers of ocean. Diet has a full-size cast of the skull and jaws with real baleen from the Rocky Harbour specimen to explain the filter-feeding behavior of the largest animal eating some of the smallest creatures – shrimp-like krill. Evolution includes impressive casts of the 5 fossil whales that beautifully track the transition of land mammals back to sea. Genome uses interactive graphics screens to describe the genetics of whales and the ROM-led research on reading the 3 billion DNA code of the blue whale and unraveling its implication to our understanding of the successful adaptation to marine life and the survival of these endangered species. Conservation looks at the history of whaling and efforts to bring these majestic beasts back from the brink of extinction. The final component is a video of marine conservation experts passionately communicating the benefits to society of having healthy whale populations and why it matters.

4. Conclusion

The Royal Ontario Museum is Canada's largest institution of natural history and world cultures with over 1 million visitors attending per year. Collection and research strengths were originally focused in our home province and country, but are now international in scope in many different fields of art, archaeology, and nature. More than 30 galleries showcase our specimens and artifacts in permanent displays and temporary shows. A recent initiative has been to develop in-house exhibitions for traveling around the world, including "Out of the Depths: The Blue Whale Story" based on our recovery of 2 whales from Newfoundland and the awe-inspiring evolution of marine mammals and the conservation efforts to bring them back from the brink of extinction.

References

[1] ROYAL ONTARIO MUSEUM. Annual Report 2016-2017[EB/OL]. .http://www.rom.on.ca/annualreport/index.php.

[2] DICKSON L. The museum makers:the story of the Royal Ontario Museum[M]. Toronto:University of Toronto Press, 1993.

[3] CURRELLY C T. I brought the ages home[J]. 1956.

[4] DYMOND J R. History of the Royal Ontario Museum of Zoology.1940.

[5] BROWNE K. Bold visions:the architecture of the Royal Ontario Museum[J]. 2007.

[6] Royal Ontario Museum. ROM business plan:fiscal 2017-2018[EB/OL]. (2017-7-13)[2017-12-1]. http://www.rom.on.ca/sites/default/files/imce/business_plan_2017_final.pdf.

The Sergei Korolev Space Museum: Promoting Science Education in the Context of International Cooperation

Victoria Chetvertak [1]

Abstract According to the Ukrainian Information Agency, UNIAN Ukraine can boast to have about 5,000 museums, which is an inspiring figure for the country. One can really have an impression that there is a museum renaissance in Ukraine. But in fact, there exists a problem of making the museums attractive to the general public. This becomes a demand in the context of modern social development. The S.Korolev Space museum being the leading space museum in Ukraine, education and cultural center elaborate its development strategy to meet this demand. In this context, science education seems to be a convenient method for bringing the museums, especially technical ones, closer to the general public providing the positive impact. Development of international cooperation will help to adopt best world practices and contribute to the development of the museum as a science promotion center.

Keywords Science Education International Cooperation Technology Museum World's Best Practices

1. Introduction

The S. Korolev Space Museum is a technology museum in Ukraine. It was established in 1970 in the city of Zhytomyr, Ukraine where the Chief Designer of the Soviet space rockets and systems and mastermind of the state space program was born. Sergei Korolev was behind

---

① Victoria Chetvertak: Head of Foreign Relations Sector, S.Korolev Space Museum. Email: victoria.chetvertak@gmail.com.

the greatest scientific and technical achievements of humanity: the launch of the world's first artificial satellite, launches of the first automatic interplanetary probes to the Moon, Venus and Mars, the first manned spaceflights, and the first extravehicular activity performed by humans in outer space.

Today, the museum has two permanent exhibitions: the Memorial House of Academician Sergei Korolev and the exhibition Space dedicated to the world's space achievements. The unique collection includes more than 25 thousand objects, among them, being personal belongings and family memorabilia of the Korolev, original spacecraft and space artifacts including the Soyuz 27 re-entry capsule, samples of lunar rocks and soil, as well as life-size precise models. One of the museum's primary tasks and missions is to educate in science and technology, promote space achievements and space-related activities and to encourage the development of planetary thinking in its visitors. The best way to successfully implement this goal is to learn from the world's best practices in the process of international cooperation and cultural exchange.

2. Why do we move towards science popularization

According to the Ukrainian Information Agency, UNIAN Ukraine can boast to have about 5,000 museums, which is an inspiring figure for our country. One can really have an impression that there is a museum renaissance in Ukraine. But in fact, there exists a problem of making the museums attractive to the general public, making the exhibitions alive. In other words to help members of the public to extract the maximum meaning from these objects with the overall aim of persuading people that museums should form a normal feature of their lives. This becomes a demand in the context of modern social development. The S.Korolev Space museum being the leading space museum in Ukraine, education and cultural center elaborate its development strategy to meet this demand.

The museum concept is most pronounced in its activities: educational and interactive programs for the general public, standard and special guided tours, festivals, cultural events and exhibitions, which make the museum attractive for multiple visits.

While planning our work we do our best keeping in mind that a museum is a non-profit, permanent institution in the service of society and its development, open to the public, which acquires, conserves, researches, communicates and exhibits the tangible and intangible heritage of humanity and its environment for the purposes of education, study and enjoyment (http:// icom.museum/the-vision/museum-definition/). This definition of the museum is based on one of the best models of the museum, which was developed in the late 1980's. at Reinwardt

Academie (Amsterdam). This model supposes preservation (includes acquisition, conservation and management of collections / objects of cultural heritage), research, and communication (combines education and exhibition).

The recent world tendencies actively supported by a number of museums in Ukraine including ours show that the most effective strategy is to use a museum as a tool to communicate to the public. The museums began to feel the power of their influence, the importance of the impression created by the museums and its exhibits, an impression that can't be replaced with any story or a written text. Especially vividly it is pronounced in the work with children, in understanding that the museums create the first impression of the "museum" concept, which will remain with a person during the whole person's life.

In this context science education seems to be a convenient method for bringing the museums, especially technical ones, closer to general public providing positive impact such as a memorable learning experience, increase of the level of visitors' knowledge and understanding of science and technology, personal and social inspirations which can enhance inter-generation learning, and encourage trust and understanding between the public and the scientific community engaged in cutting-edge technology developments.

Discovery and knowledge construction begins with something that happens, that is, a phenomenon which calls for observation, description and inquiry. In this process, no explanation is given before the demonstration of the phenomenon, and the information is never taken for granted: each new discovery is built on the personal skills and knowledge of each visitor. This means that no process is identical to the previous or to the following one. In fact, discovery and explanation of phenomena are guided by the learners and therefore determined by their age, expectations and interests. With this approach, the Museum helps to understand how scientific research takes place, i. e. that scientific investigation involves forming of hypothesis, observation, testing, trial and error, control, repetitions. Awareness of scientific method can, moreover, contribute to the development of skills that can become useful in the course of everyday life. In this approach the museum, far from being just the place for exhibiting a series of objects, is the place where the complexity and polyvalence of science and of scientific processes can be presented, along with the cultural, social and philosophical dimensions in which objects and knowledge were born.

3. How do We do it and What May We Expect From International Cooperation

In recent years the museum is actively engaged in science promotional activities. Despite our almost 48 years of museum experience we are beginners on this path moving from mere display and presentation of our collection to creating experiences that could respond to the social evolution and our visitors' enlarged range of interests. Science, which is becoming more popular in the modern society, is gaining in popularity at our museum.

The first step we did we revived our memorial exhibition by introducing retro tours into it. History becomes closer and tangible due to the activation of different perception receptors through the application of audio and video documentary, costumes of the guides. From now on, visitors can plunge into the atmosphere of the 20th century while living through the life story of the most secret Chief Designer of the Soviet Union.

Birthday may be another great opportunity to learn something new and important. An exciting game may stimulate further discoveries.

Adopting foreign experience was another challenge. Such successful projects as European Researchers' Night and Science Picnics were much welcomed by our visitors.

For us, it is also very important to present the museum educational programs in a broader extra-museum context. Indeed, the museums demonstrate a remarkable creativity in adapting its themes or exhibits so as to interest wide target audience and provide some cognitive information at the same time. Such activities are very important because they enable the museum to work side-by-side with different museums and institutions, share information, learn from each other.

In this respect, international cooperation and experience offer promising opportunities for our museum's further development. What may we expect from international cooperation? Using the best practices from the museums worldwide can help us best meet the demands of evolving society. Presentation of the museum abroad shall encourage the sharing of experience and promotion of science in the collaborating countries, exchange of specialists and will contribute to the museum's development.

4. Conclusion

The S. Korolev Space Museum is currently introducing innovations, which include promotion of science as a major line of its strategic development. Simultaneously it develops an extensive international cooperation with an aim to introduce the world's best practices encouraging the further evolution of the museum as a science promoting-institution, which is imperative of our time.

That is why we actively support the idea of the Belt and Road Initiative Science Museum International Symposium. We believe that the opportunity for international communication provided by the symposium bringing together representatives from different countries will contribute to better understanding of the ways of possible cooperation between nations, help to develop its strategy and provide a platform for fruitful networking for mutual benefit.

References

[1] Ольга Фішук. Фішук О. україна-музейна Країна[EB/OL]. (2010-5-18)[2017-12-1]. https://www.unian.ua/culture/358865-ukrajina-muzeyna-krajina.html.

[2] Hudson, Kenneth. Museums:Treasures or Tools?[J]. Education al Facilities, 1992:61.

[3] РОЖКО В М. Національна музейна політика:засади наукової діяльності музейних інституцій [M]. Prostir. Museum, 2016.

[4] Jo-Anne Sunderland Bowe. The Creative Museum Analysis of selected best practices from Europe[EB/OL]. http://creative-museum.net/wp-content/uploads/2016/06/analysis-of-best-practices.pdf.

[5] Котвіцька, К. Показати з кращого боку:як українські музеї стають сучасними[EB/OL]. (2016-8-25) [2017-12-1]. https://zeitgeist.platfor.ma/museum-in-ua.

[6] MARIA XANTHOUDAKI, BRUNELLA TIRELLI, PATRIZIA CERUTTI, et al. Museums for science education:can we make the difference? The case of the EST, Journal of Science Communication[EB/OL]. (2007-6-6)[2017-12-1]. http://www.museoscienza.org/scuole/download/ESTarticolo_Jcom0602.pdf.

The State Darwin Museum as One of the leading Russian Natural History Museums and the Methodical Centre of the Association of Natural History Museums of Russia

Anna Kliukina[①]

Abstract The State Darwin Museum was established in 1907 and celebrates its 110th anniversary in 2017. It is the leading museum of natural science museums in Russia and a methodical centre of all Russian natural history museums under the Ministry of culture. Many Russian museums are subject to different ministries and departments. All of them are united by the Association of Natural History Museums (ICOM of Russia), created in the State Darwin Museum.

The main objective is to illustrate the general development of natural history museums in Russia using the history and achievements of the State Darwin Museum in both past and present times.

The Darwin Museum is the only museum of evolution in Russia, with a total area of about 20,000 square meters and collections of approximately 400,000 items. It also collected information about natural science museums and exhibitions in Russia and issued a reference-book about 462 museums. Every year the museum hosts scientific conferences, seminars and roundtables: over 330 in the last ten years.

The exposition of the museum occupies 5,000 square meters in the main building. There is an exhibition hall with six halls which annually host 50 - 60 temporary exhibitions. The museum provides exhibitions to other museums. Over the past ten years, it has organized 723 exhibitions. The annual attendance exceeds 650,000

---

① Anna Kliukina: Director of The State Darwin Museum. Email: Anna@darwin.museum.ru.

visitors.

The Darwin museum is rapidly developing and throughout the year it continuously offers the visitors to learn, explore and enjoy something new.

Keywords State Darwin Museum the museum of evolution natural history museums in Russia the Association of Natural History Museums of Russia

1. Introduction

The State Darwin Museum in Moscow has a rich history of continuous advancement. Now it is the leading museum of natural science museums in Russia and a methodical centre of all Russian natural history museums under Ministry of culture. The past and present practice of the State Darwin Museum can be used to study and illustrate the general development of natural history museums in Russia.

The State Darwin Museum, one of the oldest natural history museums in Russia, was founded in 1907 by Alexander Kohts, a young Darwinism teacher at the Moscow Higher Women's Courses. In 2017, the museum celebrates its 110th anniversary.

In Russia, many museums began with private collections. Alexander Kohts collected zoological collections since his childhood, and after graduating the university he was offered to read the Darwinism course at the Moscow Higher Women's Courses. He took his collection to illustrate the lectures.

This was the beginning of the state Darwin Museum.

In 1911, Alexander Kohts and his wife visited many European natural history museums. In Europe, he had the idea of creating a museum of a new type. Kohts wanted to create a museum for the common people, but he had no building and money. Alexander Kohts found remarkable like-minded people whom he always called co-founders of the museum. That was his wife Nadezhda Ladygina - Kohts who was responsible for carrying out scientific research in the museum and later became Doctor of Science.

Vasily Vatagin was an artist who later became an academician of the Academy of Arts. Kohts believed that the works of Art should bring the stuffed animals to life and add beauty to the exhibition. V. Vatagin dedicated over 40 years to the museum and today it displays a lot of his paintings and sculptures.

The fourth co-founder was Philip Fedulov, a taxidermist, who spent most of his life working at the museum. Thousands of stuffed museum pieces were created by his hands. The Darwin museum was established by this great team and yet there was no building for it. All his life Alexander Kohts called on all the leaders of the country without any result. After Alexander Kohts' death, Vera Ignatievna became the second director and she managed to get the Moscow government's decision to build a building for the Darwin Museum but she did not live to see it either. It had been constructed for 20 years.

Today, the Darwin Museum has housed in three buildings: the main building, the exhibition building and the technical building. The total area is about 20,000 square meters. The main exhibition occupies 5,000 square meters. In total, the museum stores about 400,000 items (Fig. 38).

Every year, the museum hosts 50 to 60 changing exhibitions. We try to make exhibitions on various topics so that very different people can find something interesting for themselves.

Each year over 650,000 people visit the Darwin Museum; more than 40% become permanent visitors.

More than 85% of our visitors are families and only about 13% are organized groups of school children. During a year the museum gives about 4,000 excursions.

There are QR-cards on the glass-cases. Any visitor can scan a matrix code and listen to the excursion in Russian or English using their smartphones.

The museum created various interactive complexes. Among them, there is a fully interactive exhibition "Walk the Path of Evolution" (Fig. 39).

Recently there was built an interactive centre "Cognize yourself – Discover the world" with 52 interactive complexes located in a small area.

A light and video-musical exposition "The Living Planet" is switched on every day in the central hall and three times a day during weekends.

Each year the Darwin museum holds about 11 public festivals such as the International Bird Day, the International Animal Day, Mother's Day and many others.

Every year the museum publishes 20 to 45 publications. The exposures are continuously updated. Despite the small territory, the museum created two expositions. One of them is botanical. All the plants along the "Environmental path" have labels in Russian, Latin and Braille. There are also excursions in botany including those for blind visitors.

The second exposition is paleontological with sculptures of extinct animals that used to live in the Russian territory.

Naturally, the museum conducts scientific work. The staff of the museum goes to the field expeditions. Annually 50-60 articles are published in various scientific and popular magazines. The museum holds joint conferences, seminars and round-table conferences.

The entire exposition of the museum is fully adapted for the disabled visitors. They come to us individually or in groups. Last year, the museum was visited by over 8,000 disabled people.

In 1971, the Soviet Committee of the International Council of Museums established a section of natural history museums. It united representatives of natural history museums and nature departments in local history museums on a voluntary basis, as well as representatives of scientific institutions, ministries and departments.

In 1992 the Soviet Committee of the International Council of Museums was transformed into the Russian Committee and the section of natural history museums continued its work within the framework of the Russian Committee of the International Council of Museums.

In September 1996, at a conference of natural history museums held at the State Darwin Museum, it was decided to establish an All-Russian Association of Natural History Museums at the Russian Committee of the International Council of Museums. The Charter of the Association was adopted and the board was elected. The State Darwin Museum became the science data and methodology centre of the Association. The Association unites on a voluntary basis natural history museums and representatives of the nature departments of local history museums, as well as representatives of scientific institutions, ministries and departments directly associated with natural history museology.

The main goal of the Association is to unite the creative efforts and capabilities of employees of natural history museums regardless of their subordination.

So, what does the Association do?

(1) Creates a bank of information on the activities of natural history museums in Russia. In 2008 the museum published an album with information about 462 museums.

(2) Promotes professional growth of the museum staff.

(3) Supports the exchange of experience among natural history museums including the popularization of the best practices of foreign museums.

(4) Supports the publication of scientific works of the natural history museum staff.

(5) Cooperates with the International Committee of Natural History Museums.

The Association holds conferences every two years where various topics of natural history museums are discussed. More than 1,000 people have taken part in these conferences since

1996. The Association maintains close links with the International Committee of Museums and Natural History Collections (the International Committee for Museums and Collections of Natural History of the International council of Museums, ICOM NATHIST).

On June 9-12 in 2008, the annual ICOM NATHIST conference took place at the Darwin Museum in Russia. It was attended by 50 people from 16 countries of Europe, Asia, Africa and South America.

The activity of the Association is aimed at forming a common information space and creates prerequisites for the communication of like-minded people. Not only conferences but also scientific and practical seminars to improve the professional skills of the staff of natural history museums are regularly held. Since 1999, 19 such seminars have been held. 583 specialists from 246 museums participated in their work. Methodical literature is regularly published. Annually the scientific proceedings of the State Darwin Museum are published, and among its authors are not only the employees of the state Darwin Museum but also colleagues from the natural history museums of Russia and abroad. The opportunity to communicate, share knowledge and experience, establish new professional contacts is always greatly appreciated by all participants in the events held by the Association.

2. Conclusion

As many natural history museums in Russia, the State Darwin Museum managed to become prosperous and successful thanks to dedicated staff and continuous improvement of practice such as creating interesting exhibition projects, interactive centres and comfortable environment for different categories of visitors helping them to discover the beautiful world of science and wild nature.

State Geological Museum for 2016

Akhmedshayev   Ahmadzhan   Shayakubovich [1]

Abstract Uzbekistan is a treasure of unique museums and historical monuments of Central Asia. Among the most prominent is the State Geological Museum, as a repository of the wealth of the earth's interior.

The Geological Museum of Uzbekistan, which is currently run by the State Committee of the Republic of Uzbekistan for Geology and Mineral Resources, was established in 1926.

It was based on the collections of a small museum of the former Glavgeology of Uzbekistan, as well as extensive collection materials collected by geological organizations, enterprises, institutions, well-known exploration geologists of the Republic and Central Asia for more than 90 years of geological service of Uzbekistan.

Geological Museum

Collection materials

1. Works and services carried out by the State Geological Museum

Maintenance, preservation and increase of the state museum fund of stone materials (samples of rocks, ores, ornamental stone material and rare mineralogical formations) of paleontological and unique geological materials, as well as information on natural and cultural air features in the air that have scientific, cultural and aesthetic value;

Popularization among the population of geological knowledge and the need to increase the mineral and raw materials base of the Republic and the potential of its subsoil;

Education of schoolchildren and students of interest in geology in circles "Yosh Geology",

---

[1] Akhmedshayev Ahmadzhan Shayakabovich: director of the State Geological Museum of the State Committee of the Republic of Uzbekistan for Geology of Mineral Resources. Email: muzeygeologii@umail.uz.

carrying out geological expeditions, excursions and olympiads;

Drawing up instructions for recording and storage stone geological material and paleontological remains;

Work with the press, radio and television to promote the achievements of the geological industry.

Creation of educational demonstration material using the latest information technologies, including educational, popular scientific, documentary films in cooperation with the National Agency "Uzbekkino" and the National Television and Radio Company of Uzbekistan.

The variety of exhibits of the museum enables every visitor from the child of the kindergarten to professional geologists to find here information that corresponds to their level of knowledge. Experienced guides-geologists will help you deal with any issues that arise during a tour of the museum.

2. Scientific Publications

In 2013, in the northern foothills of the Kaminsky Ridge, the remains of a proboscidean were found in the vicinity of Akhangaran. The State Geological Museum is responsible for excavation and safety of this find. After studying by experts, the remains were attributed to the oldest representative of the Elephantidae family, Archidiscodon genus.

Published: in 2014, the paper "On the first finds of ancient elephants remains (Archidiskodon) in Uzbekistan" (I.B. Turamuradov, A.Sh.Akhmedshaev, N.V. Averburg, Kh.A. Saipov) was published in the scientific journal "Geology and Mineral Resources" No. 2 (Fig. 40~Fig. 43).

In 2016, the paper "Geological Museum - its role in youth education of the Republic" (A.Sh. Akhmedshaev, Kh.A. Saipov) was published in the anniversary issue of "Geology and Mineral Resources".

On May 16, 2016, A.Sh. Akhmedshaev took part in the International Scientific and Practical Conference "Education in Museums: New Global and National Strategies in the Twenty-first Century" organized by the UNESCO Office in Samarkand dedicated to World Museum Day, and made a report "Role of Innovative Methods in Working with Visitors".

Since 2008, the "Yosh Geology" Olympiad has been annually held among schoolchildren in connection with the adoption of the Law of the Republic of Uzbekistan "On Education" and the Decree of the Cabinet of Ministers "On Material Incentives of Gifted Young People of Uzbekistan", as well as joint resolutions of Goscomgeology and Ministry of Education (Fig.

44~Fig. 46).

Purpose: To develop schoolchildren' love and interest in geology.

The Olympiads include III Rounds:

I Round - school, district.

II Round - city, region.

III Round - republic.

The experience of previous Olympiads has shown that this event gives a new impulse to solve the problem of staff shortage in geology. The popularization of mining and geological knowledge is carried out in accordance with the industry staff training program practically throughout the Republic of Uzbekistan and covers more than 20 thousand schoolchildren.

For a relatively short period of its existence, the State Geological Museum developed into not only a scientific but also a training center of Goscomgeology. It serves as a center for educational tours for schoolchildren; basic lessons for students of geological and related specialties of the system of higher and secondary special education are held here, as well as lectures on mineralogy, petrography, paleontology etc. A special educational hall for students is created and equipped in the Museum.

On the basis of the museum, various thematic exhibitions, presentations of the mineral and raw materials potential of Uzbekistan, business meetings with foreign partners and investors are held with discussion of a wide range of issues related to establishing mutually beneficial cooperation in studying the mineral resources of the Republic and developing its mineral and raw materials resources. In the modern era of the transition from the industrial society to information one, the geological museum is facing the problem to fully provide scientific information contained in geological objects. This is the most important basis for obtaining new scientific knowledge and expert assessments based on geological funds as banks of information stored on various media, both primary - natural (stone) and various secondary (paper, photo, film, magnetic , electronic and other media). It is important to note that natural media are the main and sometimes the only source of reliable information about the structure and composition of geological data. The panorama of museum halls, the system of exhibits placement, and design of demonstration materials and level of information offered by guides in the state, Russian and English languages create conditions for perceptual unity and understanding of what is seen. This allows every museum visitor to leave it with indelible impressions and pride for nature and uniqueness of mineral wealth of his native heath, for selfless work of ancient prospectors and pioneers, as well as modern discoverers of the underground storerooms.

Kyiv State Polytechnic Museum

Abstract State Polytechnic Museum is one of the biggest Ukrainian museums of technology and the biggest university museum in Ukraine. The museum is a part of National Technical University of Ukraine "Igor Sikorsky Kyiv Polytechnic Institute" and occupies the building of University's mechanical workshop built in 1902 and airplane hangar built in 1914. Our University, founded in 1898, is one of the oldest and biggest technical universities in Europe. At present, the number of students at the KPI exceeds 20,000. A large number of prominent people worked and studied at the KPI. Among them are Eugene Paton, the inventor of electric welding; Igor Sikorsky, creator of Sikorsky Helicopters; rocket scientist Sergey Korolyov.Museum exposition covers wide range of technologies of 20th century: radio, television, computing, electronics, mining and environmental engineering, machinery and instruments, railway, firearms and military communication, aviation and space exploration, as well as the history of University. Total exposition area is 2,300 square meters, the collection of 17,000 objects covers the time span from the second half of 19th century to present day,as part of its operation, museum runs several programmes for young people and organizes meetings and events with symbolic figures of 20th century. Museum's mission is to preserve and promote national cultural heritage, guide youth in choosing their career.

Keywords Technology Museum   the University Museum   Education   Research

1. Introduction

One of the oldest and biggest technical universities in Europe and Ukraine - Kyiv

---

[1]   Nataliia Pysarevska: director of the Kyiv State Polytechnic Museum. Email: tala1311@ukr.net.

Polytechnic Institute,KPI.

It was founded in 1898 on the initiative of scientific-technical circles and entrepreneurs of Russia and Ukraine in response to the needs of the industry of the Southern West of the country, which developed rapidly at the end of 20th century.

The lives and activities of the world-known scientists - V.M. Kyrpychov, D.I. Mendeleyev, I, I. Sikorsky, S.P. Tymoshenko and many others were closely connected with our university.

Cherishing the traditions of profound fundamental training and its harmonious combination with practical engineering and technical developments of graduates, KPI strives for providing high-level training of Bachelors, Masters, Doctors of Philosophy (Ph.D.), Doctors of Sciences (DSc) under the present-day conditions.

Now, more than 490 museums and collections in 396 universities and institutes are in Ukraine-museums, classrooms, laboratories, 24 botanical gardens, 7 greenhouses, 17 observatories, 2 planetariums, 2 anatomical theaters, etc.

2. State Polytechnic Museum of Ukraine at the KPI

State Polytechnic Museum of Ukraine at the KPI was founded in honor of the centenary of the Polytechnic Institute. State Polytechnic Museum was established according to the Cabinet of Ministers of Ukraine Resolution No. 360 on May 29, 1995, and is the only one polytechnic museum in Ukraine. Expositions present the history of engineering development in Ukraine and include 20th-century symbols. New exposition - I.I.Sikorsky Department of Aviation and Cosmonautics - was opened in 2008.

3. Collection of the State Polytechnic Museum

Collection of the State Polytechnic Museum of National Technical. University of Ukraine (NTUU "KPI") was primarily created as the accumulation of scattered collections of University departments and academic science institutions. There were also several departmental museums located at the University. The events of the stormy 20th century did not favor the preservation of technical exhibits. Problems faced both by the country and the University were standing quite apart from the issues of saving technical heritage. Actually, there was no proper attitude and respect to the issues of technical heritage for decades. For example, despite the value of the unique rail-gun and MESM, the first electronic computer in continental Europe, they were not transferred to any Museum. The Museum in its current form was established in 1998, thanks to the efforts of Mr. Zgurovsky – Rector of the KPI, and since then it has been supported by the University administration. In some time after its founding, the Museum became the hub for the concentration of collections that could not be maintained by their owners without outer help and so they transferred their displays to the Museum. This is just how the Aeronautics and Space Exploration Department has appeared in the Museum. The view on the scope of different historical aspects of saving technical heritage in Ukraine is given in the presentation with the focus on The State Polytechnic Museum.

The museum is special because it composes an entire complex, which includes building No. 6 of NTUU "KPI", aircraft and car garage, museum collection of the State Polytechnic Museum at NTUU "KPI", including open-air museum items placed in the surrounding area as well as area with the monuments of famous scientists and engineers. At the same time, it has the status of the State and University Museum.

The exhibition areas of the Museum and stock rooms cover 1,700 m^2 and 240 m^2 accordingly.

3.1 Electronic Computers Collection

Electronic computers collection which according to areas of use and uniqueness of some items has no analogs in the world (nearly 100 showpieces) it can boast of unique pieces related to different technologies. The early computers section displays a range of the first Ukrainian machines including one of the earliest MIR-1 (made in 1965), known for having been bought by IBM in 1967. In Ukraine, the first steps in computer science were made at the Institute of Cybernetics under the direction of the mathematician and cyberneticist Victor Glushkov.

3.2 Collection of Space Section

Collection of space section that includes items used in space and in preparation for human space flight, doesn't have analogs in Ukraine (more than 1,400 showpieces), modern nanosatellite "Polytan", created by students and scientists of KPI.

3.3 Arms Collection

Arms collection from the end of 19th to the end of 20th century doesn't have analogs in Europe (263 showpieces). The exhibition reveals the complex history of small arms as a technical device from inception to the present, focusing on the revolutionary changes in design and technology. The collection presents some of the most famous systems in the world, design school: Britain, France, Germany, USA, Czechoslovakia, the USSR, China.

3.4 Archive and Book Stock

Archive and Book Stock of the Museum contains many valuable books and technical literature, original drawings of the first KPI buildings in the late 19th century. Especially valuable are general and detailed drawings of bridges made by KPI students under the guidance of E.O. Paton etc., the facsimile edition of handwritten lectures on descriptive geometry and mathematics and other.

4. Research Work

Research work of the museum is an important area of its activity and is focused on covering the history of science and technology, exploring the creative achievements of ukrainian scientists, estimation of their work in the creation and improvement of technical systems, and founded by them schools and directions. The work includes searching of items as well as memorial things of museum value and their further exploration and justification. The museum collection includes 17,000 exhibits of science and technology of various forms of state stocktaking. There are some unique collections, i.e.

5. Visitors and Events

There are 30,000 people and 650 excursions were held for them. 95 foreign delegations visited the Museum: delegations from Russia, China, Korea, Spain, Malaysia, USA, Canada, Finland, Norway, France, Germany, Poland, Switzerland, Japan, Brazil, Greece, Ghana, which had an opportunity to get acquainted with the history of machines and places of origin of scientists and technicians of Ukraine. The number of thematic excursions prepared at the request of faculties of the University, in particular, the Department of Aircraft and Space Systems, has been increased.

The museum works with various international youth organizations (for example Enactus, JCI Ukraine), holds various festivals (V Kyiv SteamPunkFest, TEDxKPI), hackathons, scientific readings of the cycle "Outstanding designers of Ukraine"(55 readings for students and staff of the university since 2002, based on materials of readings 7 volumes of book "Outstanding designers of Ukraine" were published), supports interesting student initiatives, etc.

6. Conclusion

Purpose of the Museum is to promote the identity of the Ukrainian young people as followers of technical, design and scientific elite of Ukraine.

The main tasks of the Museum are the purposeful formation of the positive image of modem university taking into account changes in social structures, ways of life, society, information technologies as the starting base for young people to receive education and profession. The Museum tries to accomplish such task using its only special methods - through exhibits (exhibits - the media contents).

Innovative Application of Special Theater System in Museums

Li Jingxia[①]

Abstract The special theater system is widely used in the Science and Technology Museums for its unique technical advantages, and it plays an irreplaceable role in the exhibition. This article introduced the concept, current situation and demands of special theater system and tried to discuss its innovative applications.

Keywords Special Theater Current Situation Prospect Innovative Applications Sustainable Development

1. The Concept of Special Theater System

Special Theater system: a file system different from the traditional film, in a three-dimensional, interactive, multi-dimensional way to esoteric, boring scientific knowledge presented in front of the audience; Special Theater has three features: Use special screening platform to show special films on special occasions. Because of the advantages of its digital technology, it has been widely used in museums in the past twenty years, and it has played an irreplaceable role in other exhibition items.

Three features of special theater:

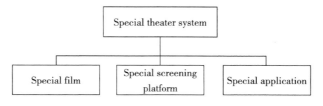

① Jingxia Li: Manager of the 3D New Culture Co., LTD, Ningbo. Email: 346849489@qq.com.

1.1   Common Special Theater System

In the museum, according to the means of expression, the common special theaters include Dream theatre, Scene theater, 4D cinema, Interactive XD cinema, Dome theater, IMAX theater, flight theater, VR theater and so on. Its biggest feature is to disseminate scientific content and display innovative technology, making the film more impacting and attractive, and maximizing the audience's sense of experience and presence.

2. Demands for Special Theater System in Museums

2.1   The Demands for Updated New Technology

In the process of disseminating knowledge, science popularization venues often disseminate through some conventional technical means, the content is relatively single, and exhibits are relatively isolated without interdisciplinary links. For the public, the new media technology more and more become the focus of public attention, the traditional technology has been unable to meet the people to open up horizons. For popular science venues, as a place to inspire people's interest in science and to enlighten innovative wisdom, it is necessary to have the most forward-looking innovative functions to maximize the significance of its social existence.

But, the special theater is a combination of a variety of new media technology, continues to integrate into new technology, and can conform to the characteristics of the times. Compared to the traditional popular science exhibits, it can convey a variety of deep, fragmented knowledge in a comprehensive way through various high-tech innovation means, and sublimate the display function of the exhibits, the high degree of integration with the humanities, stage art making it more in line with the aesthetic needs of the masses and visit.

2.2   Demand for Sustainable Development

Since the promulgation of the National Science Quality Action Plan, science and technology museums have sprung up like bamboo shoots, and in the course of rapid development of science and technology museum career, there are also some new problems and new contradictions, such as sustainable development. So how to make the science and technology museum "forever youth"? We need innovation, which requires a lot of money, but the funds are limited everywhere, if we can't keep up with them, we will have to stagnate the maintenance and renewal of exhibits. Special Theater is a kind of large input, multi-function and comprehensive means of display, for example, the dream theater is an investment, but the site can be multi-functional. It can be used as a dream theater performance, as well as a children's stage performance, and can also carry out popular science experiments or lecture hall, etc. Moreover, the content of the program can also be purchased regularly, often new, so as to achieve the goal of sustainable development.

2.3 Operating Demands Under Free Opening

Exhibit updates is an important means of sustainable development of the Science and Technology Museum. Usually, exhibit update rate is required to be about 10% per year. Large science and technology museums, such as China Science and Technology Museum, Shanghai Science and Technology Museum, etc., have their own R & D team and development site, to protect the annual update of more than 10% of the exhibits. However, most science and technology museums funding is limited, and can't achieve the annual update rate. If the museum sets up a special theater, with the corresponding charges, the museum's maintenance update can be updated to some extent.

2.4 The Demands of Audience Diversion

For a large science and technology museum, the number of exhibits is about 300-600 pieces, the average daily visit number is about 4,000, and the number of instant visitors is about 2,500, which greatly exceeds the carrying capacity of the Science and Technology Museum; this one-time special film that can accommodate 60-100 large exhibition items, for example, 1-3 pieces arranged in the museum, and can significantly deplete the number of visitors in the Science and Technology Museum. It is an important means to divert the visitors to the Science and Technology Museum.

3. The Current Situation and Market Prospects of Special Theater Development

3.1 The Current Situation of the Special Theater Development in China

Since 1959, the advent of China's first three-dimensional film, the special theater has experienced 58 years of development. Among them, a wide variety of movie forms are endless, or giant screen, or sensory class, or dynamic class, with the visual enjoyment that cannot be exceeded by conventional movies. Although the Chinese market started relatively late, comparatively speaking, the momentum of rapid development and huge market demand should not be underestimated. Especially after the World Expo, the domestic market was further stimulated, in addition, the public demand for specialty theater greatly promoted the special theater industry's development. As of the end of 2016, China had about 34,800 3D screens, over 70 dome-screen cinemas and more than 100 theme parks with specialty video. It is understood that China Film's one-time subscription of 100 IMAX movie theaters in 2011 shows that the special television market is huge.

The proportion of Special Theaters in the domestic science and technology museums is also rising year by year. Special Theater programs of science and technology museums above the provincial level are characterized by the large quantity and the large investment, and some science and technology museums have accounted for more than 20% of the total investment. In particular, the introduction of some large-scale projects such as flight cinema, dark ride, dome theater, etc. However, due to its large site requirements and high investment costs, it is difficult to popularize small and medium-sized science and technology museums. In order to better adapt to small and medium-sized science and technology museums, it is necessary to rely on some innovative technologies and innovative means such as VR and AR to make large-scale specialty theater more suitable for market demand.

The following is a review of the use of some special theaters in science and technology museums:

Table 1

Science Museum	Opening Time	Special Theater
Shanghai Science and Technology Museum	2005	IMAX three-dimensional giant screen theater, IMAX dome theater, Four-dimensional theater, Space theater, Dream theater
Guangdong Science and Technology Museum	2008	3D giant screen theater, Dome theater, 4D theater, Virtual navigation dynamic theater, Dream theater
China Science and Technology New Museum	2009	Dome theater, Giant theater, 4D theater, Motion theater
Linyi Science and Technology Museum	2010	4D theater, dynamic loop theater, Dome theater, Interactive three-dimensional theater, Dynamic three-dimensional theater, 3D panoramic sound theater, Science theater (Laser Theater), Dream Theater
Shanxi Science and Technology Museum	2013	Dome theater, 4D triple theater, Dream theater
Liaoning Province Science and Technology Museum	2015	Dream theater, Dome theater, 4D theater, Motion theater, IMAX theater

3.2   Prospects of the Special Theater Market Development in China

For special theater customized market, the main application areas are in science and technology museums, museums, theme parks, and according to "Tourism Green Book 2009" released by the Chinese Academy of Social Sciences, China will enter a new period of development of a large-scale theme park from 2010 onwards. The continuous expansion of the size of the market of science museums venues and theme parks will drive the continued expansion of the special video market size, especially for the provincial and municipal science and technology museums. Special Theater as a new product is very attractive to visitors, and has a leading role in the development of science popularization and tourism culture, so, the market demand is more exuberant.

From the supply point of view, a few years ago, a large part of specialty films required by domestic theme parks, major science and technology museums and museums were purchased from abroad. There were fewer Special Theater providers in China, who did not fundamentally solve the problem of domestic market demand; in recent years, some domestic enterprises engaged in Special Theater have risen abruptly, to a large extent, they have mastered the full range of technologies from device production to film production and have shown promising prospects in certain fields.

4. Innovative Application of Special Theater in Science and Technology Museum

A large part of the innovation and application of special theater depends on the market. Because of its high technology content, excellent audio-visual effect, and a strong sense of experience, it is favored by the major science and technology museums, has become the standard configuration of the science and technology museum, and is also one of the popular science projects that the majority of viewers are striving to experience first. The introduction of many technologies such as AR Technology, Virtual Reality, Holographic Phantom, Real-time Interaction and VR Technology has provided a new breakthrough for the development of special theater and has formed a new field that has produced an interactive theater, Dream Theater, Dark Ride and other high-end special theater forms.

For example, Liaoning Science and Technology Museum Dream Theater - "Cretaceous

Journey"(Fig. 47) is a typical case of the AR technology, it is a comprehensive special film with the traditional science play combined with multimedia interpretation. Through the use of high-tech means such as holographic ghosting, real scene &virtual scene, live-action & virtual image and multi-level picture interaction, it vividly interprets the cretaceous growth environment, animal and plant species and the extinction of dinosaurs.

Audiences without wearing stereo glasses can see lifelike three-dimensional images, multilevel image shuttling, the real-life actor's magical appearance and disappearance, the cartoon image and the audience's real-time interaction, and they produce a virtual, real and illusory special effects so that the audience can't find out the mystery.

The theater takes full advantage of the special effects of Special Theater to break the space, time and other objective constraints, it is a brand-new attempt of the traditional popular science drama, which comprehensively displays the obscure and fragmented knowledge points, systematic communication and artistic deduction, so that the audience exposure among them to learn scientific knowledge and feel the science and culture.

The Traditional Ride program is another special form of the video that incorporates the latest technology. It is highly sought after in domestic theme parks because of its superb experience, some of the science museums also introduced this project, for example, Hangzhou Low Carbon Museum, "Global Warming", science and Technology Museum, Foshan, "Dream" and so on, is also very popular in science and technology museums, but not popular in small and medium science museums, for three main reasons: a large venue, big investment, and high operating costs. First of all, in the science and technology museum, such projects require about 2000 square meters space, and 2000 to 30 million yuan in investment, which is not affordable for general science and technology museums (Science and Technology Museum Special Theater usually covers an area of 200-500 square meters, and about 200 to 600 million investment). Second, the high operating costs, including: ① maintenance costs. ② equipment replacement costs. ③ environmental update costs. ④ program update costs. ⑤ personnel costs, maintenance costs and secondary design costs have greatly increased the cost of science and technology museum operations.

In order to adapt to small and medium science and technology museums, in line with the public taste, proceed from the following considerations: First, reduce the cost of making big scenes. Narrow our visual world to the front. In the technical means, the popular VR head display replaces multiple large scenes, the physical space that the human eyes can't see is achieved through VR glasses, moreover, seven degrees of freedom dynamic platform is equipped, and in combination with virtual imaging technology, panoramic VR, scene effects and other high-tech means, 360-degree panoramic experience is realized, which we call VR Ride (Fig. 48). The advantage is that the venue is greatly reduced; at the same time, the content of the program can be changed at any time without limitation of the venue and the scenery; furthermore, the audience can experience the operation by themselves and only one technician can operate, greatly reducing staff costs.

The table below analyzes the Traditional Ride and VR Ride in Science and Technology Museum:

Table 2

Project Name	Venue(m²)	Investment	Second Investment	Staff
Traditional Ride	About 2000	20 -30 millions	Equipment maintenance update 5%=1-1.5 million	4-7 persons
VR Ride	50	1 -3 million	Equipment maintenance update 5%=50-150 ten thousand yuan	1-2 persons

The figure above shows that the combination of traditional ride and VR ride makes the two complement each other and solves the bottleneck of space, investment and its later operations, and then, we will explore the education mode of science and technology museum and promote the sustainable development of science and technology museum, bringing new highlights for the Science and Technology Museum. Of course, the development of popular and interesting film also needs to follow the pace to make it more consistent with the purpose of the Science and Technology Museum to disseminate scientific knowledge.

5. Conclusion

In short, the innovative application of digital technology in the Special Theater is one of the mainstream trends in the future. It not only brings a brand new audiovisual experience but also overcomes the problem of sustainable development of traditional Special Theater in science and technology museums. Therefore, the Science and Technology Museum must occupy the latest cutting-edge technology in order to make more and more public go into the Science and Technology Museum.

The History and Development of China Science & Technology Museum

Yin Hao [1]

Abstract As the unique comprehensive museum of science & technology at national level in China, China Science and Technology Museum (CSTM) is a large-scale science popularization facility for the implementation of the national strategy of invigorating the country through science and education, strengthening the comprehensive national power of the country by relying on talented people and innovation-driven development, as well as for the enhancement of the scientific literacy among the general public. Since Deng Xiaoping agreed on its establishment in 1978, CSTM has gone through three phases in its construction with Phase I works completed and opened in 1988, Phase II works in 2000 and the new complex in 2009. Each phase of works truly represents a historical leap. In recent years, we have made the positive exploration and gained relatively good results around such aspects as strengthening the demonstration role in exhibition and education of complex, serving for the construction of Modern Science & Technology Museums System with Chinese Characteristics, and promoting the development of the science & technology museums in China. With the approaching of its 30th anniversary in the year of 2018, CSTM will surely undertake the new mission and extend the new thinking to contribute more to improving the scientific literacy among the general public.

Keywords China Science & Technology Museum (CSTM) History Development

1. The Development History of CSTM

The preparatory construction of CSTM can be traced back to 1958, and it was suspended

---

[1] Yin Hao, Director-General of China Science and Technology Museum, E-mail: yinhao@cstm.org.cn.

for various reasons. Until 20 years later, Deng Xiaoping reiterated the viewpoint of "Science and technology are productive forces" at the National Science Conference in March 1978, and China thus ushered in the springtime for science. Against this background, 83 famous scientists such as Mao Yisheng and Wang Daheng jointly proposed the resumption of the construction of CSTM at the conference, and this proposal was approved by Deng Xiaoping in November of the same year. Since then, CSTM embarked on a new journey of museum building venture, and mainly experienced three projects:

Since 1979, after ten years of hardships, CSTM, the first "Science Center" in China, was opened to the public on September 22, 1988 (Phase I) with the building area of 20,000 square meters and the exhibition hall area of 5,000 square meters.

CSTM's unique exhibition education contents and methods were gradually well received by the more public. To ride on the momentum, the initiation of Phase II works was fulfilled in November 1996 and completed and opened to the public on April 29, 2000, with the building area of 23,000 square meters and the exhibition hall area of 15,000 square meters.

With the promulgation of the *Law of the People's Republic of China on Science & Technology Popularization*, the public demands for science popularization increased day by day. In this case, the project initiation for the new museum of CSTM was fulfilled in April 2005, and officially opened on September 16, 2009, after four years of hard work.

The total investment of the new museum reached RMB 2.03 billion, with the building area of 102,000 square meters and the exhibition hall area of 40,000 square meters. The new museum is provided with five themed exhibition halls in terms of "Science Park," "Light of Huaxia," "Exploration & Discovery," "Science & Technology and Life," "Challenges & Future," including more than 1,000 pieces (sets) of exhibition items. There is also one short-term exhibition hall provided with a public space display area and four special effects theaters that comprise dome cinema, dynamic theater and 4 D theatre. Moreover, there are a number of laboratories, classrooms, science lecture halls and multi-purpose halls. During the eight years since the new museum is opened to the public, 27,206,700 people have been received by the end of October 2017, surpassing the accumulative tourist arrivals of 21 million person-times equivalent to that during Phase I and II in the last 20 years. From 2010 to 2016, the average annual of tourist arrivals reached 3.3 million person-times. In 2016, the tourist arrivals reached 3.83 million person-times. In the summer vacation time of 2017, the tourist arrivals reached 1.75 million person-times, inclusive of 55,866 person-times on the single day of August 12, setting a new record of tourist arrivals on a single day.

The construction and development of CSTM have always received the kind attention of state leaders. In 1958, Premier Zhou Enlai approved the construction of CSTM. In November 1978, Deng Xiaoping approved the construction of CSTM and wrote an inscription for the foundation-laying of Phase I works in 1984. In April 2000, President Jiang Zemin wrote an inscription for the completion and opening of Phase II works, that is, "Carrying forward the scientific spirit, popularizing the scientific knowledge, and disseminating scientific ideas and scientific methods." In May 2004, President Hu Jintao inspected CSTM and celebrated festivals with children. In May 2010, He once again came to the new museum for the celebration of festivals with children and adolescents. In September 2009, Vice–President Xi Jinping came to Beijing home court activity of the National Popular Science Day. In May 2010, he once again came to visit the new museum and attended activities on the Children's Day.

2. Summary of Major Work Performance

In recent years, based on the goal of "Popularizing scientific knowledge, promoting scientific spirit, spreading scientific ideas and advocating scientific methods", CSTM has focused on promoting some of the work by upholding the concept of "Experiencing science, enlightening innovation, serving the public, and promoting harmony", and centering on the construction of modern science & technology museum system with Chinese characteristics, as well as the enhancement of scientific and technological literacy for all:

2.1 Strengthening the Demonstration Role of the Physical Science & Technology Museum in Exhibition and Education

(1) Actively promoting the transformation of exhibition halls: As early as 2011, the exhibition area of "Meteorological Tour" was launched. In 2016, the renovation of two exhibition halls that include "Space Exploration" and "Bridge of Information" was launched and completed. From February to September 2017, the entire exhibition hall of "Light of Huaxia" was closed for upgrading and transformation. At present, the transformations of Children's Science Park and other exhibition halls are underway.

(2) Laying stress on the brand building of science and technology education: In 2016, 42,000 educational activities were conducted throughout the year, with a total audience of

about 2.26 million person-times. To focus on the needs of primary and secondary students, the new museum introduced "Consecutive lectures on senior high school entrance examination", "Customizing your trip to CSTM" and a series of other activities. To focus on the reinforced combination of museum and school, the new museum launched "The first lesson" and other brand activities. The new museum held more than 180 sessions of "CSTM Lecture Hall" that attracted more than 70,000 audiences, and also plotted the "Moonlight of Spring River" and other key scientific performance projects. During May to October 2017, the original large-scale interactive sci-fi fairy tale named *Pippi's Mars Dream* was successfully put on the stage within CSTM, Hong Kong Space Popular Science Exhibition, Shanxi and other places. There were a total of 60 sessions, with an audience of 23,400.

(3) Deeply tapping the potentialities of a short-term exhibition: A total of 75 short-term exhibitions were organized since the establishment of the new museum. There are four main modes: i.e. firstly, independent R&D, namely, "Internet Exhibition" "Psychological Exhibition" and so on; Secondly, collaborative R&D by means of major alliance and great collaboration, namely, "Aerospace Exhibition" "Salt Story" and so on; Thirdly, holding exhibitions by crowdfunding, namely, "Virtual Reality" and "Unmanned Power" and so on; Fourthly, introduction of international high-level exhibitions, namely, "Einstein Exhibition" organized by Switzerland, "Science Tunnel 3.0" exhibition organized by Germany-based Max Planck Society and so on. The current "Ancient Greek Technology and Art Exhibition" will be exchanged with Greece on November 3.

(4) Giving play to the role of popular science film & TV and improving the originality of film and TV: In the past 8 years, a total of 96 special effects films have been screened, with the audience of 3.9 million person-times. Three original productions of 4D films, five 3D films and one full dome film were completed, Moreover, the original production of Balance Artifact and other dozens of popular science films was completed, and some of them were broadcasted in CCTV.

2.2 Serving the Construction of Modern Science & Technology Museum System With Chinese Characteristics

In response to the requirement for "Promoting the equalization of fundamental public services" proposed at the 18th National Congress of the Communist Party of China in November 2012, China Association for Science and Technology (CAST) timely proposed to build the modern science and technology museum system with Chinese characteristics based on the integration of previous work. To be specific, constructing physical science & technology museums at the place where conditions are not yet available, implementing itinerant exhibition of mobile science museum within a county, carrying out Popular Science Caravans at townships and remote areas, allocating science & technology museums in rural middle schools, and developing Internet-based digital science & technology museum website. As a state-level museum, CSTM has mainly undertaken the following tasks in the construction of service system:

(1) China Mobile Science & Technology Museum: R&D was launched in 2010, and officially put into operation in 2013, taking the county public, especially primary and secondary students as the main service targets. As of the end of 2016, there had been a total of 295 sets of mobile science & technology museum exhibitions, a total of 1747 stations were visited during the itinerant exhibition, and accumulative benefited tourists of 67.57 million person-times. The expected goal in terms of "basically covering counties (cities) within 4 years" was consummated. In September 2017, the second round of nationwide itinerant exhibition was officially launched with the distribution of 69 sets. As of the end of October 2017, a total of 2,216 stations were visited during the itinerant exhibition, and the number of visitors reached 83.82 million. The popularization of the educational idea, content and form of science & technology museum within county-level cities was preliminarily realized.

(2) Popular science caravan: Started in 2000, with primary and secondary school students in rural areas and rural residents taken as main service targets. As of the end of October 2017, an accumulative of 1,445 vehicles had been allocated, 192,000 activities had been implemented, with a mileage of 33,771,000 km and the total number of beneficiaries of 210 million. Its mobility and flexibility can well meet the needs for popular science among the general public at grass-roots, and it was thus kindly called as the "light cavalry of popular science."

(3) Science & technology museums in rural middle schools: Started in 2012, young students and the general public in rural areas and the surrounding areas were taken as the main service targets, in an effort to solve the "last mile" bottleneck in popular science. As of the end of 2016, a total of 293 such museums were built, benefiting more than 1.37 million youngsters. By the end of 2017, the number of quantity possessed will reach 539. This project is considered as a kind of beneficial exploration dedicated to the cause of "Accurate poverty alleviation via the power of technology" by means of an organic integration between three organization resources at the government, business and foundation level.

(4) China Digital Science & Technology Museum: Started in 2005 and its current positioning is to create a public-oriented popular science website that serves as a hub for the construction of national science & technology resources sharing service platform and modern science & technology museum system. As of the end of October 2017, the website had an average daily PV of more than 2.93 million, ALEXA China Rank reached the maximum of 76, and recently stabled at around 100. The total amount of official resources reached 9.95TB. "Science Talk Show" "Science Enlightenment" and several original columns have been successively launched. The construction of the digital museum has promoted the co-construction and sharing of high-quality popular science resources, and it is building a science & technology museum that can "never be closed."

2.3   Serving the Cause Development of Science & Technology Museum

(1) Strengthening the construction of industry standardization & normalization: After 4 years of preparations, the Technical Standardization Committee of the Chinese Popular Science Service was established in June 2017 after the approval from the China National Standard Commission. At the same time, the revision of *Standards for Science & Technology Museum Construction* and the studies & compilation of the *Evaluation Standards for Environmentally-friendly Science & Technology Museum Building* are underway, so as to set standards and norms for the industry development.

(2) Strengthening industry exchanges and discussions: Relying on the Science & Technology Museum Development Foundation, China National Association of Natural Science Museums (CANSM), special committee under CANSM, special committee for special effects theatre at science museum venue and so on, CSTM organizes the selection & recognition of science & technology museum development award, nationwide science & technology counselor contest, training courses for nationwide science & technology museum curators, forum for science & technology museum curators and so on. In addition, CSTM also organizes the compilation of the magazine named *Journal of Natural Science Museum Research*, and the translation of *Dimension*, a magazine of the United States.

3. International Exchanges and Cooperation of CSTM

3.1   Basic Situation

CSTM has constantly been attaching importance to international exchanges and cooperation. As early as 1982, even before the completion of venues, the "Ancient Chinese Traditional Technology Exhibition" was plotted and exhibited at Ontario Science Center, Canada. Until now, this exhibition has been touring 23 cities in 13 countries and regions, including the USA, Switzerland, UK, Germany, Belgium, Italy, Thailand, Singapore, Holland and Greece, with a total number of 6,572,000 visitors. In the meantime, as early as 1983, Canada's "Ontario Science Center Exhibition" was introduced.

In recent years, CSTM is trying its best to deepen international exchanges and cooperation through inspection, attendance, training and itinerant exhibition. Since the opening of this new museum, as of the end of October 2017, a total of 40-item & 118 person-times bilateral exchanges or exchanges between Hong Kong, Macao and Taiwan were conducted. CSTM has participated in 31 annual meetings of international organizations or international conferences for 173 person-times.

At present, CSTM has joined five international organizations such as cinema, World Science Center Summit International Program Committee, the International Committee for Museums Association, Grand Screen Net work of Science and Technology Centers Association and Asia Pacific Technology Center Association. Among them, as one of the founding member of ASPAC, CSTM hosted the first annual meeting after the founding of this association (1997) in 1998, once again hosted the 16th annual meeting of ASPAC in 2016. A total of 164 overseas delegates and 356 delegates from mainland China attended this conference, and good results were achieved.

3.2 Future Plan

In order to implement The Belt and Road Initiative proposed by Chairman President Xi Jinping, further strengthen international exchanges and cooperation, and contribute to the building of human destiny community, CSTM has proposed the following considerations in four aspects:

First, CSTM plans to intensify the efforts of implementing the global itinerant exhibition of "Light of Huaxia" series thematic exhibitions, promote and introduce the "Ancient Chinese Traditional Technology Exhibition" "China Ancient Machinery Exhibition" and other existing exhibitions. Furthermore, new exhibitions such as "Charm of Mortise and Tenon" and "Be Marco Polo for one day" will be launched.

Second, CSTM will promote the operation mode of China Mobile Science & Technology

Museum and popular science caravan in Central Asia and Southeast Asia, and promote the normalization of the itinerant exhibition. During the meeting, CSTM has signed a letter of intent for cooperation with the Ministry of Education of Myanmar with regard to the itinerant exhibitions of mobile science & technology museum in Yangon and other places.

Third, CSTM plans to launch a number of high-quality science communication services by focusing on countries and regions along The Belt and Road Initiative and integrating such modes as "exhibitions + activities + lectures + popular science films & TV programs +art creation+ Internet + virtual reality + training".

Fourth, CSTM plans to organize "One Conference" (International Forum for Promoting Citizens' Scientific Literacy), "One Exhibition" (China Hi-Tech Achievements Exhibition), and "One Tournament" (Global Scientific Performance Invitational Tournament) during the World Conference on Science Literacy 2018" organized by China Association for Science and Technology in 2018, with the purpose of creating a grand international science center exchange event.

As President Xi Jinping portrayed like this, The Belt and Road Initiative construction aims to transcend the civilization gap through cultural exchanges, transcend the clash of civilizations through mutual reference among civilizations, transcend the superiority of civilization through civilization coexistence, and promote mutual understanding, respect and trust among all nations." CSTM is very willing to take this international seminar as a good start for enhancing exchanges and cooperation with popular science venues within countries and regions along The Belt and Road Initiative, vigorously carry forward the Silk Road spirit with "Peaceful cooperation, openness & inclusiveness, mutual learning & reference, mutual benefit & win-win cooperation" as its core, work together with various museums for pushing forward substantive progress in all the cooperation proposed by the seminar, and jointly promoting the win-win cooperation and common prosperity of popular science venues within countries and regions along The Belt and Road" Initiative.

Foreign Cooperation in the Development of Beijing Elk Ecological Research Center

Song Yuan [①]

Abstract Elk is a plain wetland animal unique in China. A hundred years ago, it was orphaned overseas because of the decline of the country, and not extinct due to the rescue by a British Duke. Beijing Elk Ecological Experiment Center was set up in 1985 for the reintroduction of elk to China, and over 32 years after its foundation, the development of the Center has been basically conducted around elk. In recent years, by actively cooperating with overseas scholars and institutions, it has studied the road to development of elk and elk culture and promoted the concept of ecological civilization construction. By going out, introduction, and going out again, the development of the elk center cannot be achieved without the assistance of all parties, and under the guidance of The Belt and Road Initiative policy, the elk center will see a better future of development.

Keywords Beijing Elk Ecological Research Center   elk   history   development

Beijing Elk Ecological Research Center (Elk Park) is also called Beijing Elk Museum and Beijing Biodiversity Conservation Center. Elk Park, a research institute focusing on elk protection, is a science communication institute for elk protection, ecology conservation, and environmental conservation, and is also an outdoor museum which displays elk and elk's habitat. Elk Park was established in1985 for the reintroduction of elk. From 1995, Elk Park was opened to the public and concentrates on research and science communication in wildlife. During the 32 years development, Elk Park has made great achievement in elk protection and diffusion,

① Song Yuan: section head of science communication of the Beijing Elk Ecological Research Center. Email: yl_s2000@aliyun.com.

wetland ecosystem recovery, biodiversity research, and science communication in ecosystem and wildlife protection. With the rising of the idea of the ecological civilization construction and the implementation of Belt and Road Policy, the science communication work in Elk Park improved, and the cooperation with foreign countries increased. Cooperation leads to improvement. Now, Elk Park has increasing science communicating equipment, exhibition, and activities which are connected with ecological civilization construction.

1. The Relationship Between Elk and China, the UK, and France

Similar to the Giant Panda, the elk is a special species in China. It used to live in the wetland of the plain in the east of China. It was used as the symbol of the power of the emperor and lived in captive for sacrifice and hunting since the Shang and Zhou dynasties. In China, there are two idioms which referred particularly to the elk. Because the habitat of Elk was ruined by wars and natural disasters during the long history, after Qing dynasty, the elk was extinct in mostly China, except one group which was still alive in the Beijing Nanyuan Royal Hunting Garden.

In 1865, the French Père David came to China. As a natural historian, he found elk in the Beijing Nanyuan Royal Hunting Garden. He bought two specimens of elk and transported them to the Paris Natural and History Museum for identification. Since that, the elk was scientifically found and well known in Europe. It belongs to the deer family.

After Père David introduced the elk to Europe, European people imported some elks to the European zoos. When elk was extinct in China in 1,900 because of war and flooding, European zoos still kept some alive. Unfortunately, these elks had very low reproduction rate, which might lead to the extinction of the whole species. Around 1,900, the 11th Duck of Bedford from England bought all of the elks from the European zoos, totally 18, since he was interested in Deer Family. The 11th Duck of Bedford kept these elks in the Woburn Abbey. Since then, the elk was saved from extinction.

During the Second World War, to keep elk safe, the Duck of Bedford family sent elk to other countries. However, the elk was not sent back to China during the war. Until 1985, after 85 years of the elk extinction, the 14th Duck of Bedford signed the agreement with the Chinese Government and Beijing Government. Elk was reintroduced back to Nanhaizi, Beijing, China, where it was extinct in 1,900.

2. Development of Elk Park

Elk Park was established in 1985 for the reintroduction of Elk. Since 1985, Elk Park has kept a very good relationship with the family of the Duck of Bedford in elk protection cooperation. After the reintroduction, the Duck of Bedford sent a professor who was famous in elk to China to help Elk Park for the arrangements. At the beginning of the establishment, the duty of Elk Park was elk protection in China. From 1995, Elk Park was opened to the public as a science communication institute. In 1999, Elk Park got the permission to be Beijing Elk Museum. After then, Elk Park was branded as Beijing Science Communication Base, National Science Communication Base, Outdoor School Base, Beijing Environmental Education Base, Beijing Ecological Civilization Education Base, and so on. As time passed by, the science communication work in Elk Park developed well.

2.1 Science Communication Equipment

As the only outdoor museum in Beijing, Elk Park displays a wetland ecosystem to the visitors, which is the habitat of elk. Coordinated with the wetland habitat, Elk Park built some science communication equipment for wildlife protection and environmental conservation disseminating. Some equipment uses pictographic form sculpture and humorous words to let visitors know the principle that there is no killing without buying and selling, such as "the end of deforestation" "the bird's nest", and "the shark's fin". The equipment which is named "the Earth Ark" (Fig. 49) tells people why the ecosystems in the earth need to be protected. The Domino of Extinct Animals (Fig. 50) tells how human behavior leads to extinct and what we could do to prevent it. In Elk Park, visitors could also found a series of sculptures called "Chinese Traditional Poems and Drawing in Wildlife Protection" and a wall which is covered by all kinds of the Chinese character "deer", which is called "Deer Wall". These two pieces of equipment show the traditional Chinese culture in wildlife protection and deer. After visiting, visitors said that the science communication equipment in Elk Park was interesting, original, and knowledgeable. Elk Park is separated from other parks and zoos because of the unique equipment.

2.2 Displays and Exhibitions

In 2001, Elk Park opened the first indoor display called "the Legend of Elk" (Fig. 51).

Since this display, Elk Park initiated a new way in science communication. In 2010, Elk Park cooperated with a German private museum to get nearly 3,600 pieces of deer specimens. By using these deer specimens, in 2014, Elk Park had a totally new display which was named "the Deer of All Lands" (Fig. 52) opened, which was successful. This display showed knowledge about the deer family, the influence of human on deer, and the relationship between deer and human. In 2015, the exhibitions were opened for the 30th anniversary year of elk reintroduction, which was named "Show of Deer Antler" and "the Achievement of Elk Park in 30 Years". These two exhibitions enriched the Elk Park science communication content, and they also increased the level of the ecological civilization education in the park. As more and more displays and exhibitions released to the public, the number of visitors to Elk Park was keeping increasing. Totally more than ten million visitors visited these displays and exhibitions, and most of them indicated that these displays and exhibitions are useful and helpful.

2.3 Activities

Science communication activities are also important in Elk Park. Elk Park has its own science communication education system since exchange activities these years. Now, Elk Park has designed some series of science courses and games which are considered the superiority of Elk Park in ecological civilization. Elk Park organized a large number of activities for outdoor school courses and social practices for students. For example, the courses, which include the "Spring and Summer Phonological Album in Elk Park", "Play in the Wild", and "Elk Park Ecological Tour Map", is designed especially for the primary and middle school students. These courses try to lead the students to feel nature in the class and let them know the idea about the harmonious coexistence of humans and nature. Students who attended these courses said that the courses were designed originally and they learned these ideas from the courses. Another famous activity in Elk Park is science communication drama. Since drama is an easy way for the public to accept, Elk Park produced some dramas which had an idea about wildlife protection or Elk knowledge. Now Elk Park has totally about 5 dramas which can be played to the public. In 2016, and 2017, the activity called the "Exhibition of Science Communication Drama" was set to release to the public. Each year there was three performances in the theatre and about 5 dramas were played in one performance. Visitors to these performances not only include school students, but also parent-child families and community residents. Visitors felt released when they watched the drama, and at the same time, they learned the idea and knowledge about wildlife protection from the drama. Furthermore, Elk Park has another series of activity called "Lectures in Natural Story". In this series, lectures, courses, games, and handwork are combined together to let the participators feel the idea of ecological civilization.

Also, Elk Park joined in social science communication activities in many ways. In recent years, the science communicators in Elk Park gave lectures, courses, drama, and games to the public all over China. Totally more than 10 million people from 20 provinces had involved into the Elk Park science communication activities. By these activities, more and more people know why nature need to be protected, how people could do for helping the wild, and the importance of the ecosystem biodiversity.

3. Development and Cooperation

With the development of Elk Park, the chances of cooperation between foreign countries and Elk Park increased, and also Elk Park attracted more and more foreign visitors. In 2015, Elk Park organized the meeting "First International Symposium on the Conservation, Protection, and Management of Elk and on Biodiversity". This meeting was arranged in the year of the 30th anniversary of the reintroduction of Elk. The conventioneers were from 9 countries, including the UK, Germany, Japan, France, Poland, Israel, Czech, Slovakia, and China. This meeting increased the communication between these countries in Elk protection achievement and opened the curtain for the long-term mechanism for international cooperation for Elk Park. Now, Elk Park keeps the long-term cooperation relationship with Australia, the UK, Germany, and so on. All of these cooperation gives Elk Park chances to develop. By Cooperating with these countries, more and more foreigners know Elk and the level of science communication of Elk Park increases.

In the future, Elk Park will enhance the cooperation between other countries to export its own ecological civilization idea and import the technique and experience for the better way to do science communication.

On the Development and Research of Exhibition and Educational Activities of Chongqing Science and Technology Museum

Zhang Jie[①]

Abstract "The Belt and Road Initiative" is a major initiative that the Chinese government has proposed according to the characteristics of the times and the global trend. The cooperation and exchange of science education, as the important content of The Belt and Road Initiative, is an important field to enhance the level of cooperation along the culture and education of the counties, and also an effective way to promote the cultural integration of the countries along the BRI. It is also a part of the science education about the construction and development of science and technology museum. By taking Chongqing Science and Technology Museum as an example, this paper introduces its development history, exhibition and education activities and lessons, discusses the future development difficulties and challenges of the venues and proposes some countermeasures and suggestions. It is hoped to promote the understanding, cognition and mutual trust among countries along The Belt and Road Initiative in the construction and development of science and technology museums.

Keywords Science and technology museum   history   exhibits   educational activities

The Science and Technology Museum, as the infrastructure for the development of the country by science and education, an important symbol of urban civilization, and a place for public acceptance of popular science education, receives more and more attention from all walks

---

① Zhang Jie: Staff of the Chongqing Science and Technology Museum. Email: 182262914@qq.com.

of life. With the deepening of people's understanding of the science and technology museum and the lack of improvement in the concept of construction, the flourishing development of the science and technology museum, which originated from Western civilization, has formed, the culture of science and technology, in the practice of combining Chinese activities of popular science.

1. Concept and Function Orientation of the Science and Technology Museum

The science and technology museum is a place for science education, which shows the way and function of education through visual stimulation and interactive experience. It arouses people's interest in exploring and learning, and gradually changes ideas to form a scientific world outlook and humanistic values. Therefore, it is also the second classroom, tourist hotspot, civic training and education base, advanced science and technology extension center, science and technology service center, academic exchange center at home and abroad, etc.

2. Construction of Chongqing Science and Technology Museum

2.1 Construction Guiding Ideology of Chongqing Science and Technology Museum

To adhere to the people-oriented tenet, take the "innovation and harmony" as the concept, and take the "life, society and innovation" as the theme, in order to stimulate scientific interest and enlighten scientific innovation for educational purposes. Create a situation for the public to learn science from practice. Through interaction, participation, experience and other educational methods, the public will be guided into the process of exploring and discovering science. Improve the public scientific and cultural quality, build an innovative city and construct a harmonious society service. Actively create a platform to experience the charm of science, a hall to enlighten the innovative ideas, a window to display scientific and technological achievements, and a position to carry out popular science education.

2.2 The General Survey of Chongqing Science and Technology Museum

Chongqing Science and Technology Museum is one of the key projects of the ten major social and cultural undertakings established by the municipal party committee and the government of Chongqing. It is an institution directly under the Chongqing Association for science and technology, a modern, comprehensive, multi-functional, large-scale popular science education venue for the public, and also the basic science popularization facility to implement the strategy of "Rejuvenating Chongqing through Science and Education" and to improve the scientific and cultural literacy of citizens.

It is located in the core area of the Chongqing Jiangbeizui Central Business District (CBD) at the intersection of the Yangtze River and the Jialing River. It was laid a foundation on January 7, 2006, constructed on October the same year, and opened in September 9, 2009. So far, it has been opened for 8 years. This museum covers an area of 37 Mu and a building area of 48. 3 thousand square meters (of which the exhibition area is 30 thousand square meters), with a total investment of 567 million yuan (including 400 million yuan for construction and installation projects and 167 million yuan for the exhibition projects). Till up to August 2017, Chongqing Science and Technology Museum has received more than 12. 4 million visitors.

3. An Overview of the Exhibition and Educational Activities of Chongqing Science and Technology Museum

Exhibition and science education activities are the driving force and source for the sustainable development of the science and technology museum, as well as the core competitiveness of the pavilion development. Here is a brief overview of the exhibition and activities development of Chongqing Science and Technology Museum.

3.1 The Present Situation of Exhibition Features and Design and Development of Chongqing Science and Technology Museum

The exhibition concept is the soul of the science and technology museum. The exhibition of Science and Technology Museum is not simply the display of exhibits, but through exhibitions, inspires people's thoughts and awakens audience awareness. The content construction of Chongqing Science and Technology Museum is divided into the permanent exhibition, short-term exhibition, science popularization training experiment, science popularization activity, popular science film and television. Among them, the exhibits are mainly composed of multimedia demonstration type, interactive participation type, original object and model display type.

3.1.1 The principles of the design of exhibition items in Chongqing Science and Technology Museum

(1) Pay attention to the combination. The content of the exhibition is focused on the combination of science and technology with economic and social development, and the combination of independent innovation with reference.

(2) Highlight interactions. The exhibits are based on interaction, participation and experience, and a variety of forms are used to experience the magic and beauty of science and technology and to create a suitable learning situation.

(3) Embody characteristics. By adopting the design method of theme expansion, the organic connection between the contents of the storyline and the knowledge chain is constructed, which embodies the own characteristics of the science and technology museum.

3.1.2 Layout and upgrading of the exhibition content of Chongqing Science and Technology Museum

(1) The layout of the exhibition of the permanent exhibition

Around the permanent exhibition "Life, Society and Innovation", we have set up 6 thematic exhibition halls for life, disaster prevention, transportation, defense, aerospace technology and basic science, as well as 2 thematic exhibition halls of Children Science Park and Industrial Light. Exhibits include many fields, including military, aerospace, microelectronics, virtual reality, life sciences, basic science and so on. The number of exhibits is over 400.

(2) Short-term exhibition content planning

The short-term exhibition is rich and complementary to the contents of the permanent exhibition, which has the characteristics of a distinct theme, relatively independent content and short display time. The short-term exhibitions will focus on important technological events, major scientific and technological achievements, the latest technological progress and important scientific and technological activities. It helps to increase the flow of people in the Science and Technology Museum and expand the influence of the Science and Technology Museum.

(3) Other content construction planning

The experiment of science popularization training is the expansion and deepening of the exhibition education. It mainly includes training, experiment, science lecture, report and youth science and technology competition, science and technology club, winter and summer camp activities.

The exchange of popular science is the expansion and enrichment of the exhibition function. It mainly includes scientific and cultural exchanges, such as the science forum, the scientific salon, the debate on science and technology, and the meeting of scientists and the public.

Popular science and television are one of the characteristic educational forms of the Science and Technology Museum. 4D motion film, giant screen film and other popular film and television projects are set up.

(4) To prepare and start the upgrading and upgrading project of the exhibition items of the stadiums and stadiums

With the loss of interests of the new pavilions, the flow of visitors in the stadium has dropped sharply, and it is imminent to renew and rebuild the exhibits. After a long-term plan, Chongqing Science and Technology Museum started the 5 years renovation project at the beginning of 2015, and gradually renovated the exhibition area of the exhibition hall, so as to achieve the frequent exhibition hall and satisfy the visiting demand of loyal fans. The preparatory work of the early stage transformation is manifested in the following aspects:

① During the operation of the venue, we should observe the exhibits dynamically, compare and analyze the exhibits that are prone to damage, with low participation and damage resistant and popular exhibits, so as to provide theoretical and data basis for the renewal and renovation of the exhibits. ② We should strengthen the accumulation of frontier information in science and technology development, pay attention to current hot topics, understand the latest scientific and technological research achievements, investigate and collect hot products that audiences are interested in, and update timely exhibits. ③ Collect the relevant information of the production units and collate them. Understand the direction of the production units, such as technology, credit or exhibits stability, and so on, to provide a solid backing for the design and transformation of the exhibits.

3.2 The Design Principles and Typical Popular Science Activities of Chongqing Science and Technology Museum

The educational activities of science and technology museum is a series of scientific, interesting and participatory activities aimed at improving the scientific quality of the public based on the scientific interest of the public and the scientific interest of the public.

3.2.1 The principles of the design of educational activities

The design of educational activities follows the principle of "putting people first", giving full play to the public's main body consciousness, mobilizing their enthusiasm for participation in activities, and easily experiencing and exploring the secrets and principles of science in activities.

According to different objects, it is required for the development of activities:

(1) The pertinence of age;

(2) In accordance with the general requirements of Outline of the National Scheme for Scientific Literacy (2006–2010–2020), the interactivity and interest are highlighted.

(3) To embody the educational value of the activities and to reflect the close relationship with the social life.

3.2.2　Typical educational activities with audience favorite

Through learning science education activities of domestic and foreign outstanding science and technology museums, and constant practice and summary, Chongqing Science and Technology Museum has formed a lot of fixed brands science activities. According to different objects, the following activities with the museum features are listed as follows:

(1) Science, Technology and Humanities Forum

Object of activity: Citizens

To transmit scientific and academic ideas to the public and to spread the knowledge of science technology and humanities, the Chongqing Science and Technology Museum held a great forum of "Science and Technology, Humanities" in 2010. It is a large public lecture which is free and open to the people. It is offers the two-way communication between the elite and the general public, such as scientists, experts and scholars, social celebrities and other industries. By the end of September 2017, 63 stages have been successfully held, with an audience of about 60,000.

(2) The comprehensive practical activities of the cooperation between museum and school

Object of activity: Teenagers

This activity is oriented to meet the needs of the school. It aims to enlighten curiosity, cultivate imagination and inspire creativity. Relying on the exhibition resources of science and technology museum, develop a multi-level and diversified curriculum system with the characteristic education mode of the museum to meet the different needs of students in Grades 1~9. The contents include special visits, theme activities, science experiments, popular science dramas and scientific small production, etc.

The project "Guide to Comprehensive Practice Activities of Chongqing Science and Technology Museum and School" has established a clear "menu" science popularization service mode and mechanism for the school, and effectively connected the bridge between the museum and school to understand each other and interact with each other. The implementation of this project is based on the orientation of the teachers of science and technology instructors, and the students who serve the compulsory education stage.

(3) Light on parent-child Science Time

The object of activity: Family of parents and children

This project is a family of interactive science activities. It lets parents and children spend a fun parent-child times through the experiment, handmade activities and other ways. It can promote mutual communication, enlighten scientific interest, and encourage children to become lifelong learners.

4. Difficulties, Challenges and Countermeasures in the Development of Chongqing Science and Technology Museum

4.1 Difficulties and challenges of the development of the museum

After 8 years of development, Chongqing Science and Technology Museum has accumulated some experience in the educational activities of the popular science exhibition, but still faces some new difficulties and challenges.

(1) The development of the exhibits and permanent exhibition are not enough. The renewal speed of the exhibits is slowly, and the ability of independent research and development is insufficient.

(2) The development of science popularization activities is not enough, and the characteristics of the stadiums are not prominent.

(3) The knowledge reserve of the scientific and technical instructors is not enough.

4.2 Countermeasures and Suggestions For the Development of the Museum

(1) In the area of exhibition development, we should speed up the transformation process of "imitation→imitative innovation→self innovation", "exhibition form design innovation→exhibition exhibits technology innovation→exhibition principle content innovation" and "exhibition innovation→exhibition area innovation→the overall innovation of the exhibition in science and technology museum". Besides, we should attach great importance to the short-term thematic exhibitions, and introduce the short term exhibitions that are popular among the audiences. It will be an effective way to increase the audience and expand the social

impact of the Science and Technology Museum.

(2) In the development of science education activities, although the "Museum and School Coperation" project has a good demonstration effect in the city and the national science and technology museums, the influence gradually vanished with the passage of time and fading of freshness. Therefore, the people of the science and technology museum still need to continue to strengthen their own learning and the research and interpretation of national policies, looking for new innovation activities, and strive to create more new models of scientific education with the characteristics of the science and technology museum.

(3) In the training of talents, we should focus on strengthening the training of professionals in the existing internal museum, especially in science and engineering, science and technology history, social development history and so on, and strive to cultivate versatile and versatile professionals. Through the way of "making up a missed lesson", we should enhance employee's business learning and academic exchanges, form a strong learning atmosphere, build learning oriented science and technology museum, and lay a foundation for the sustainable development of venues.

5. Conclusion

The Belt and Road Initiative advocates to jointly build and share principles, and aims to build a community of destiny, interests and responsibility. Upholding the ideas of equality and mutual benefit, solidarity and mutual trust, tolerance and mutual learning, cooperation and win-win is the direction and important guarantee for the future development of The Belt and Road Initiative construction. It is an important part of cultural exchanges and cooperation in science and technology museum. Each country has different characteristics, experiences and comparative advantages. Through deepening the exchange and cooperation of scientific education and learning from each other, it is beneficial to the overall promotion of the level of science education and the scientific quality of citizens of the countries along the BRI. It is believed that under the guidance of The Belt and Road Initiative, with the deepening exchanges and cooperation of the countries along the BRI, co-construction of the silk road to science communication will be realized in the undertaking of the science and technology museum.

References

[1] LI XIAOLIANG. The Belt and Road Initiative science and education cooperation and exchanges [J]. China Science & Technology Education, 2017. 07:24–27.

[2] ZENG CHUANNING. On the thinking about the system construction of the characteristic science and technology museum. [J]. Science & Technology Association Forum, 2014. 11:25–27.

[3] ZHENG NIAN. Research on the status and development countermeasures of national science and technology museum [J]. Science Popularization 2010 (6): 68–74.

[4] China Science and Technology Museum. Report on the theoretical research on educational activities of China Science and Technology Museum [R]. 2009.

[5] ZHANG JIE, ZHU HAIGEN. Thinking about the design and evaluation of the educational activities of the combination of the science and technology museum and the primary and secondary schools [M]. Scientific Education Papers about Combination of the Museum with the School, 2012, 89–95.

The Idea and Practice of Science Popularization of Chongqing Natural History Museum Under the Belt and Road Initiative

Zhao Di [①]

Abstract Through the introduction to development overview, exhibition design, educational interaction and problems and challenges of Chongqing Natural History Museum, this paper expound the concept and practice of science popularized by Chongqing Natural History Museum during the implementation of the strategy of The Belt and Road Initiative and the opening and construction of the new museum.

Keywords Chongqing Natural History Museum   The Belt and Road Initiative   Idea   Practice

In the process of implementing the national strategy of The Belt and Road Initiative, in order to enhance public understanding of the tremendous achievements made by China in the fields of natural history and science from a scientific perspective, the strategy of The Belt and Road Initiative enjoys popular support, and the Chongqing Natural History Museum, with nearly ninety years of profound historical accumulation, carried out a large number of theoretical exploration and practice activities, and achieved some success. In this process, we also encountered some new problems and challenges. The development of Chongqing Natural History Museum and the problems encountered are discussed as follows:

1. Development Profile

Located at the intersection of The Belt and Road Initiative strategy and the Yangtze River

---

① Zhao Di: Staff of, Chongqing Natural History Museum. Email: 18384378@qq.com.

Economic Belt, Chongqing is a central hub city of The Belt and Road Initiative strategy and plays an important strategic supporting role. Economic and trade cooperation and cultural assistance. As one of the three national first-level museums in Chongqing, Chongqing Natural History Museum is also the only comprehensive museum of natural sciences in Chongqing that interprets the concept of The Belt and Road Initiative. Let the public learn a great deal of scientific knowledge during the visit, cultivate scientific thinking, scientific methods and scientific spirit. As a popular science ambassador of The Belt and Road Initiative, efforts should be made to improve the quality of science and technology for all and boost the smooth implementation of The Belt and Road Initiative strategy.

The new museum of Chongqing Natural History Museum (Fig. 53) is located in the foothills of Jinyun Mountain, with a total area of 216 mu and a total construction area of 30,000 square meters. The exhibition area is about 20,000 square meters. Its design is based on the "leading west, domestic first-class" pursuit. In the first year of opening, the total reception volume of the audience was 3.07 million, with operating income of 14 million yuan.

Its predecessor was the "Western China Academy of Sciences" founded by Lu Zuofu, general manager of Minsheng Company. It is the only private academy of sciences in Southwest China, and also the first one in China. Its purpose is "studying practical science and assisting in the development of economic and cultural undertakings in western China". During the War of Resistance Against Japan, the Western Academy of Sciences of China liaised with more than a dozen national academic institutions that relocated to Beibei in Chongqing to jointly set up a museum in western China. In 1953, it was merged into the Southwest Museum, later renamed the Chongqing Museum, and in 1991, independently established as Chongqing Natural History Museum. On November 9, 2015, the new museum of Chongqing Natural History Museum was officially opened to the public for free. In 2017, it was named the third batch of national first-level museums by China Museum Association. Basic information data after the opening of the new museum is as shown in Table 1:

In the course of the development of Chongqing Natural History Museum, its achievements are mainly as follows: In 1941, Yang Zhongjian wrote and published the first Chinese scientific monograph on dinosaurs in China's Western Academy of Sciences. The monograph named "A complete osteology of Lufengosaurus huenei Young (gen. et sp. nov.) from Lufeng, Yunnan, China". In the same year, he carried out the first public exhibition of the skeleton of Lufengosaurus hueneiin in the Western China Academy of Sciences Central Geological Survey, which is also China's first unearthed dinosaur fossil skeleton; in the summer of 1943,

Li Shanbang, founder of modern Chinese seismology, an important pioneer of geophysical prospecting in China developed Ni-style horizontal seismograph. This is the first mechanical record-type horizontal seismograph developed by China itself. In 2017, "Earth, Creature and Humanity - Chongqing Natural History Museum Basic Exhibition" of Chongqing Natural History Museum won the "National Top Ten Exhibition Quality Awards" in the "Top Ten Exhibition Promotion Activities of the 14th National Museum" sponsored by Chinese Museums Association and China Cultural Relics Newsreceived.

Table 1　Basic data table of the opening of Chongqing Natural History Museum

Annual audience reception volume		Number of exhibitions		Number of activities		Exhibition area		Research results	
3.25 million		16		28		30,000 square meters		78 articles	
Permanent exhibition	Temporary exhibition	Permanent exhibition	Temporary exhibition	Campus Activities	Countryside activities	Exhibition area	Workspace	discourse	scholarly monograph
3.07 million	180,000	6	10	21	7	20,000 square meters	10,000 square meters	75	3

Description: The research results include 27 English papers, including 8 SCI papers, and one English monograph with250,000 words

2. Exhibition Design Overview

The permanent exhibition theme of Chongqing Natural History Museum is "Earth, Creature and Humanity - Basic Exhibition of Chongqing Natural History Museum". In addition to a few of its exhibits for the museum collection, the vast majority are new treasures. More than 8,000 exhibits were collected extensively on seven continents of the world, carefully selected and arranged according to the exhibition outline design, proactively associated with the theme of the exhibition, and formed the core content support through the combination of exhibits and exhibits.

The new museum area of Chongqing Natural History Museum is divided into three layers around the theme of the exhibition. Dinosaur Hall, Behring Hall, Chongqing Hall, Earth Hall, Evolution Hall and Environment Hall are distributed throughout the exhibition. The content systems of the six exhibition halls are both focused, echoing each other, based on the material, and people-oriented, with a systematic and orderly combination, showing the evolution of the Earth, biological evolution, biological diversity and relations among resources, environment and human activities, covering the contents of geology, biology, ecology, paleontology, paleo-anthropology and other disciplines. Under the guidance of the new concept of the Earth, the Earth Hall will use the latest theoretical achievements to reveal the progress of earth science in many dimensions and at various levels. The Evolution Hall will clarify the evolutionary mechanism and explain the evolutionary principles. The Dinosaur Hall reconstructs dinosaur spectacles of the world with informative specimens and scientific layouts to arouse the interest of the public in exploring the dinosaurs; the Behring Hall highlights the geographical distribution of creatures in a global perspective and interprets the relationships among species. The Environment Agency reviews the relationship between man and the earth in the course of human civilization and expands the future. The broad perspective of development leads to the thinking of building a community of human destiny. The Chongqing Hall displays the changes to local natural history, the interdependent wisdom of people and the environment, strengthens the scientific concept of "clear waters and green mountains can bring us prosperity and wealth", and advocates the theme of urban ecological civilization development. Throughout the exhibition, the exhibits serve as a support point for exhibition contents. Through the clues of cocooning and interlocking narratives, the exhibition uses the differentiated expression of visual lines and the modeling art of alien space to realize the exhibition-style in which science and art are integrated.

Taking the Dinosaur Hall (Fig. 54) as an example, in contrast with the simple display of other dinosaur exhibitions in China on the site of dinosaur burial and ecological restoration, the museum visually subdivided the entire space of the exhibition hall into three levels, from the hall entrance arranged in the basement of the simulated field excavation site to the ground floor of the fossil repair room and scholarly bookstore scene, to the dinosaur evolutionary tree and the dinosaur life scene recovery, from dinosaur burial state to the anatomy of the local structure, and two layers of dinosaurs on the ground accompanied by ancient animals and plants. Present concepts of presentation through dynamic linear axes, reveal the biological behavior of dinosaurs and the biodiversity of the dinosaurs, and provide inspiring, participatory answers to public scientific questions. It not only shapes the independence of cell space but also expands the connection with other exhibition areas.

3. Overview of Educational Interaction

In order to establish Chongqing Natural History Museum as a platform for dissemination and sharing of knowledge, culture and education and scientific knowledge, the Chongqing New Natural History Museum opened its doors and worked hard to establish activities in various forms with educational institutions at all levels in Chongqing. The activities organized in recent years are:

3.1 Participating in the Work Projects Carried out by Chongqing Municipality Cultural Relics Bureau to Improve the Youth Education Pilots of the Museum

By setting up four groups of school ages, kindergarten, lower primary school, primary school and junior high school, as architecture, it designed 22 youth science education curricula and 12 popular science books. It signed a cooperation agreement with 15 elementary and junior high schools such as Chongqing No.1 Middle School and Chongqing Zhuangyuanbei Primary School and entered the schools to carry out more than 20 theme education lectures including "Bookmarks of Nature", "Have a Good Trip - Story of Asiaticus", "Education of Sand Painting of Adventures of Little Dinosaurs" and "My Hometown". Over 20,000 young people were covered.

3.2 Theme Exhibitions and Popular Science lectures Including World Earth Day, International Museum Day, Chongqing Science Week, and National Science Popularization Day

It organized the special exhibitions such as "Treasure the Earth resources to change the mode of development - to promote ecological civilization and build a beautiful China", "Enter into Africa", "Stories about Dinosaurs". The "Ichthyosauria in Ancient Oceans of the Triassic", "Elves of Nature", "Carbon Dioxide and Low Carbon" and "Origin and Evolution of the Giant Panda" were held in schools, troops and communities including Chongqing No.1 Middle School, Elementary School of Southwest China Normal University, Beibei District Fire Brigade and Beibei Chaoyang Community. Visitors were more than 30,000 people.

3.3 "Dream Classroom Activities" Organized by Publicity Department of Chongqing Municipal Committee of CPC

It popularized popular science to villages and towns and distributed more than one thousand copies of publicity materials to primary school students in Jiaping Town Center, Jiangjin City. With topography and natural environments of Jiangjin incorporated in the promotional materials, it introduced the main local flora, fauna and unearthed paleontology fossils. The obscure scientific knowledge was popularized to children's familiar living environment to stimulate children's pride in their hometown and even the nation.

3.4　Special School "Natural Science Park" Project in Beibei District

It donated to build a science education park for Beibei District Special Children's School. Aiming at the characteristics and difficulties of special children in cognition, using the means of sound, light and touch, we should refine the science communication and popular science knowledge so that the special needs children can transcend the physiological inconvenience and enjoy the same recognition as the general public for the natural sciences.

3.5　Popular Small Exhibitions to the Townships

In the surrounding counties of Chongqing, 157 exhibitions such as "Flying Flowers", "Low-Carbon Families", "Debris Flow Prevention", "Insects Around", "Dinosaur Experiences", "Bizarre Stones" and "Dinosaur on Stamp" (Fig. 55) were held, and the number of viewers reached more than 129,800.

4. Problems and Challenges

In the long-term science popularization practice, Chongqing Nature Museum insists on protecting the natural environment as the science popularization concept to protect the continuation of human civilization and has also achieved certain achievements. However, we also clearly recognize that there are still many deficiencies in comparison with the vision and planning of The Belt and Road Initiative. How to solve these problems is a long-term challenge to our work. For example, during the process of building popular science education brand, the guarantee of special funds, the cultivation of diplomatic talents of science and technology, the contradiction between the intellectual property protection of brands and the exchange and cooperation of popular science museums need to be tackled one by one. In terms of exhibitions and exhibitions, Chongqing Natural History Museum also needs to do a great deal of research and practice on the core idea of exhibition design, the consistent storyline of exhibition contents, the localization principle of collection and the authentic experience of audiences.

Therefore, strengthening exchanges and cooperation in The Belt and Road Initiative popular science and technology venue will not only share the resources of all parties in such areas as exhibitions and scientific education, and will jointly establish a Silk Road for science communication. Moreover, it provides natural science museums of different cultural backgrounds and themes with opportunities for mutual understanding and in-depth exchange to make up for the cognitive limitations of a single museum and form a powerful alliance.

References

[1] Liu Xiao. The Belt and Road Initiative foreign communication research [D]. Xiangtan University, 2016.

[2] Hou Dechu, Zhao Guozhong. Patriotic industrialist Lu Zuofu and Western China Academy of Sciences [J]. Journal of Sichuan Normal University (Social Science Edition), 2000 (1).

[3] Hu Changjian. Development overview of Chongqing Cultural Relics Museum over the past 60 years [J]. China Museum, 1996 (2).

[4] Li Shanbang. Principle, design and manufacture of Ni-style horizontal seismograph [J]. Journal of Geophysical Research, 1945 (3).

Science Popularization, Mutual Benefit, Coordination and Sharing

—— Development of Hebei Science and Technology Museum

Xu Jing[①]

Abstract Since its opening in 2006, Hebei Science and Technology Museum has gradually become an important place for people around to improve their scientific and cultural qualities and leisure and entertainment. In addition to the permanent exhibition update constantly, the rich popular science activities are also an important aspect of attracting the public. Since 2013, China Association for Science and Technology has launched a nationwide tour of science and technology museums in China. Hebei Science and Technology Museum undertook a number of sets of mobile science and technology museums, and carried out propaganda in the counties, not only to expanding the Hebei Museum's influence, but also driving the enthusiasm of the people in the counties for learning science, setting off the upsurge of learning science and loving science. From the actual condition of Hebei Science and Technology Museum, this article discusses the current development and future prospects of the museum and would like to take The Belt and Road Initiative as an opportunity for the development of science popularization, mutual benefit and reciprocity. Each region will share excellent resources.

Keywords Science and technology museum science activity development

1. Introduction

In September of 2013, President Xi Jinping proposed to build the new Silk Road Economic

---

① Xu Jing: staff of the Hebei Science and Technology Museum. Email: xujingkjg@163.com.

Belt and the 21st Century Maritime Silk Road, emphasizing the need to create a mutually beneficial collective of interests and a shared destiny. The Belt and Road Initiative runs through Eurasia, east to the Asia Pacific economic circle, west to the European economic circle. Whether to develop the economy, to improve people's lives, or to deal with the crisis, and to accelerate the adjustment, many of the countries along BRI have shared interests with our country. And as the main position of popular science education of China, the science and technology museum has played a unique role in the popularization of scientific knowledge, the scientific method and the spirit of science, and the science and technology museum is also the main ground for the national scientific literacy education. It is the responsibility of each of our popular-science people to strengthen the communication and cooperation between the countries along The Belt and Road Initiative and the popular science and technology venues in the region, and to promote the development and progress of the venues, making it an important force for local economic and social development.

2. Overview of Hebei Science and Technology Museum

Hebei Science and Technology Museum is made up of old museums and new museums. Since the opening of the new museum in 2006, the old museum has become an office area. In order to adapt to the new situation of social development, the new findings of Hebei science popularization work have been made to meet the new requirements of the public to learn scientific and technological knowledge, and to give full play to the education advantage and important role of the science and technology museum. Hebei Provincial Committee and Provincial Government invested in the construction of the new museum of science and technology in Hebei province in 1998. In March of 2006, it was officially opened and mainly composed of the exhibition area, the movie theater district, and the conference lounge. The permanent exhibition hall includes four exhibition areas, which involve math, machinery, life and so on. In 2010, the hands-on park on the third floor was upgraded to a safety exhibition area, the application exhibition area on the second floor was updated in 2013 and constructed into a small children exhibition area, and in 2015, the third floor was updated into a robot exhibition area. In September 2017, the children exhibition area on the second floor was updated and is now entering the exhibition arrangement stage.

Hebei Science and Technology Museum, as a provincial building, has a certain guiding role in the construction of small and medium science and technology museums in the country. Hebei Science and Technology Museum has carried out abundant exhibition activities since the opening of the museum in 2006, attracting crowds of people to visit the science and technology museum, to understand the scientific knowledge and to feel the power of technology. According to the incomplete statistics of the museum, since the opening of the new museum in 2006, the permanent exhibition has been open for more than 300 days, with an annual audience of 200,000, more than 600 performances in the sky and sky shows, and more than 500 games in 4D cinema. Since it was opened free in June of 2015, there have been 400,000 people and 98 groups from all walks of life. It has performed more than 70 popular science performances every year and held nearly 20 science popularization presentations. There are 93,266 people in the museum during the summer of 2017, an increase of 165% over last year. More and more people are choosing to use the Hebei Science and Technology Museum as an important place to improve their scientific literacy. In addition, it attracts more people's participation through popular science experiments, science shows and popular science dramas. Popular science exhibitions, such as "World of great view", and "How to deal with chronic diseases in scientific health", have received nearly 150,000 visitors from all walks of life. The museum has also composed more than 10 works of popular science drama and popular science experiments have been in the school for about 10 times a year and have organized nearly 10 scientific activities. These activities promoted the science popularization work, school education and the combination of traditional culture effectively.

Hebei Association for Science and Technology bought a caravan in November of 2002 and set up a popular science caravan working team of Hebei Science and Technology Museum. The exhibits of the caravan are available for hands-on participation, and there are dozens of popular science exhibition signs featuring "Advocating for Science, Opposing Superstition", "Aerospace Knowledge", "Science and Technology, Civilization and Better-off Life" and so on, which can be diversified according to different regions and groups. Since 2003, the science caravan has participated in the large-scale series of exhibition activities and special events, such as "Three Rural Areas", "Popular Science Activity Day" and "Entering Communities Four Times" of the science and technology, culture and hygiene organized by relevant departments and the Hebei Association for Science and Technology. The science caravan exhibition ranges from Chengde in the north, Handan in the south, Qinhuangdao in the east, to Zhangjiakou in the west. So far, the caravan has crossed 11 districts and cities and dozens of counties in Hebei Province, with more than 15,000 kilometers, and more than 200,000 visitors.

3. Characteristics of Hebei Science and Technology Museum

3.1 Initial Results of Informationization Construction

In 2015, Hebei Science and Technology Museum increased its efforts in creating a large number of excellent scientific experiments and popular science dramas, and put them on the website of the science and technology museum through the Internet, which can be viewed by visitors and made into a series of CD-ROMs of "Cognitive Science" and "Cognitive Science II". In 2016, the Internet + public service science popularization platform of Hebei Science and Technology Museum was completed. The platform is based on information technology and Internet media, mainly including: optimizing the website construction of Hebei Science and Technology Museum, opening the WeChat offical account and mobile phone client, and strengthening the construction, application and standard management of new media. It is the implementation of "Internet + popular science" in Hebei Science and Technology Museum and efforts to promote the new ways of the effective combination of science popularization work with the resources online and offline. Hebei Science and Technology Museum has also won the honorary title of "Excellent Base of Science Popularization Education 2015".

3.2 The Original Science Popularization Activity is Magnificent

Hebei Science and Technology Museum is hosting social charity events about 30 times a year, including the science and technology museum activities in the campus, free astronomy training classes, astronomical outdoor observations, public seminars, and so on, which are the most popular science topics, such as "Fluctuation in time and space - gravitational waves", "Splendid China manned spaceflight", "The maker was in the maker's time", left-behind children's public welfare activities, environmental-themed activities, etc.

From the first national original microcap-drama series in 2013, many of the original scripts and the science shows have won many awards, just like "pm2.5 who is the killer", "The tornado" and "The death of the haze". In addition, the original scientific experiments "The quiet ball," "Playing with the dry ice", "The magic feast", "The friction around" and other works also brought science and laughter to the children during the holidays. These activities were deeply loved by the audience.

3.3　Mobile Science and Technology Museum has reached high tide for Science

Since 2013, with the support of the Ministry of Finance, the China Association for Science and Technology has launched the China Mobile Science and Technology Museum exhibition. Hebei Science and Technology Museum also undertook the exhibition work. In 2015, 8 mobile science and technology museums were carried out in 18 districts and counties. The total number of visitors reached 335,600 person-times, they were widely welcomed by grassroots people, especially the youth, and had achieved a good social benefit. In 2016, the mobile science and technology museums in our province have been increased to 10 sets. They have visited 42 stations in the province and received more than 750,000 social audiences. So far in 2017, there have been 17 mobile science and technology museums, which have shown 90 stations and are expected to increase by the end of the year.

4. Current Situation and Development Plan

In 2015, the central government spent 350 million yuan of subsidy funds, promoting 92 science and technology museums to be open to the public for free before since May 16. Hebei Science and Technology Museum is one of the first free open venues. Since the free opening, there has been a sharp increase in the number of visitors, especially during the holiday season, which has caused great pressure on the operation of exhibition hall. How to provide more efficient services in the case of surging audiences is the main problem for Hebei Science and Technology Museum.

4.1　Science Communication is not Enough

Despite 11 prefecture-level cities and 22 county-level cities in Hebei Province, there is only one science and technology museum. Most of the visitors to the museum are people and school groups around, and occasionally travel agencies. There are many people in this city who don't know what Hebei Science and Technology Museum is doing, not to mention other places, which means that the Science Communication of the museum is not enough.

4.2　Visit Experience Needs to be Improved

The first thing that the science and technology museum has to focus on is the public's experience of visiting, including the experience before, during and after a visit. Hebei Science and Technology Museum has paid more attention to the experience of public visits, such as the presentation of exhibits, the experiment of performing science, and the interaction of the audience, etc, which are more neglected in the pre-visit and post-visit experience. The pre-visit experience includes information about the science and technology museum obtained through various means, including the relevant information that should be obtained when the public reach the science and technology museum but have not yet entered the museum. Hebei Science and Technology Museum does not do enough. The post-visit experience includes offline interaction, visits, and experiences. The website of Hebei Science and Technology Museum was changed in May 2016, and the comparison was made in the form of pictures and words. Now richer content has been added to the website, such as "Amusement Park Roaming", "Special Effects Cinema" and "Work Display". The "Work Display" module is a popular science popularization experiment and popular science lecture directed and performed by Hebei Science and Technology Museum offsite, which can make the public again feel the charm of science without being able to watch.

4.3 The Construction of the Professionals Needs Further Improvement

The Education Department of Hebei Science and Technology Museum is responsible for the explanation of the exhibition hall, the science experiment, coming into visit the campus and explanation and performance in some of the mobile science and technology museums. This is the most important department in the science and technology museum, and the quality of the staff can represent the soft power of the museum. There are two obvious problems with this team. Firstly, the starting point of personal quality is not high, the Education Department has 24 people, including 50% with the bachelor's degree, 1 with the master degree, and remaining 50% with the college degree. The second is that the majors are not match with for the work, and the majors of literature, management and economics occupy more than half. It shows that the professionals of Hebei Science and Technology Museum not only needs the promotion of quality but also cultivates the comprehensive personnel. It has been 12 years since the new museum was opened in 2006. In the age of the big bang, the technology is changing, Hebei Science and Technology Museum is falling behind in scale and facilities. Therefore, Hebei Science and Technology Museum is actively planning to build a new museum, so that it can radiate a wider area and bring more and better popular science activities to more people, which will make Hebei Science and Technology Museum a brand.

5. Conclusion

The Belt and Road Initiative strategy proposed by China means that China's production elements, especially high-quality excess capacity, will be delivered to the developing countries and regions along the BRI through policy communication, road connectivity, unimpeded trade, currency circulation, and people-to-people exchanges. Common sense is to facilitate exchanges and dialogues between different civilizations and religions and promote education, cultural exchanges, development of tourism, etc. Education communication has led the way. The New Silk Road University Alliance has been joined by 124 universities in more than 30 countries and regions of the five continents, and various forms of cooperation with the universities in the alliance have been conducted in terms of interschool exchanges, human training, research cooperation, and the construction of think tank. The silk road economic belt, the big data cloud service innovation academy and the Chinese western technological innovation port are all building up a scientific think tank. The International Hantang College and the Chinese Calligraphy Institute have been building intensive research centers in Turkmenistan, Russia, Poland and other countries. Popular science venues can also build a popular science community to play a bigger role in The Belt and Road Initiative. For example, we can strengthen scientific and technological cooperation, jointly build a joint laboratory and an international technology transfer center, enhance exchanges between scientific and technological personnel, cooperate in major scientific and technological breakthroughs, and jointly improve the capacity of scientific and technological innovation. There are endless opportunities for cooperation between different countries and regions. During the four years of The Belt and Road Initiative, the world witnessed a lot of achievements of education, technology, and innovation in the countries along the line. The popular science venue also wants to carry out The Belt and Road Initiative innovation banner to contribute to the development of countries and regions along the line, linking the "Chinese Dream" and the "Dream of Technological Take-off".

Development of Inner Mongolia Science and Technology Museum

Guo Yu [1]

Abstract Inner Mongolia Science and Technology Museum is a project built by the Government of Inner Mongolia Autonomous Region on behalf and is one of the key people's livelihood projects during the 12th Five-Year Plan of the autonomous region. It was constructed in August 2010, completed by the end of 2013, and officially opened on September 20, 2016. The Science and Technology Museum is a large science infrastructure for improving the science quality of all people in the autonomous region and is also the only comprehensive science venue in the autonomous region.

Keywords Science and Technology Museum history development

1. Overview

The old museum of Inner Mongolia Autonomous Region Science and Technology Museum (Abbreviated as Inner Mongolia Science and Technology Museum) was officially opened on December 5, 1983, and one of the earliest science and technology museums established in China. The new museum of Inner Mongolia Science and Technology Museum is a new comprehensive popular science venue built during the 12th Five-Year Plan Period of the Autonomous Region. The project was approved in October 2009 and the foundation was laid on August 18, 2010, with the total construction area of 48,300 square meters, exhibition education area of 28,830 square meters and the total construction investment of 608.3 million yuan. The main functions include Exhibition education, public service, business research, management security and so on. It was opened free to the outside on September 20, 2016.

---

[1] Guo Yu: staff of the Inner Mongolia Science and Technology Museum. Email: 276774029@qq.com.

The shape of the new museum means "the rising sun", and its modeling indicates saddle Hadad sand dunes and other geographical features and connotations. The dome cinema, with the background of green grass, seems like a round of rising sun from a distance, giving the prairie a visual vision of the sun that will rise without falling.

Inner Mongolia Science and Technology Museum regards scientific and technological innovation and scientific popularization as the two wings to realize the development of innovation, and takes the promotion of scientific spirit, the dissemination of scientific thought, the advocacy of scientific methods and the popularization of scientific knowledge as its own responsibility. By taking "the world vision, features of the times, characteristics of Inner Mongolia, Innovation and development" as the construction goal, it has been built into a modern science and technology venue with national characteristics, First class in the western China, and domestically Leading.

The permanent exhibition of Inner Mongolia Science and Technology Museum is composed of five exhibition halls and indoor iconic exhibition items of public space, with the idea of "Exploration, Innovation and Future", concerning nine themes, i.e. "Exploration and Discovery", "Creation and Experience", "Earth and Home", "Attractive Ocean", "Life and Health", "Technology and Future", "Universe and Space", "Intelligent Space" and "Children's Paradise". A total of 457 exhibits are displayed, and with interaction and experience as the main way of display, the exhibits embody scientific, intellectual and interesting features. In addition, there is also a digital stereoscopic large screen cinema, a digital ball screen cinema, a 4D dynamic cinema, a special exhibition hall, a science report hall, etc.

2. Characteristic Activity

In Inner Mongolia Science and Technology Museum there is a unique exhibition area - Mongolian Medicine Theme Exhibition Area. With the support of the related agencies of Mongolian Medicine of the Inner Mongolia Medicine, in combination with phantom imaging, physical interaction and other domestically and internationally most advanced display means, this exhibition area in the Inner Mongolia Science and Technology Museum, lasting three years, focused on the display of the development of Mongolian medicine and theoretical knowledge, diagnostic methods, Mongolian medicine knowledge, Mongolian medicine characteristic therapy and other details, and introduced the essence of Mongolian medicine. Mongolian medicine has three methods: inspection, inquiry and palpation. The Mongolian medicine characteristic therapy also includes moxibustion therapy, sour horse milk treatment, orthopedic surgery, etc. In addition, the Inner Mongolia Science and Technology Museum also houses a Mongolian medicine acupuncture bronze figure, a bronze person of Buddha treatment. Its original version is now collected in the Inner Mongolia Medical College affiliated to the Mongolian Medical Museum, the bronze person is 61 cm high, weighing 21 kg, and the body is marked with 611 acupuncture points, with accurate acupoint tagging. The body indicates acupuncture points, moxibustion points, bloodletting points and so on. In the word of mouth medical teaching age, it not only instructs Mongolian doctors to use therapy to treat diseases but also is one of the most valuable cultural relics in the study of Mongolian medicine acupuncture and moxibustion.

This year, we start to set up the Science Museum Laboratory, based on the concept of STEAM education internationally, and establish the science laboratories concerning energy, robotics, mathematics and other topics, aiming at guiding teenagers to choose research topics, do scientific experiments, and cultivate their creative problem-solving ability. Moreover, our museum is still actively studying and formulating a school-based science curriculum. The goal of the course is to make rational and full use of science and technology museum resources to enhance the scientific literacy of adolescents. The choice of course content follows the science and technology museum resources as the foundation, aims to develop students' scientific literacy, based on the students' existing knowledge and experience, meets the students' cognitive characteristics and interests, and selects the basic principle to reflect the modern science and technology and embody the content of Science, Technology and Society.

In terms of education activities, our museum has carried out many attempts, which received the good response. In January 2017, "Eagle Plan 2017" Youth Growth and Personal Experience Camp, sponsored by the Inner Mongolia Science and Technology Museum and Inner Mongolia TV News Channel, and undertaken by Inner Mongolia Ruixue Education, camped in the Charm Ocean Exhibition Hall of the Inner Mongolia Science and Technology Museum. The children spent the night in the Charm Ocean Exhibition Hall, experiencing a different kind of popular science tour. In March 2017, the Inner Mongolia Science and Technology Museum conducted a three-month school association and set up a joint partnership with 18 primary schools in Hohhot. By upgrading the educational function of the Science and Technology Museum to achieve a win-win situation, the Science and Technology Museum will become the second classroom for primary and secondary school students. Museum-School Cooperation mainly includes theme visits, exhibition hall theme activities, watching science shows and popular science films and so on, with good response from students. On July 10, 2017, the Inner Mongolia Science and Technology Museum began its five-day summer camp, with a total of 6 sessions, with 150 small camp members in this period. From Hubei, Shandong, Zhejiang, Jiangsu and Hunan provinces respectively, the children learned about the ethnic knowledge of Inner Mongolia and gained unprecedented experience in science and technology through the summer camp. They harvested and became happy in science and technology.

3. Difficulties Encountered in the Development of Museum and Prospects

Inner Mongolia Science and Technology Museum has just been opened for a year. Within this year, we found a lot of shortcomings and also encountered many difficulties and plights. Inner Mongolia Science and Technology Museum, located in central and western China, enjoys fewer resources compared with those in developed areas. Inner Mongolia Science and Technology Museum started late although we are not backward in hardware facilities, and are relatively backward in terms of software conditions. The ability to design and develop popular science curricula and exhibits is still very weak. Furthermore, because Inner Mongolia Science and Technology Museum is carrying out a series of transformation and construction, it is time to learn from other science and technology museums and it is especially important to strengthen the cooperation and exchange visits between museums. On the other hand, there are many exhibition items with local characteristics in Inner Mongolia Science and Technology Museum. The scientific knowledge displayed by these exhibition items should also be disseminated through the platforms of other regional science and technology museums. Strengthening cooperation and exchange visits between venues can be used for mutual reference and complementing each other.

4. Suggestions on BRIAMIS

First, science and technology museums in various regions can carry out The Belt and Road Initiative thematic exhibitions, while science and technology museums located in the Silk Road economic belt can combine their own characteristics, history and culture to enrich the contents of the thematic exhibitions. For example, Inner Mongolia has 4,221 kilometers of border with Russia and Mongolia. There are historical routes connecting the Ancient Silk Road,

Ancient Tea Road, Ancient Salt Road and other routes connecting the Mainland with Russia and Mongolia. When related exhibitions are staged, communication with the relevant scientific and technological institutions in Russia and Mongolia can also be performed. We should enrich and perfect this special exhibition together and promote the common development of science and technology museums.

Second, regularly hold science and technology museum exchanges, share each other the educational activities between science and technology museums, and take example from advanced concepts from abroad. After a further step, we can link up with the science and technology museums of other countries, learn from all kinds of people, and especially communicate and exchange with the countries that are more mature in science and technology education.

Thirdly, define the popularizing population of science in underdeveloped areas. Within the one year of opening the museum, many people think that the museum is just a playground for children, and fewer people came to the museum to learn scientific knowledge on their own initiative. Many parents take their children to play, in fact, this is also because parents do not understand the function and role of science and technology museum. We should let adults have the scientific idea, scientific spirit and scientific thinking, and then enable them to guide children. After all, parents plays a pivotal role in the process of education of children.

It is believed that this seminar can promote the development and progress of the museums in various regions and make them an important force in the local economic and social development. By providing "mutual benefit of science popularization and sharing of resources", we should share the resources of all parties in exhibitions and displays, science education and other aspects, and jointly build the Silk Road of science communication.

Making Use of Ningxia's Important Regional Characteristics to Build Up the Activities of Science and Technology Dissemination with Their Own Characteristics

Chai Jishan[①]

Abstract The Belt and Road Initiative strategy is the major strategy made by the Party Central Committee and the State Council according to the profound changes to global situation and overall planning of the international and domestic situations, and is of great significance to creating a new pattern of all-round opening to the outside world in our country, promoting the process of the great rejuvenation of the Chinese nation, and promoting the world peace and development. Joint construction of The Belt and Road Initiative is an important move of our country to comply with the world multi-polarization and economic globalization, cultural diversity and social informatization, an active exploration in the new mode of "China Solutions, Global Governance", and a new positive energy added to make community of human interest and fate community. As a key area of western China development and the important fulcrum of the silk road economic belt, Ningxia strengthens the communication and cooperation of popular science venues between the countries and regions along The Belt and Road Initiative, and promotes the development and progress of venues to make it becomes the important force in the local economic and social development.

Keywords The Belt and Road Initiative　Popular science venues　Communication and cooperation development

① Chai Jishan: senior economist of the Ningxia Science and Technology Museum. E-mail:3134923407@qq.com.

1. Introduction

Located in the upper reaches of the Yellow River in the northwest of the motherland, Ningxia is bounded by Shaanxi Province in the east, by Inner Mongolia Autonomous Region in the west and north, and by Gansu Province in the south, with a total area of 664,000 square kilometers. Since ancient times, the ancient Silk Road is the necessary place and commercial town. Since the reform and opening to the outside world, Ningxia has, through the State Council's "Strategy for the Development of the Western Region", "Some Opinions on Further Promoting the Economic and Social Development of Ningxia", "Planning for an Open Economic Pilot Zone in the Ningxia Mainland" and "Strategic Plan for the Development of Ningxia's Space Development", taken a series of strategic measures, while ensuring the great economic, social and cultural development of Ningxia, laying a solid foundation for the current strategy of deepening and comprehensive docking with the "Silk Road Economic Belt". To realize the two Centenary Goals and the dream of the great rejuvenation of the Chinese nation, it is necessary to support the power of science and technology and power of independent innovation. Implementation of innovation-driven development strategies is centered on the promotion of science, technology and innovation and economic and social development.

2. Pay Attention to Functional Positioning, Create First-class Science and Technology Venues, and Provide the Best Popular Science Resources and Popular Science Services for the Public

2.1 Give Full Play to Regional Strengths and Build Unique Venues

Ningxia Science and Technology Museum is the key dedication project, a gifted presented to the 50th anniversary of the establishment of the autonomous region, with the investment of 250 million yuan, the total area of 3.88 hectares, floorage of nearly 30,000 square meters, the permanent exhibition hall of 16,101 square meters, opened in September 2008. Over the years, by adhering to the idea of "Experience science, inspire innovation; serve the general public and promote harmony", Ningxia Science and Technology Museum has given full play to the role of comprehensive publicity and education venues throughout the region, integrated the construction of venues with Hui nationality culture and Ningxia regional characteristics, and made due contributions to promoting national unity and improving the scientific quality of the public in minority areas.

2.2 Construction of Venues is Progressing Steadily and Functions are improving Constantly

In accordance with the standards for the construction of the Chinese Science and Technology Museum, the new Science and Technology Museum in Ningxia has one ordinal hall and 13 exhibition areas. It has nearly 500 exhibits and exhibition items, more than 1,500 specimens of minerals, animals, plants, paleontology and 4D special effects theaters, as well as two special theaters, including science and technology exhibitions, training, production and screening of science and technology films, science and technology reports, and youth science and technology activities.

2.3 Scientific Planning, New Exhibits

In view of the actual operation and public appeal, from the building characteristics of the venues and exhibition of exhibits, Ningxia Science and Technology Museum employs experts from the industry of the National Science and Technology Museum, through repeated calculation and demonstration, to formulate a five-year plan for the renovation of Ningxia Science and Technology Exhibition Hall. According to the plan, relying on the Central Financial Free of Charge Special Funds for Science and Technology Museums, so far in 2013, a total of 62 million yuan has been invested to upgrade the exhibition hall, realizing the permanent exhibition of new exhibits, so that the capacity of science and technology museum venue services is significantly enhanced, with more prominent role as a position.

3. More Measures, "Small Museum and Big Popular Science" Pattern Formed

Based on the current situation, in popularizing scientific knowledge, advocating scientific methods, disseminating scientific thought and promoting scientific spirit, Ningxia Science and Technology Museum actively carries out the principle of large-scale cooperation, gives full play to the advantages of popular science resources, and expands the coverage of popular science, having formed the "Small Museum and Great Popular Science" pattern.

3.1 Innovation in the Concept of Exhibition and Education and the Ability and level of Exhibition and Teaching is Constantly Improving

Combining the advantages of the region and the characteristics of the venues, Ningxia Science and Technology Museum, while upgrading the hardware facilities such as the exhibition hall, carries out a series of Popular Science Education Activities, Scientific Fun Experiment, Popular Science Drama Performance and other performing activities for young people all year round.

3.2 Active efforts to Explore and Innovate, and development of Scientific and Technological Education for Young People in depth and breadth

Ningxia Science and Technology Museum (Ningxia Youth Science and Technology Activity Centre) is responsible for the popular science education of students in the region, in order to innovate the Youth Science and Technology Education Model, promote the youth science and technology activities to develop in-depth and vigorously have made fruitful achievements. Over the years, we have organized and conducted the Youth Science Festival, the Youth Science and Technology Innovation Competition, the robotic competition, the "big hand pulling small hands into the campus", the activities of scientific investigation and experience among young people, the Tomorrows of Small Scientists, the Scientific Image Festival and other comprehensive educational activities. Holding the "little doctor" training activities during the winter and summer; in 2016, the first Ningxia Youth Science Festival was successfully held, which lasted 36 days. Four levels were linked, covering 22 counties(districts) in five cities and over 300 scientific and technological education activities for young people.

4. Significant Progress has Been Made in Building the Modern Science and Technology Museum System

Ningxia Science and Technology Museum insisted on the important spirit of implementing President Xi Jinping's "Science and technology innovation and science popularization are the two wings of innovation development", to achieve that popular science education service has connotation, exhibition education activities have its own features, youth science and technology education activities have bright spots, science and technology museum system construction has its own innovation, focusing on youth science and technology innovation ability training and the improvement of public science quality, having formed a modern science and technology museum system with the physical science and technology museum, digital science museum, mobile science and technology museum and popular science caravan as the main content.

4.1   The Permanent Exhibition Hall is Open for free and the Audience are Increasing

Since the opening of the Science and Technology Museum has been free in 2014, the number of visitors to the permanent exhibition hall has increased by a great amount, with 550,000 visitors received throughout the year, an increase of 10 percent over the previous period.

4.2   The Momentum of the Development of Mobile Science and Technology Museums has Been Encouraging and Effective

In the national scale, the Mobile Science and Technology Museum took the lead in achieving the full coverage of Ningxia, and popular science benefited the people's livelihood. Ningxia, as the country's first mobile science and technology museum pilot, has been operating in the region, with the footprints in 22 counties of 5 cities, the popular science benefiting more than 1 million people in the old, small, border and poor areas.

4.3   Promote the Construction of Digital Science and Technology Museums in the Province and Regions, and Set up a Network Popular Science Sharing Platform, "Internet + Popular Science" Achieving Significant Results

In 2011, with the support of the network, we built the digital science and technology museum in Ningxia Science and Technology Museum, which was established by China Digital Science and Technology Museum as the first stage construction pilot, and constructed the Ningxia site of secondary sub-station of China Digital Science and Technology Museum. In 2013, a patented Ningxia Digital Science and Technology Museum was built, and since 2014, a digital interactive and integrated platform has been jointly established with enterprises to maximize the benefits of popular science resources to the people and enhance the scientific quality of the public. Advance Internet + popular science actively and make use of all kinds of new media to strengthen the function of popular science education and enlarge the benefit area of popular science.

5. Deepen the System Reform of Ningxia Science and Technology Association and Promote the Sustainable and Healthy Development of Science and Technology Museums

5.1 Focus on Personnel Training and Provide Strong Support for the Construction and Development of Science and Technology Museums

Ningxia Science and Technology Museum has established and improved a supervision, examination and reward mechanism, linking work performance with wage treatment, job appointment and evaluation, etc., so as to effectively stimulate the enthusiasm and creativity of the staff of the museum. By using the "come in, go out" approach, expand inter-museum cooperation, strengthen exchange and learning, and improve professional quality and management level of personnel. We will train our youth business backbone, create practical training opportunities and work out career development plans for them. In the past 5 years, middle managers have completed half of the new and old alternations, and the young cadres who have grown up quickly have injected new blood into the management team of the science and technology museum, introduced new ideas and new motivation, and effectively promoted the innovative sustainable development of the work of the science and technology museum.

5.2 Forming Popular Science Boutiques and Improving Quality of Popular Science

With the national and local attention to popular science, the public is focusing on the construction science and technology museums with the main function of spreading scientific knowledge and enlightening scientific thought. Facing the new historical opportunity of science and technology museum development, Ningxia Science and Technology Museum will, in deepening the system reform of Ningxia Association for Science and Technology system, aim at building a modern science and technology museum system with China characteristics, to transform the design concept, the focus of science and technology development and the way to change popular science education, to create popular science boutiques, to improve the quality of science and technology, and to give full play to Ningxia Science and Technology Museum's important role in promoting innovation drive development in the region, and continuously meeting the public demand for improving the quality of citizen science Literacy.

6. Opportunities and Challenges for the Development of Science and Technology Museum

With the development of science and technology museum undertakings, the educational function of Science and Technology Museums is improving and expanding. Faced with the increasing public demand for science and technology education and the urgent need of the public to provide accurate science and technology services, the science and technology museums face bottlenecks and new challenges in innovation, especially after the free opening of science and technology museums free of charge, they are confronted with the proliferation of visitors, the increase in operating costs, the increasing pressure on the operation of science and technology museums, the increasing damage to exhibits, the long maintenance cycle of exhibits, the slow updating and other adverse factors. Therefore, in the design of the exhibits, we should conduct in-depth research on exhibition items suitable for the actual and development of the museum, not only ensure the interactive participation and experience operation but also ensure the safety of the audience, easy maintenance of the exhibits and other comprehensive factors. In terms of the permanent exhibition of new exhibits, Ningxia Science and Technology Museum shall explore and experiment profitably. In accordance with the requirements of the "Outline of Action Plan for the National Science Literacy", Ningxia Science and Technology Museum puts emphasis on popular science exhibition education and popular science education activities for young people, relies on standardized construction and the development of the quality of staff and workers, continuously strengthens cooperation and innovation in public science and technology education, highlights the social benefits of popular science and technology education, and constantly meets the needs of the public for popular science, realizing innovation and sustainable development. We will continue to make innovations in the concept of exhibition education, enhance the level of exhibition education, increase the capacity of exhibition education, and give full play to the important role of the science and technology museum in improving the scientific quality of citizens.

7. Countermeasures and Recommendations for the Construction and Development of the Science and Technology Museum

President Xi Jinping stressed in the report of 19th National Congress of the Communist

Party of China that "Innovation is the first driving force for leading the development and the strategic support for building a modern economic system. It will provide strong support for speeding up the construction of an innovative country and realizing the construction of a strong country in science and technology, quality, spaceflight, network, transportation, digitalization and intelligent society". By focusing on the construction of The Belt and Road Initiative, we should pay equal attention to the importance of bringing in and going out, follow the principle of "common building and sharing", strengthen the capacity for innovation and open cooperation, and form an open pattern of linkage between China and foreign countries and mutual aid between the East and West. Innovative national construction cannot be separated from the improvement of the scientific quality of the whole people, the Science and Technology Museum should grasp this rare historical development opportunity tightly, keep studying the concept of innovation construction and development, and play a important role of the Science and Technology Museum in publicity and education.

The Science and Technology Museum should focus on popularizing scientific knowledge for the general public, especially young people, disseminating scientific ideas and methods, promoting the scientific spirit, enlightening wisdom, cultivating innovative abilities and making positive contributions to the training of innovative scientific and technological talents. The Science and Technology Museum should pay attention to the popularization and application of experiential inquiry learning in the venues, focus on the deeper excavation of the exhibition resources, combine with the latest popular science education ideas at home and abroad, carry out various forms of popular science education activities, and maintain the vitality and vigor of the Science and Technology Museum. It is necessary to speed up the construction of the popularization of science and technology, improve the construction of the digital science and technology museum, with digital information technology as platform and carrier, by centering on the physical science and technology museum, make use of the "Internet + Science and Technology Museum" Platform and digital popular science facilities of the science and technology museum, and gradually realize the informatization construction and development in the work of the digital science and technology museum, mobile science and technology museum, and the youth education of science and technology, expanding the coverage of the Science and Technology Museum and promoting the scientific management of free opening of the Science and Technology Museum.

Analysis of the Development of Qinghai Science and Technology Museum

Zhai Yong[1]

Abstract Qinghai Science and Technology Museum was founded in 1987, and the new one was opened in October 2011, which is the largest and well-appointed science museum. Making full advantage of existing resources, Qinghai Science and Technology Museum has achieved good social benefits by enriching the content and form of exhibition and education activities. Moreover, with the increasing popularity of science and improvement fame of the museum, some problems in operation have been encountered gradually, which have affected the function of the Science and Technology Museum to a certain extent. By giving the introduction to the basic situation and education activities of Qinghai Science and Technology Museums, this paper analyzes the current situation of its operations, including the implementation content of education activities, achievements, existing problems, etc.

Keywords Qinghai Province   Science and Technology Museum   Education   Difficulties

1. Brief Introduction

Qinghai Science and Technology Museum was founded in 1987, and the new one was opened on October 242,011, which is located No.74 Wusi West Road, Chengxi District, Xi'ning City, Qinghai Province. With the total investment of 430 million, it covers 3.67 hectares and has the total floor area of 33,179 square meters. The exhibition hall, with an area of 14,000 square meters, is divided into 7 theme exhibition areas, 3 special effects theaters, one preface hall, one youth science studio and one education and training center. There are more than 280 permanent

---

[1]   Zhai Yong: deputy director of the Qinghai Science and Technology Museum. Email: 459970338@qq.com.

exhibits, of which more than 90% are accessible to the public, covering basic disciplines, environment, life, energy, transportation, safety, information, aviation and aerospace, as well as the unique natural environment and high primordial ecology of Qinghai and key scientific and technological content in social and economic development. As the largest and most well-equipped popular science museum in Qinghai Province, the Science and Technology Museum plays an important role in popularizing scientific knowledge, spreading scientific ideas, carrying forward scientific spirit and advocating scientific methods for the general public, and it is one of the main implementing units in Qinghai Province to carry out the "Outline of the Action Plan for National Science Literacy". As of the middle of September 2017, more than 4.482 million people had been received, of whom 1.976 million were young people, accounting for 44% of the total, effectively promoting the spread speed and breadth of science popularization activities. The museum always takes the lead as a position to popularize the knowledge of science and technology.

2. Implementation of Popularizing Science and Technology

2.1 Relying on Major Holidays to Carry Out Thematic Science Popularization Activities

In recent years, the museum has held the activities more than 10 times per year on varied holidays, such as legal holidays, public holidays, important anniversaries (New Year's Day, winter vacation, China's New Year, May Day, International Children's Day, summer vacation, National Day, Christmas Day, Science Week, Popular Science Day), and good social benefits have been achieved. Since 2015, the museum has made a series of popular science activities under the theme of "Popularizing Science, Helping Ecological Civilization", expanding the depth and breadth of educational activities and increasing the brand effect. At present, the large-scale theme science and technology activities held by the museum have formed an important influence to the public, and become a fixed platform and window for the public to visit and learn the scientific and technological knowledge in holidays.

2.2 Focus on the youth and plan the theme activities of popularizing science

The museum has always regarded improving the scientific quality of young people as the top priority of popularizing science activity and has carried out colorful and targeted youth activities every year. Firstly, through the form of "Holding activities in the campus", we would organize popularizing science activities regularly on campus to enrich students' life in spare time, and promote the integration of education on and out of school. Secondly, carry out various kinds of science and technology camp activities for youth, such as seeking science camp activity in winter and summer vacation, outdoor survivability training camp, science camp in colleges and universities, etc., and organize a series of visits, study tours, camping activities, and activity location including Distillery, Guide Science Base, Delingha Observatory, Kruk Lake, alien ruins and top universities in Beijing/Shanghai/Guangdong. Thirdly, strengthen cooperation between the museum and schools, and carry out new and interesting "science and technology classroom" teaching activities. Since the opening of the new museum, by relying on the museum's resources and taking the steady progress of museum and school cooperation as the premise, we have strengthened the research and development of courseware and improved the construction of curriculum system. Try our best to create a first-class brand of science education. Up to now, the museum has independently developed 11 series 1185 courseware, and we have established a cooperative relationship with 9 primary schools in Xi'ning, and 82088 students from 13 schools have taken science courses in the studio, which inspired them to learn science and love science.

2.3 Carry out Mobile Science Exhibition Tours to Bring Benefits to the Public

Since June 2011, the museum took the activity of China Mobile Science and Technology Museum as an opportunity to continuously integrate high-quality science resources into remote areas. It has effectively improved the current situation of relatively weak resources of popularizing science at the grass-roots level. At the end of 2015, the mobile science and technology museum successfully completed the goal of full county coverage in the province. In 2016, the second round of tour was started, and an attempt was made to extend from county to township. Up to now, the mobile science and technology museum tour exhibition has been carried out at 64 stations in all villages and towns, cities and counties of the province, with a total journey of nearly 70,000 kilometers and 758,300 people were covered. After several years of exploration and practice, the mobile science and technology museum tour has become a major brand of science popularization work in our province. It has been widely praised in the province and has been recognized by the industry, having reached the national first-class level. In addition, the museum has also actively carried out "Grass-Roots Science Popularization Activities" approximating the life of the masses, closely integrating them with such themes as

"Science Popularization Winter" and "Science Popularization and the Construction of Three Districts". The dissemination of scientific knowledge to the grassroots through the popular form of activities have played a very good role in communication and guidance.

3. Problems and Difficulties

3.1 Insufficient Funds and the High Cost of Operation and Maintenance

For many years, the museum has made unremitting efforts to improve its popularity in and out of the province and also has been recognized and favored by the public. However, the increasing number of visitors has led to problems such as high damage rate of exhibition items and accelerated aging, so the operating and maintenance expenses have been increasing year by year, further increasing the expenditure of funds as the renovation and upgrading of some exhibition areas since 2016.

3.2 Shortage of Talent and Floating Workers

For a long time, the museum has always had the problem of large flow of people and lack of professional talents. The reasons are as follows: Firstly, the higher proportion of casual workers. The difference in income, security, sense of belonging between full time and part time workers is still very large. The factors such as low wages, long working hours, heavy workload, and no reunion with their families during holidays further lead to frequent turnover of staff. It affects the stability of the construction of the talent team. Secondly, job trait. Qinghai has a vast territory, scattered population distribution and low scientific quality per capita. As the main scientific popularization unit in Qinghai Province, it is necessary to go deep into agricultural and pastoral areas to carry out popularizing activities, besides that, the special geographical conditions, such as high altitude and cold, further increase the difficulty of popularizing science. Hard working conditions and periodic going to the countryside are one of the factors contributing to the greater turnover of people. Thirdly, lacking professional talents. At the present stage, the talents of the museum are mainly people with the bachelor degree and the college degree, most of them are at art and science majors, and there is no qualified person with the relevant major in science and technology museum. Together with the above-mentioned reasons, the team has been lacking of scientific and stability.

3.3 Counselors (teachers) of the Museum do not Have Independent Titles

At present, science and technology counselors have a trend of younger age. On the one hand, they are enthusiastic, active and dare to innovate, on the other hand, they also have a high demand for professional titles. The work of science and technology counselors is educational in nature. For a long time, the demarcation of counselor "educators" is not clear, and it is more prominent at the level of professional title evaluation. For nationwide, apart from educational series, museum staff series and engineer series, there is no independent professional title evaluation system and unified assessment standard for science and technology counselors, which cause the failure of most technology counselors to participate in the evaluation of professional titles. This affects their enthusiasm for work and is not conducive to their career development and promotion. In addition, casual workers cannot participate in the evaluation of professional titles, resulting in that the income situation is not optimistic and work enthusiasm and initiative is relatively deficient, indirectly leading to a decline in work efficiency and exacerbated brain drain.

4. Suggestions on Communication and Cooperation of Museums Along The Belt and Road Initiative

Qinghai has been the route of the Silk Road since ancient times and has an irreplaceable important position in The Belt and Road Initiative. BRISMIS serves as an international, mutually beneficial and shared scientific gathering for the promotion of international and domestic science popularization museums. It is of great significance to the construction and development of popular science museum in China, especially along the route of The Belt and Road Initiative. The museum hopes to take this as an opportunity to integrate into the construction of The Belt and Road Initiative and to promote the development of science popularization in the countries and regions along the route together with the domestic and foreign science museum.

Firstly, hoping that the organizer can build a platform of equality, cooperation, exchange and sharing so as to provide the museum with channels for mutual exchange and learning.

Secondly, paying attention to the current situation and development needs of museums in the areas along the route, understanding the difficulties existing in the development in the economically underdeveloped areas, and providing corresponding policy support.

Thirdly, combing the typical practices of outstanding museums in the management, operation, talent training, exhibition, science education and other aspects, and aiming at promoting learning exchanges between different museums, in order to narrow the gap for joint development.

At last, it is hoped that BRISMIS will become a link for the sharing of resources at home and abroad, for common development, for promoting the construction of popular science museums in the underdeveloped areas of the western region, and for facilitating balanced development of science popularization in China.

Sharing Recourses, Building a Popular Science Position

—— The History of Innovation and Development of Shaoxing Science & Technology Museum

Gu Yaogen, Ding Yaodong [①]

Abstract Under the background of times framework of The Belt and Road Initiative, by setting the goal as "Gathering Popularity, Creating a Brand, Building High Level Science and Technology Museum in Prefecture-Level Cities", through innovating management, exploring and going ahead, over three years since its opening, Shaoxing Science & Technology Museum has got the social confirmation from all walks of life in promotion of popular science knowledge, cultivation of creating-science consciousness and improvement of the quality of whole people.

Keywords Shaoxing Science & Technology Museum   popular science position   The Belt and Road Initiative   development history

1. Introduction

President Xi Jinping, stressed on the national "Three meetings of science and technology", that scientific and technological innovation and scientific popularization are the two wings to realize the development of innovation, and scientific popularization and scientific and technological innovation should be put in coequal and significant position. Shaoxing Science and Technology Museum further implements the spirit of the 19th National Congress of the Communist Party of China, closely relies on the support of the Party Committees and

---

① Gu Yaogen, Office director of Shaoxing Science and Technology Museum, Zhejiang, Email:35490827@qq.com.

governments at all levels and the care from all aspects of society, targets at creating "Gathering Popularity, Creating a Brand, Building High Level Science and Technology Museum in Prefecture-Level Cities", unifies the strength of cadres and masses, digs potential advantages, and realizes the construction record of museum construction and the opening in the same year, offering the best quality service, the best facilities, and the best resources to the general public. Over the three years since its opening, Shaoxing Science and Technology Museum has attained significant achievements in the promotion of popular science knowledge, cultivation of scientific innovation consciousness, and improvement of the quality of the whole people through innovation management and exploration, and has been appraised by all aspects of the society due to its contributions to the construction of The Belt and Road Initiative in Shaoxing. On August 17, 2017, Shang Yong, ex party secretary, executive vice chairman and first secretary of the secretariat of China Association for Science and Technology, expressed his appreciation for the rich and novel exhibits and science courses when he visited Shaoxing Science and Technology Museum.

2. Overview of Shaoxing Science and Technology Museum

Shaoxing Science and Technology Museum, located in No. 528, Yangjiang West Road, Jinghu New Area, Shaoxing City, was listed in the key projects of "Three-year Construction Plan of Shaoxing City" in 2009, construction initiated in March 2011, civil works handover completed in May 2014, and the museum opened on December 29, 2014, with the total land area of $55,860m^2$ (83.75 mu), a total construction area of $31,000 \ m^2$, including the overground area of $25,000m^2$, underground area of $6,000 \ m^2$, and building height of 24 m, with a total investment of 250 million yuan of Construction (Fig. 56). The new museum is divided into north and south zones. The north zone is the Rainbow Children Playground and the Popular Science Theater Zone. The south zone is a sequence hall and a temporary exhibition area on the first floor, earth and life, exploration and discovery exhibition area on the second floor, and the technology and life exhibition area on the third floor. There are also four public exhibition areas, such as Shaoxing Academician, Youth Education, Anti-drug Education, Shaoxing Water and Life, and a science and technology education practice base for primary and secondary school students and an outdoor small science and botanical garden.

3. Achievements and Characteristics of Shaoxing Science and Technology Museum in Recent Years

3.1 Strengthen the Position, Demonstrate the Power of Popular Science

3.1.1 The permanent exhibition hall keeps pace with the times

Focusing on the audience's needs for popular science and innovation of exhibits (exhibition items), maximizing the social value of the exhibition hall and presenting it to the general public in a novel way. For example, in the technology and life exhibition hall, the internal structure of the new energy vehicle is presented by means of the structural section view. Moreover, through cooperation between the academy and local government, i.e. the cooperation with the Chinese Academy of Sciences Institute of Vertebrate Paleontology and Paleoanthropology, the Chinese Academy of Sciences was responsible for planning and design of the content of the earth and life hall, the science and technology museum completed the project implementation, and such cooperation is the first in the country. Recently, AR equipment was added to the exhibition hall to display the prehistoric life to attract more young people to visit and experience in modern ways.

3.1.2 The free opening is popular

Through active efforts, Shaoxing Science and Technology Museum was listed in the first pilot unit for free opening in the national science and technology museums, and on May 16, 2015, some free-of-charge people-benefiting policies were implemented, including a permanent exhibition area, popular science lectures, and other supporting services. After the free opening, it greatly enriched the public's technological and cultural life, promoted the scientific literacy of the citizens of Shaoxing, and strengthened the exertion of social benefits. Over the past three years, the number of visitors has exceeded 1.5 million, and the youth audience occupies about 75.2%, with an average annual audience of over 500,000.

3.1.3 Short - term exhibition of new exhibits

Short-term exhibitions can effectively keep the novelty of the science and technology museum. According to the requirements of the audience, combined with the latest scientific and technological progress and achievements at home and abroad, the hot topics in society, influential science and technology characters and events, Shaoxing Science and Technology Museum enriches the content of short-term exhibitions. We have made full use of the exhibition hall of 2,200 m^2 on the first floor, and introduced large and medium-sized short-term exhibitions in multiple channels. In the past three years, we have held a short exhibition 15 times, attracting more than 500,000 people to stop and visit. During "Sailing Downwind – Exhibition of Ancient Ship Models" held in May, we undertook "Shaoxing Ancient Bridge along The Belt and Road Initiative" exhibition, which fully shows the role of Shaoxing ancient bridges in the construction of The Belt and Road Initiative. One ticket was difficult to get in the short-term exhibition of "Crossing Darwin Planet" VR/AR new media science popularization held in July this year.

3.2 Highlight the Characteristic, Arouse Popular Science Vitality

3.2.1 Multiple good "dramas" staged in brand activities

With the theme of "loving science, playing with science and showing science", more than 50 popular science brand activities with linkage and social participation were created. For example June 1st Science Carnival, Popular Science Camp, Science Experiment Show, Popular Science Puppet Show and other theme activities. Falling in Love with Science and Technology Museum, Big Horn and Small Hall, Experts in Exhibition Hall and other series of activities were mainly implemented, Shaoxing campus scientific talent contest was held for three consecutive years, and through active exploration, "Campus Scientific Talent" competition was upgraded to the provincial "Science Players" teenager science talent challenge. Together with the provincial Juvenile Science and Technology Activity Center, and the provincial TV Station, we will hold the second Zhejiang "Scientific Players" Juvenile Science and Technology Talent Challenge this year. Through regular activities, the science and education base will remain active. We also make active efforts to bring the "Popular Science Caravan" into the communities, schools, and the countryside, letting the people in remote mountainous areas also feel the endless charm of science and technology in their home. The "Popular Science Caravan" traveled more than 3,000 kilometers in 2017, with its footprints around Shaoxing, letting the light of science shine more and more people.

3.2.2 Teaching through games for medals

Informatics reinforcing training on students and tutors of informatics was conducted, according to their own level of primary and middle school students, training courses were arranged for junior and senior classes, and while giving special training to players preparing for the National Youth Informatics Olympic League, knowledge of informatics was popularized for primary and middle school students in the city. Moreover, we organized all the outstanding informatics teachers in the city to carry out the discussion, and invited the senior experts inside and outside the city to impart the experience, combining the research with teaching, effectively promoting the information technology in the teaching application. Since 2011, Shaoxing has accumulatively obtained 5 gold medals and 2 silver medals in the international informatics Olympic contests, ranking the first in the country, Shaoxing students won the optimal results, 4 gold medals, 7 silver medals and 9 bronze medals, in the 34th National Informatics Competitions this year, and nearly 60 students of Shaoxing having been recommended for admission to Tsinghua University and Peking University through the informatics competitions. From 2015 to present, more than 50 scientific and technological competitions have been held, organized and participated in, and the youth science and education atmosphere is strong. In 2016, in the "Fourth National Science Performance Competition", our integrated achievements ranked first in the province, and all the six programs were given an award, including three national second prizes. In 2017, the original micro science play "Removing from Hero by Means of Cups of Wine" won the first prize in the Fifth National Science Performance Competition, showing the good spirit of science and technology museum.

3.2.3 Talents introduced through cooperation between the academy and local government

On the date where the new museum was opened, the Popular Science Education Base for Chinese Academy of Sciences Institute of Vertebrate Paleontology and Paleoanthropology was established. In October 2015, the expert team led by Zhang Miman, academician of Chinese Academy of Sciences, and academician of Royal Swedish Academy of Sciences, signed an agreement to establish an academician workstation. This is the first-of-its-kind in the industry of the national science and technology museum, representing the fruitful cooperation. Over the past three years, academician experts came to Shaoxing Science and Technology Museum frequently, carrying out extensive public welfare popular science activities. Hao Yue, Cao Chunxiao, Zhang Miman, Else Marie Friis and other famous experts and scholars were invited successively to carry out the "Science Lecture Hall", more than 10 times a year.

3.3 Highlight the Base and Create a Popular Science Card

3.3.1 Exhibition and teaching linkage reflects charm

In January 2015, Shaoxing Education Bureau and Shaoxing Association for Science and Technology jointly issued a document to establish the education practice base at Shaoxing Science and Technology Museum for primary and secondary school students in Shaoxing City. The base is open to Grade 4, 5 and 6 students in primary schools and first year and second-year students in secondary schools, with the opening hours from Wednesday to Friday. One day is arranged for the activity "Unity of Exhibition and Teaching", half a day for the "Science Dream Workshop" to carry out scientific practice activities, and half a day for the exhibition hall to carry out popular science activities. The "Science Dream Workshop" is provided with a total of 16 classrooms, providing more than 10 scientific practice projects, and can accommodate more than 400 students for listening to lectures, including the science classroom, informatics classroom, model airplane exploration box, modeling exploration box, robot experience room, 3D printing workshop, 3D innovation laboratory, thinking training room, creative best spell room, psychological discovery room, Science Dream Workshop studio, etc. Such a science and technology practice place is first in the national science and technology museums.

3.3.2 Museum-school cooperation expands to improve the quality

Since the operation of the practice base, the school and society have responded enthusiastically. At the beginning of 2015, a notice was issued for the establishment of the base, the construction was officially started in March and completed in early September. The new "Scientific Dream Workshop" was officially opened on October 21. For more than two years after its opening, the scope has been further expanded, more than 90,000 students from 39 schools in Neiyuecheng District, Keqiao District and Shangyu District were received. Through the establishment of outside-school science and technology education practice base, more students participated in science education, the effective connection of outside activities of science and technology with the school curricula were affected, and as a supplement of school education, it has become the "second classroom" for science and technology education among primary and middle school students of the city.

3.3.3 Innovation to create a brand column

Through cooperation with Shaoxing TV Station, the Scientific Dream Workshop, a juvenile science column, very popular with primary and secondary school teachers and students and parents, was launched. Firstly, the novel show means, it promoted a science and technology museum brand and displayed the science and technology museum environment facilities, exhibits and exhibition items in a stereoscopic manner. Secondly, in the form of developing small members and volunteers, awareness of activities was raised to reinforce the practice base of science and technology for primary and secondary school students. The column was divided into four parts: all the scientific experiments in Scientific Exploration Museum was designed and developed by the science and technology counselors of the Juvenile Activity Department of Shaoxing

Science and Technology Museum, and the hosts and the small guests worked together to jointly explore the magic and mystery of science; the "Small Flying Horse Club", through a host entering the classroom, lets the children learn the scientific knowledge in the questions and answers; the "Rainbow Paradise" combines personal games and team games, which is conducive to the cultivation of a parent-child relationship and increases the enjoyment of the column. The "Fantastic Idea Theatre" is a platform to showcase scientific and technological inventions, science and art programs, and scientific and technological innovations, aiming to enhance students' awareness of innovation.

3.3.4 Catching up accurately to lift and upgrade the level

We will further improve the practice base to provide better scientific practice for students in the whole city. First, the curriculum development keeps pace with the times, and more than 30 science practice curricula have been developed, which are taught on a trial basis every week to improve the overall teaching level. Second, the teaching environment is improving continuously, and the "Science Dream Workshop" improves some classrooms. Thirdly, an annual youth science popularization program is formulated, and the education department is cooperated to issue a document to organize the adolescents in the whole city youth to participate in various competitions. Fourthly, in Cooperation with the schools for comprehensive introduction, the competent principals of the schools are organized in the early year to hold a practice base working advance meeting at the science and technology museum to discuss the base science and technology practice program, deliberate over the matters needing attention in handover with the schools, and further improve the mechanism to promote effective operation of the base. Moreover, all efforts should be made to create the practice base as a business card of Shaoxing City and national innovation popularization education.

4. Difficulties Encountered in the Development of and Suggestions for the Exchange and Cooperation of The Belt and Road Initiative

4.1 Current Difficulties of Our Museum

While conducting the characteristic exhibition and teaching activities on the high-quality basis, we have developed all kinds of youth science and technology innovation education practice, forming a good atmosphere of popular science education, leading the similar science and technology industry in the prefectures. However, in the actual operation, there are problems restricting the further development of science and technology museum. Firstly, the new museum was completed earlier than the mature period of Jinghu New Area. Currently, the museum is far from the residentially concentrated area, and the surrounding facilities are not yet perfect, such as lack of catering support, inconvenient public transportation and inadequate parking lots. Secondly, the strength of staffing is insufficient. According to the "Standard for Construction of Science and Technology Museum", staffing of our museum should reach 155 people, while the total staff actually approved by the staffing office is 111 people only (48 people with institution staffing, 63 without such staffing), as compared with staff of 140 and 145 respectively in Zhejiang Science and Technology Museum (30,452 m^2) and Hangzhou Low Carbon Science and Technology Museum (33,656 m^2).The annual total income of a person without such staffing is 40,000 yuan, lower than that in a similar museum in the province, most employees must continue to work on two-day weekends, legal holidays, and the conflict in the rest time and low treatment lead to turnover rate of irregular employees being as high as 120% since the opening, seriously influencing the sustainable development of the undertakings of our museum. Fourthly, there is no separate title appraisal system, the institution personnel in the museum complying with professional title promotion conditions need to participate in the title appraisal in the education system and museology system, and due to various reasons, the review pass rate is low, thus becoming a stumbling block to hinder the promotion of institution personnel. This is not conducive to the stimulation of motivation.

4.2 Working Hard to Solve Development Problems and Promote Innovative Development

The development of science and technology museum is an important foundation for social welfare service and citizen science quality construction. Firstly, we should strive for the leadership and support from the municipal Party Committee and the municipal government, include the goals and tasks of development of science and technology museum and the construction of scientific quality of citizens into the local development plan, make all efforts to resolve the predicaments of our museum such as staff shortages, and unmatched surrounding facilities. Secondly, combine science and technology museum youth science and technology education with school education, and cooperate closely with the Chinese Association for Science and Technology, education and other relevant institution to conduct the construction of the science and technology education practice base of primary and middle school students of Shaoxing City. Third, we need to establish and improve the system of incentive assessment, innovate the organization and management mechanism for science popularization and the education work of youth science and technology, and give spiritual and material rewards to the advanced individuals emerging in the work. Fourth, strengthen the construction of staff team, achieve the normalization and institutionalization of learning and education, deeply carry out the theme education "stay true to the mission, keep in mind the mission", and lead the development of the museum through party construction. Fifth, strengthen the construction of volunteer service team, improve the system of volunteer recruitment, training, assessment and management, give full play to the roles of the three volunteer teams - students, experts and citizens, and establish a sound working mechanism of carrying out popular science volunteer service together with universities around, effectively reducing the operating costs, and promoting the smooth development of public service culture in the science and technology museum.

4.3 Suggestions on Exchange and Cooperation of The Belt and Road Initiative Popular Science and Technology Museums

BRISMIS 2017 was put forward to inject a strong impetus into the construction of The Belt and Road Initiative. Shaoxing acts as an important hub city in The Belt and Road Initiative, and this is not only an opportunity but also a challenge to our museum. We are willing to work together with peer venues to join the "friend circle" of The Belt and Road Initiative popular science venue exchanges and cooperation to promote the development of popular science undertaking. Here, we put forward some humble opinions.

4.3.1 Make up the letdown

Imbalance and inadequacy still exist in regional popular science education and youth science and technology innovation, and science and technology education cause, for example, science popularization in remote mountainous areas and rural public must be further intensified, youth science and technology innovation ability cultivation still must be further strengthened, "Museum-School cooperation" practice still need further promotion, and inter-museum cooperation still requires further deepening. The display content of domestic popular science and technology museums is uniform and stereotyped to a certain extent, the exhibition arrangements in some museums are slightly tough, and communication and exchange with famous science and technology museums in foreign countries should be further strengthened.

4.3.2 Build a platform

The Beijing Declaration proposes to create a The Belt and Road Initiative international collaborative innovation platform with high efficiency and information sharing. In the opinion of the author, the effective means to promote the exchange and cooperation of The Belt and Road Initiative popular science and technology museum is to establish an information communication and data sharing platform, for example, under the leadership of CANSM, with the collaboration of professional committees and working committees, a number of regional associations for exchange, communication, win-win and sharing will be set up in divided areas along The Belt and Road Initiative to push forward the development of popular science and technology museum in the areas.

5. Conclusion

General Secretary President Xi Jinping pointed out that all organizations of China Association for Science and Technology should "adhere to the service for the scientific and technical workers, service for innovation drive and development, service for improvement of the scientific quality of all people, and service for scientific decision-making of the party and the government". Shaoxing Science and Technology Museum, as a directly affiliated institution of the municipal association for science and technology, has always kept in mind the responsibility mission, and has made great efforts to strengthen the construction of popular science positions, to give full play to the important role of science popularization and the science and technology education of adolescents. Under the new normal, Shaoxing Science and Technology Museum will continue to excavate the potential, focus on the future, exert the radiation effect of science popularization, and will continue to promote the popularization of popular science with activities, enhancing the popularity and vitality of the science and technology museum. Moreover, we will further improve the construction of digital science and technology museum, present the new glamour of popular science activities by means of "Internet +", and drive the new development of science popularization, to contribute the popular science power to the construction of Shaoxing The Belt and Road Initiative Important Hub City.

Development of Wuhan Science and Technology Museum

Wang Ruili [1]

Abstract Wuhan Science and Technology Museum has experienced large construction for three times in the last 27 years since it was opened. The exhibition ideas keep pace with the times. The construction of Science and Technology Museum fits the needs of the public. The activities of the popularization of scientific knowledge are rich and colorful. The visiting form of participation, interaction and experience is well received by the public. The forms of the popularization of scientific knowledge will be continuously expanded and improved in the future according to the present situation of the museum and the needs of the public.

Keywords Science and Technology Museum History Development

1. Introduction

Wuhan Science and Technology Museum has gone through 27 years. It consists of two museums, one is old and the other is new. The old museum, at No. 104, Zhaojiatiao, Jiang'an District, was opened to the public since 1990. The new museum is at No. 91, Riverside Avenue, Jiang'an District. Its opening ceremony was held on Dec. 28, 2015. The two parts of the museum provide service for the public as youth bases of science education at the same time (Fig. 57).

2. The Construction of Wuhan Science and Technology Museum

2.1 The First Building of Wuhan Science and Technology Museum

After the National Science Conference in 1978, many Science and Technology Museums

---

① Wang Ruili: Museum assistant of Wuhan Science and Technology Museum. Email: 381397238@qq.com.

in some cities and provinces were put into construction with the upsurge of the construction of the Science and Technology Museums in China. Wuhan Science and Technology Museum experienced the unusual 15 years from the official opening on Mar. 18, 1990 to opening to public again on Dec. 30, 2006, after the reconstruction and expansion.

The construction of Wuhan Science and Technology Museum was listed as one of the 20 main projects for Wuhan people by Wuhan Municipal Party Committee of CPC and Wuhan Municipal Government during the 7th Five-Year Plan. The site of Wuhan Science and Technology Museum, which covered an area of 58.08 mu, was set in Houhu Township, Jiang'an District, in 1984. The foundation stone laying ceremony of Wuhan Science and Technology Museum was held on Aug. 18, 1986.

The building of Wuhan Science and Technology Museum was completed in 1989. It covered an area of 26,623 square meters and the floorage was 13,374 square meters. The building scale was at the top of all the domestic science and technology museums in provincial capital cities. The opening ceremony of Wuhan Science and Technology Museum &The First Wuhan Science and Technology Expo was held on Mar. 18, 1990.

Wuhan Science and Technology Museum insisted on the direction of popular science education under the condition of insufficient funds after it was opened to the public. Thirty- four thematic exhibitions and 11 large itinerant exhibitions were held by the Wuhan Science and Technology Museum including The First Science and Technology Expo, Large "Living" Dinosaurs & Rare Aquatic Animals Exhibition, Popular Knowledge of Change Space Science Exhibition, Large Earthquake Relief Picture Exhibition—We Are on the Scene, Adhere to Scientific Development and Build Ecological Civilization, Series of Activities of International Year of Astronomy 2009 & Total Solar Eclipse Observation, 60 Years of Brilliant Science and Technology, Close to Robots, A Million Citizens of Wuhan Learn Science—Series of Activities of The First Health Wuhan of Scientific Cancer Prevention. Healthy Life, Exoplanets, A Journey of Biodiversity Exhibition--Listen to the Sounds of the Earth, Popular Science Exhibition of Food Safety Theme and Cosmic Exploration Exhibition. Meanwhile, Wuhan Science and Technology Museum has done a lot in improving the scientific quality of juveniles by actively expanding the working range to actively carry out all kinds of youth scientific and technological activities and host large youth popular science competitions like Wuhan Youth Scientific and Technological Innovation Competition.

In the 21st Century, the Science and Technology Museums in China have ushered in unprecedented opportunities for development because the Central Committee of the Party and the State Council paid great attention to the construction of popular science museums and the public's enthusiasm for learning scientific knowledge was rising. Wuhan Science and Technology Museum was no longer able to meet the public's requirement to learn scientific knowledge for its small exhibition hall, simple content and a single form of popular science activities. Because of the appeal from all walks of life in Wuhan and the attention paid by the city leaders, the main building of Wuhan Science and Technology Museum began to be reconstructed and expanded in Aug. 2005 after 15 years' operation.

During the construction period, the Wuhan Science and Technology Museum, on the one hand, prepared exhibits and on the other hand, sent popular science activities to schools, communities, towns and barracks by adopting the way of "going out" with the combination of the popular science resources of the Science and Technology Museum and the Caravan of Popular Science. It received more than 220 thousand visitors in total in less than two years.

2.2 Wuhan Science and Technology Museum Has Opened Again After Being Reconstructed and Expanded

Wuhan Science and Technology Museum has opened to the public again after being reconstructed and expanded on Dec. 30, 2006. The exhibition area was enlarged, popular science exhibits were updated and the function of science popularization was upgraded, followed by a four dimensional cinema, the largest popular science gallery in Wuhan, an electronic screen for the popularization of science, a celestial museum with the supporting facility of astronomical knowledge exhibition area (Fig. 58, Fig. 59). The overall image and educational function of Wuhan Science and Technology Museum were significantly improved. The main functions include science and technology exhibitions and education, academic exchanges, science and technology training, youth science and technology experiments, popular science films, popular science galleries, and an electronic screen for the popularization of science. The exhibition area was enlarged from 1,600 square meters to 6,400 square meters and the floorage was enlarged up to 15,435 square meters. There are over 260 (sets) scientific exhibits of which over 90% can be operated by hand. The exhibits cover the fields of mathematics, science and life science. They show and introduce the application of technologies and the scientific knowledge in our daily life, such as industry, agriculture, medical & health, meteorology and environmental protection, and show and introduce modern science and technology, especially high and new technology and the relevant industries, such as satellite and space technology, computer & information technology, bioengineering, genetic engineering, new materials & new energy, and laser. With the improvement of hardware facilities, Wuhan Science and Technology Museum has a more clear purpose and educational idea. The two forms of education become prominent in exhibitions and education. The first is the participatory of the exhibitions. Visitors can feel the pleasure of science and technology and learn scientific knowledge in the experience by directly participating in operating the exhibits. The second is the thought of science. It is more concerned with the public's mastery of scientific ideas and methods. It receives more than 250 thousand visitors every year.

2.3   New Wuhan Science and Technology Museum Was Built and Opened

With the development of society and the progress of science and technology, people are more enthusiastic about learning scientific knowledge. The emphasis on the work of science popularization of science and technology museums gradually turns from popularizing scientific and technological knowledge to the public to improving the scientific literacy of the public. The objective condition of the old Wuhan Science and Technology Museum, including the hardware facilities, can not meet the demand for scientific popularization then. Six academicians of the Chinese Academy of Sciences (CAS) and the Chinese Academy of Engineering who worked in Wuhan, Yang Shuzi, Li Peigen, Li Deren, Ye Zhaohui, Zhao Zisen, Deng Xiuxin, jointly offered a proposal to Wuhan Municipal Party Committee and Wuhan Municipal Government to build a new Wuhan Science and Technology Museum in Jul. 2010 [1] .Wuhan Municipal Party Committee and Wuhan Municipal Government made a decision to transform the passenger building of Wuhan Port into a science and technology museum in Aug. 2010. The construction of the new museum was listed in the 12th Five-Year Plan of our city and was written in the Government Work Report twice in a row.

The New Wuhan Science and Technology Museum is located on the beautiful Marshland in Hankou. The foundation laying ceremony was held on Nov. 20, 2011, and the museum

---

[1]   Four-Show concept-link up the scientist behind the exhibits and the era which the scientist lived in, and the context of scientific development, scientific discoveries and humanistic spirit, stories about scientists, science in life and other elements and show them to visitors. Their interest in scientific inquiry can be cultivated by experiencing science. The concept was put forward by Xu Shanyan, who is the chief expert of the expert group of the construction of the New Wuhan Science and Technology Museum, the honorary chairman of China Natural Science Museum Association and a doctoral supervisor of Tsinghua University.

was opened to the public on Dec. 28, 2015, after the construction was completed. The whole floorage is around 30 thousand square meters. The total investment of the reconstruction project of the main building and the exhibition project was over 500 million yuan. The museum is a super-large science education venue which is multi-functional, comprehensive and intelligent and also an important component of the group of the Jianghan Chaozong (The Changjiang River and Hanjiang River Join to the Sea) cultural tourism scenic spots which is a key project Wuhan City puts forth the effort to (Fig. 60). The new museum had received about 2.9 million visitors by Sept. 2017.

The top-level design of the New Wuhan Science and Technology Museum was mainly created by domestic masters of popular science education. The design of the museum shows the collective wisdom of many scientists. The exhibition concept was set up on the basis of Four-Show Principle[2] , that is, showing objects, figures, spirit and wisdom, linking up the scientific discoveries, humanistic spirit and stories about scientists. The exhibition forms were designed by famous domestic and international design companies. Over 40% of the exhibits are innovative ones. Besides, some classical exhibits were selectively introduced from foreign countries.

The permanent exhibition of the museum begins with Questions about the Heaven by Qu Yuan, who was a famous poet in ancient China, to show the spirit of seeking knowledge and exploration of human beings since ancient times. There is the Plate of Nature consisting of exhibition halls of the universe, life and water and the Plate of Creation consisting of exhibition halls of light, information and traffic. Besides, there is an exhibition hall of mathematics and an exhibition hall for children. The number of the exhibits is over 600. The exhibition not only combines natural science and engineering science but also highlights the distinctive local characteristics. The visiting form of participation, interaction and experience is well received by the public.

2.3.1 Unique Geographical Location

The New Wuhan Science and Technology Museum is located in Wuhan central urban area and enjoys the superior regional environment. It can be described to be Three Close and Three Old. The new museum is close to a street, the Changjiang River and the Marshland and in the Old Town, near to the old station and the old wharf. The new museum next to the river is one of the core view spots of the Jianghan Chaozong (The Changjiang River and Hanjiang River Join to the Sea) cultural tourism scenic spot (applying for national AAAAA grade scenic spot). The unique location means that the public has high expectation for the hardware and software facilities of the new museum. It is an opportunity and also a challenge for the Wuhan Science and Technology Museum.

2.3.2 Distinctive Buildings

The New Wuhan Science and Technology Museum is one of the largest Science museums in China that was rebuilt from the old buildings. The museum still holds the architectural shape of a ship of Wuhan Passenger Port which was once the landmark of the River City. The new museum embodies the new concept of environmental protection and scientific development with the injected science and technology elements. The ship of popular science will sail far. The historical building will radiate new vitality.

2.3.3 Exhibitions with Shining Points

First, the new museum has realized the full integration of science and technology, history and culture. The buildings and the location of the new museum and the contents exhibited in the new museum with historical and cultural connotations including Meeting a Bosom Friend at the Guqin Platform, Ancient Human Wisdom, show that the new museum has the clear feature of the integration of science and technology, history and culture. Second, the contents exhibited have distinctive local characteristics. The local industrial characteristics are shown in the Traffic Exhibition Hall and Information Exhibition Hall. The characteristics of regional resources are fully shown in Water Exhibition Hall. Third, the things exhibited in the new museum show the characteristics of innovation, strong interaction and combination of local features. With a unique and graceful shape, the exhibition project of Questions about the Heaven whose idea was from Questions about the Heaven by Qu Yuan is a key project of the new museum. Visitors can interact with it by answering questions. It is unique in China.

2.3.4 High-end Intelligent Construction

The new museum is equipped with a leading intelligent system in China. The system has three great functions of public service, communication and operation management. It consists of 15 subsystems which involve security, ticket business, visitors guide & public service, RFID management system, the website of the museum, intelligent office, exhibition hall management and others (Fig. 62~Fig. 64).

3. Diversified Forms of Popular Science Activities

3.1 Wuhan Selection Competition of National Youth Science and Technology Innovation Competition Was Organized

Wuhan Science and Technology Museum successfully hosted from the 23rd to 32nd Wuhan

Selection Competition of National Youth Science and Technology Innovation Competition. Nearly 50 thousand students participated in each competition in Wuhan.

3.2 Summer (Winter) Camps of Popular Science and Training Courses

Summer (Winter) camps of Magical Science and Dreams Help Me Grow are held in summer (winter) holidays to help youngsters learn science easily and have happy holidays. Over 10 varieties of Summer (Winter) camp activities are organized every year. Wuhan Science and Technology Museum holds training courses for the specific population. Training Courses for Wuhan Primary and Middle Schools Scientific Tutors are held all year round for the tutors of National Youth Science and Technology Innovation Competition. It also offers Vocal Training Class and Thinking Training Class for Primary and Secondary School Students.

3.3 Popular Science Dramas

Wuhan Science and Technology Museum launched successively popular science dramas written, directed and performed by itself, such as A Talk between Household Electric Appliances, Magic Experimenter, in 2009 and it launched popular science dramas like Physical Properties of Different Articles, Shizuka's Birthday Party, Escape from Mars, SpongeBob SquarePants's Undersea Concert and "Shine" the Sky, scientific experiment drama The Power of Gas, and scientific shows Balloon Adventures and Trove of Mystery. The interesting, meaningful and fresh forms of popular science are well-received.

3.4 Combination of Museum and Schools

Based on the principle of close cooperation, resource sharing and common development, Wuhan Science and Technology Museum started the combination of museum and schools and then signed co-construction agreements with 11 primary and middle schools in Jiang'an District, Jianghan District and Hanyang District. They have carried out rich and colorful popular science activities.

3.5 Wuhan Scientist Speech Group of Popular Science·Chinese Academy of Sciences

Wuhan Scientist Speech Group of Popular Science was established in Mar. 2010. The speech group consists of academicians and experts from CAS Wuhan Branch and some universities. Meanwhile, the activity of A Million Wuhan Citizens Learns Science—Academicians and Experts Enter Campus whose main content was sending popular science reports to schools was officially launched. A platform for the students and scientists to communicate face to face was set up through the activities. The platform has helped students expand their scope of knowledge and scientific visual field. The number of people who have listened to the 23 reports is up to 8,200.

3.6 Mobile Science and Technology Museum

Wuhan Science and Technology Museum transformed and updated caravan exhibits in 2009 and carried out caravan Four Entering Activity (enter squares, schools, communities and units). The Mobile Science and Technology Museum carried out more than 50 tour exhibitions a year and gave public not only scientific knowledge but also happiness (Fig. 65, Fig. 66).

3.7 Theme Exhibitions

Both the Old and New Science and Technology Museum stick to holding theme exhibitions. There is a temporary exhibition hall covering an area of 1630 square meters in the New Science and Technology Museum. Some large theme exhibitions including Popular Science Exhibition of Wild Animal Specimens, Sea Elves-Jellyfish Theme Exhibition, Future Fabric, Marine Rights & Interests and Military and The Beauty of South Sea-Theme Exhibition of Marine Ecology and Protection have been held since it was opened. The contents of temporary exhibitions are close to people's lives and the exhibitions play an important role in improving the scientific quality of the whole people. The theme exhibitions had served over 10 million people until Sept. 2017.

3.8 Wuhan Public Science Forum

The New Wuhan Science and Technology Museum focused on creating the brand of Wuhan Public Science Forum when it was opened. The reports including The Characteristics and Responsibilities of Popular Science Times, Common Sense of Photography of Reading and Walking in Great Wuhan, Observation and Shooting of Solar Eclipse, Appreciate the Glamour of Science and Technology and Move Forward Hand in Hand to the Future, Wall-E, April 22 World Earth Day-Public Lecture on Garbage Classification, have been made. The forum has provided rich popular science feasts for the citizens.

4. The Expectation of the Wuhan Science and Technology Museum

The mayor of Wuhan investigated and surveyed the Wuhan Science and Technology Museum on May 16, 2016. The mayor expressed the idea clearly, "The old and the new museums complement each other in the aspect of exhibitions and popular science activities and form differential development". He also put forward the requirement that exhibitions should expand the scope of knowledge and display or introduce the latest scientific and technological achievements based on stable and theme exhibitions to look forward to the future direction of scientific development. He advised Wuhan Science and Technology Association and Wuhan Education Bureau to make the work program to turn the old museum into a scientific experience center and a scientific inquiry room.

The differential development is an inevitable choice to run two museums at the same time. The New Wuhan Science and Technology Museum focuses on exhibitions, displays and education with the combination of the knowledge of natural science and that of engineering science. There is Plate of Nature consisted of exhibition halls of the universe, life and water and Plate of Creation consisted of exhibition halls of light, information and traffic. Besides, there is an exhibition hall of mathematics, an exhibition hall for children and a temporary theme exhibition hall. The Old Wuhan Science and Technology Museum focuses on hands-on experiments and inquiry learning. It has been preliminarily determined to build ten innovative theme scientific inquiry rooms of STEAM, robots, sensors, little scientists, model sports, astronomy, mechanical plays, mathematics, optics and innovation education, a teaching base of a community popular science university, a popular science cinema and so on. Each of the two museums has its own emphasis. They complement each other and promote the work of science and technology in different aspects to provide better services for the public.

5. Suggestions on the Exchange and Cooperation under The Belt and Road Initiative

More and more common views have been reached between countries and regions since The Belt and Road initiative began to be implemented. There are huge differences in Natural environment, social policy, economy, culture and so on in the countries along The Belt and Road

Initiative. Each of the countries is faced with complex development challenges. The countries and regions along The Belt and Road Initiative share the concept of development, development model and development results and finally realize a win-win development of science and technology museums.

5.1 Online Model

With the help of emerging media like the Internet, mobile terminal, and the model of Internet Plus, the public science museums in the countries and regions along The Belt and Road Initiative can help to make up what the others lack and share the concept of operating museums and forms of carrying out popular science activities in the field of public science.

5.2 Off-line Model

The public science museums in the countries and regions along The Belt and Road Initiative can exchange and learn from each other through diversified activities such as academic forums, popular science exhibitions and summer camps in order to achieve good results of science popularization.

References

[1] Brief Introduction to Wuhan Science and Technology Museum by Wuhan Association for Science and Technology[J]. 50 Years of Wuhan Association for Science and Technology. Wuhan Publishing House, 2012.
[2] WANG GANG. The Group of Academicians and Doctoral Tutors behind the revival of Wuhan Science and Technology Museum[J]. Focus News, Edition 5, Yangzi River Daily, 2016.

Highlight the Key Point, Cover the Whole Province, Carry Out Activities to Popularize Scientific Knowledge

—— Cases of Yunnan Science and Technology Museum

Xiang Wenyi [1]

Abstract In the respect of popularizing scientific knowledge, Yunnan Science and Technology Museum spares no efforts to bring science popularization resources to the whole province by means of permanent exhibition such as "Scientific Exploration" and "Experience Science" and China Moving Science and Technology Museum Yunnan Itinerant Exhibition as the platform, by cooperating with schools, agencies and enterprises, and takes familiar science knowledge around us as entry point, showing frontier technology to audience, inspiring teenager's science spirit. Our staff can bear hardships anytime at work, with the purpose of bringing science and technology exhibits closely to students from border areas especially those minority students, general public and cadres etc., stimulate public's enthusiasm towards science, enhance scientific culture quality of the whole province, and furthermore, strengthen our patriotic emotion and a sense of national pride.

Keywords Science popularization   border areas   frontier technology

1. Introduction of Yunnan Science and Technology Museum

As a popular science education base awarded by China Association for Science and Technology and Yunnan Province Government, Yunnan Science and Technology Museum

① Xiang Wenyi: Deputy head of Experience and Education Department of Yunnan Science and Technology Museum E-mail:15860043@qq.com.

is an important position for rejuvenating Yunnan through science and education and implementing Outline of the Action plan for Yunnan's Science Literacy(2016–2020), a provincial demonstration science and technology museum, and a public welfare education institution for carrying out the science popularization and science popularization and education and training activities geared to the needs of adolescents and the public. In 2011–2012, Yunnan Science and Technology Museum carried on the transformation of the permanent exhibition capability, renaming the original "Science Park" to "Scientific Exploration" Popular Science Exhibition. In September 2012, it was free and open to the public again, and as of December 2017, the audience received has broken through 1.2 million person-times. With exhibition hall area of 3,550 square meters, 121 interactive exhibits, 175 models, with the history of science and technology as the main line, it has been divided into four themed exhibition areas, "Scientific Enlightenment", "From the Laboratory to Industrialization", "From the Particles to the Universe" and "Low Carbon and Development", and five characteristic exhibition areas, "Children's Heaven and Earth", "Television and Scientific Lands", "Dream Factory", "Health Clinics", and "Military Technology Heights", providing free interesting scientific experience for the public.

2. The Background of Science Popularization Activities.

2.1   The Situation in the Province

Yunnan is a populous province and the largest ethnic group in the country, with 15 ethnic minorities. Moreover, located in the southwest border of China, Yunnan is a mountain plateau terrain, mountainous area accounts for 94% of the province's land area, with big elevation drop, and the topography is complex, such natural conditions leading to the traffic inconvenience and information block for people in the mountain areas. For ethnic minorities and remote mountainous areas, science resources are seriously inadequate. In order to improve the situation, and strive to improve the scientific literacy of masses in our province, especially ethnic minority people in mountainous areas, the science and technology museum insists on raising the banner of "Science Popularization" in Yunnan Province, is committed to free public science popularization activities, and spares no effort to benefit the popular science resources to the public in the province.

2.2   Regional Situation

Yunnan borders on Vietnam, Laos and Myanmar, with 4,060 kilometers of border lines, and is a province in China with most neighbors. The specific territory and history reflect the friendly relations of Yunnan, China, with its neighboring countries. Originating from China, Lancang River - Mekong River connecting the six countries, geographically, brings China, Laos, Vietnam, Myanmar, Thailand and Cambodia closely together, just as President Xi Jinping pointed out: "In terms of the geographical position, natural environment and the relationships, the periphery has very important significance to our country". Yunnan with neighboring countries have formed the regional community, fateful community, a community of interests, and strategic community, and these fully demonstrate the irreplaceable important role of Yunnan in the restoration of relations with neighboring countries, and in construction of national strategies such as The Belt and Road Initiative and "Marine Silk Road".

"The basic principle of our country's diplomacy with neighboring countries is to have a good neighbor, a safe and rich neighbor. Highlight the concept of affinity, sincerity, mutual benefit and inclusiveness. To develop good-neighborly and friendly relations with neighboring countries is a consistent policy of China's neighborhood diplomacy. We should uphold good-neighborly friendship and help each other. Talking about equality, affection, meeting and walking; we should do more popular and warm things to make the neighboring countries more friendly, closer, more approachable and more supportive, and enhance our affinity, charisma and influence. To cooperate sincerely to treat peripheral countries, weave the more close network of common interests, promote the interests of both sides fusion to a higher level, and let the surrounding countries benefit from the development of our country, and our country also gain benefit and common development from neighboring countries".

3. Principles of the Design of Science Popularization Activities

(1) Based on the various popular science plays, science shows and scientific experiments in our museum.

(2) Integrate with the pace of the times and represent the achievements of the development of science and technology.

(3) Organize popular science activities around social hot topics.

(4) The supporting activities should not be similar to the conventional exhibits and the "Mobile Science Museum".

(5) The activity plan must be scientific, operable and interesting.

4. Way of activities

(1) By making good use of two platforms of my museum ("Scientific Exploration" Popular Science Exhibition and "Experience Science" China Mobile Science and Technology Museum Yunnan Tour), our museum science counselors carry out a variety of popular science activities.

(2) Joint social resources, explore groups or individuals who are enthusiastic about science popularization, and jointly research and develop science popularization activities such as popular science lectures, fun activities and cutting-edge technology experience.

Table 1

Place	Content	Form	Purpose
Activities within the museum	Scale tube, bamboo dragonfly, Kaleidoscope, national batik, rubbing, reproduction paper, small machine tool, air car, float and sink, leaf pulse bookmark	Use the given material to complete the related activities	We carry out various kinds of science popularization activities, and play the purpose of "promoting the spirit of science, spreading scientific ideas, advocating scientific methods, popularizing scientific knowledge and improving the quality of science". In the process of making, the students put forward the idea to solve the problems in the face of difficulties and problems, by the ingenious use of tools, solve difficulties and problems encountered, etc., and form the habit of comprehensive thinking, sparking interest in "exploration"
	Air cannon, small robot performance, popular science play, popular science and television broadcast	Look through your eyes, use your brain, discover problems, and propose solutions	
Activities outside the museum	Astronomical observations Cosmic theatre Science education institution	Joint social science organizations. 1. Observation with the astronomical telescope; 2. The ball screen plays popular science and television; 3. Sharing of science education resources	
	Yunnan Open University Primary school affiliated to Yunnan Normal University Kunming Second Kindergarten	Joint activities with theme activities. For example lifelong learning activity week, school science festival/technology week	
	Kunming city Shilin County Science and Technology Association Yuxi City Science and Technology Association	Popular science activities are carried out in the national science and technology week, the popular science day and the national fitness activities. Popular science play performance, 3D printing demonstration, robot performance, etc	
	China Mobile Technology Museum	At the start of the exhibition tour of Yunnan county level, we carry out supporting science activities, such as "dream of galloping", aerial photography experience, VR technology experience, etc	
	Bilingual popularization in minority areas	Joint action. We came to the frontier minority areas to carry out the frontier science and technology demonstration, and the popular science caravan activity	

Table 2

Content carried out	Annual benefit indicator	Completion situation
Scientific exploration of popular science exhibition featured activities	More than 50 times	64 sites for the whole year
Scientific exploration of popular science exhibition featured activities, audience participation	Over 15,000 person-times	The number of participants was 17,620
Scientific exploration of popular science exhibition, free opening days	More than 280 days	Open for 312 days throughout the year
Scientific exploration of popular science exhibition, total visits of audience	Over 200,000 person-times	321669 person-times throughout the year
Scientific exploration of popular science, exhibits integrity rate	More than 95%	Staying above 95%
Scientific exploration of popular science exhibition, audience return rate	More than 20%	In 2017, the number of visitors to the exhibition is over 50%
The number of key activities organized by major popular science activities	Up to 8 and above	12 activities carried out
The number of people involved in major science activities	The number of visitors was up to 30,000	48,600 person-times
The number of popular science resources displayed in the exhibition	35 to 49 items	45 items
The types and quantities of popular science publicity materials issued by major science popularization activities	2	5
The number of participating units of major popular science activities	The number of participating units, learning associations and communities reached 15 and above	21 participating
The number of media coverage of activities in major popular science activities	Number of 20 and above	35 reports
China mobile science and technology museum 13 sets of exhibition tours, 52 sites, completion rate	The number of exhibition sites is not less than 52	Starting work of 53 county-level sites completed
The proportion of local primary and secondary school students covered in the audience of China mobile technology museum	The coverage of primary and middle school students reached 70%	For each site, a document is jointly issued by the local education bureau, which requires schools to organize the visiting experience of primary and secondary school students, and the coverage of primary and secondary school students is over 70%
Audience participation in China mobile science and technology museum's experiments and performance activities	More than 41,600 people attended the event	The number of participants was 53,000
Perfectness ratio of exhibits in China mobile science and technology museum	The perfectness ratio is over 90%	Staying above 95%
The proportion of old and poor counties in China mobile science and technology museum exhibition sites	70% are old and poor counties	Complete

Content carried out	Annual benefit indicator	Completion situation
China mobile science and technology museum to promote the municipal science exhibition matching funds input	90% of the counties have the corresponding investment	There is a corresponding investment in all regions
The leading group of China mobile science and technology museum and the national team leader of the People's Republic of China take the lead in visit and experience	100% lead the tour experience tour	All the local leaders of the county and all the people in charge of the leadership take the lead in visit and experience
China mobile science and technology museum supporting science activities	In the 90% city-county tour, there are some supporting activities	Complete
The favorable rate in the audience visit of China mobile science and technology museum	90% favorable rate	99% favorable rate
China mobile science and technology museum (county-level cities) are willing to conduct the second round of the exhibition during the 13th Five-year Plan period	100% willing	100% willing
Rural high school science and technology museums successfully completed and on time	80% successful completion and timely implementation	Complete
Perfectness ratio of exhibits in rural high school science and technology museum	≥ 85% perfectness rate	The perfectness rate is over 90%
Rural high school science and technology museum construction school is located in the national poverty-stricken county	90% of them are in poverty-stricken counties	Complete
Training rate of science and technology counselor of rural high school science and technology center	90% of the schools participated	All schools participate
The total number of visitors to the rural high school science and technology museum	≥ 150,000 person-times	172,839 person-times
Audience favorable rate of rural high school science and technology museum	80%–90% favorable rate	95% favorable rate
Bilingual popularization in minority areas	Joint action more than 2 times. In the frontier minority areas, we carried out frontier science and technology demonstration and popular science caravan	6 times jointly carried out

5. Science Popularization Activities: From Exploration to Insistence, Finally Forming a Brand

5.1 Hands-on Experience and Production Activities

Through the preliminary investigation and preparation, and combining with the actual the situation of the museum, we set up a practical production activity which is in accordance with the development of our pavilion.

Plan Rules for the Production of Kaleidoscope.

Activity theme: experience feeling, handmade.

Activity object: nine-year compulsory education students.

The number of participants: 20–30 people.

Activity time: 2 hours.

Introduction

Kaleidoscope is an optical toy, and as long as you look in the eye, a beautiful "flower" will appear. Turn it a little, and there will be another pattern of flowers. After continuous rotation, the pattern is also changing, so-called "kaleidoscope". The kaleidoscope was originated in Scotland in the 19th century and invented by a physicist who studied optics. Two or three years later, it was spread to China and Japan almost at the same time. When many Chinese toys entered Japan in the early 19th century, they included kaleidoscope. At that time, as an optical game, fresh and interesting, kaleidoscope became a signature toy in a candy store to attract children.

How did the pattern of the flower tube come? It was reflected off the mirror. It is a prism composed of three glass mirrors, at one end, we put some fragments of glass, these fragments will show symmetrical patterns after reflection by the three mirror, and look like blooming flowers.

Activity

1. Activity preparation

(1) Preparation for materials

Mirror, cardboard, adhesive, scissors.

(2) Preparation for grouping

3 to 5 people in each group, the materials assigned to each group.

2. Activity process

(1) It is cut according to the cut line (black solid line);

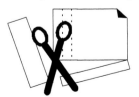

(2)Tear off the protection film of the mirror;

(3)Use double-sided adhesive to paste the three-sided mirror into the position of the cardboard, with the mirror reflector upward;

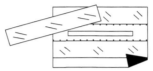

(4) It is folded according to the dotted line, and pasted with double-sided adhesive to form an equilateral triangle, with the mirror inside the kaleidoscope;

(5) With the eye observing from the other end, move the kaleidoscope slowly from the other end of the kaleidoscope to the monitor or other objects.

Activity record:

_____

_____

_____

_____

_____

_____

...

After a long-term conduction of activities, we continuously explored and learned, and successively formulated multiple hands-on plans: scales tube, bamboo dragonfly, national batiks, rubbings, recycled paper, small machine tools, air vehicle, float and sink, and veins bookmarks.

5.2   Outdoor Science Activities

5.2.1   Bilingual popularization in minority areas

In 2017, we cooperated with the national science popularization task force six times and went to the frontier minority areas to carry out cutting-edge science and technology demonstration and popular science caravan activities. There are 75 caravans in the whole province, each vehicle enters the country every year for no less than 10 times, and the number of audiences is not less than 500. In the six activities conducted jointly, 18 times of cutting-edge technology exhibition and experience were conducted, a total of 3,200,000 popular science materials were distributed, including 100,000 ethnic minority bilingual popular science materials, covering Chuxiong Prefecture, Lincang City, Banna Prefecture, Ninglang County and other places (Fig. 67~Fig. 75).

5.2.2   Astronomical observation

Yunnan Science and Technology Museum and Yunnan Association of Amateur Astronmer jointly carry out science popularization activities; in initial stage, by the use of "Mid-Autumn Festival", "Science and Technology Week", and "Popular Science Day", they conducted the activities of observation of the sun in daytime and moon at night, took all kinds of astronomical telescopes to citizens, for free observation and popularization of knowledge of astronomy; in recent years, with the development of China Mobile Science and Technology Museum Yunnan Patrol, took astronomical telescopes to remote mountain areas, explained knowledge of astronomy to the local public and told them to observe the stars.

5.2.3   "Cosmic Theater" activities

"Cosmic Theater" is a dome screen three-dimensional cosmic theatre, a standard hemisphere, 18 meters in diameter, with 15 degrees of dip angle, equipped with curtain as the screen, exquisite images of high resolution are played inside the theater, the picture is gorgeous moving, and the 3D effect is remarkable. The main films are "Mystery of Dinosaurs" and "Space Capsule", with the playback time of 10−15 minutes.

Because the outstanding effect in the pilot sites, the supporting activities of "Cosmic

Theater" have been recognized by China Association for Science and Technology and China Science and Technology Museum were arranged in subsequent exhibits, and its name was changed to Dome, with the films up to five.

5.2.4　Popular science play

Since 2008, Yunnan Science and Technology Museum has prepared the first popular science play "Beautiful Bubbles", which was presented to the public free of charge. Then, it gradually developed a "Strong Pig and His Friends," "Look Up at the Starry Sky", "Mystery of the Light", "Flying Dreams" and other popular science plays, which obtained the consistent high praise from school teachers and students in the daily performance. The original science popular play Wonderful Night in Toy Store directed and performed by itself won the first prize in the western conference of the fifth national science and technology museum counselor competition, third prize in the national finals, and first prize in the fifth national science performance competition. All the competition platforms promoted exchange and learning in the industry.

We will bring popular science plays into kindergartens and primary and secondary schools, and perform different popular science plays according to the characteristics of students of different ages (Fig. 76~Fig. 80).

5.2.5　Science show (scientific experiment)

The students are led to conduct scientific experiments, and occasionally a few words of witty statement or a few cool experiment shows can quickly reduce the distance between students and counselors, letting the student accept the education of scientific knowledge in a relaxed environment, being more conducive to the students in absorbing scientific knowledge and understanding so as to achieve the goal of science popularization.

5.2.6　Handmade

Using safe and reliable equipments, students can experience the basic working principle of machine tools and the working process of various machine tools while making small souvenirs for themselves.

5.2.7　Museum-community cooperation

We have established a good relation of cooperation with the associations for science and technology in prefectures and communities, and to reflect the leading role of provincial museums, we often participate in important science popularization activities, popular science activities in all places and Kunming, such as "Science and Technology Week", "Popular Science

Day" and regional themed popular science activities.

6. Questionnaire

In 2017, I carried out the random survey of the audience to visit or participate in the popular science activities aperiodically, having taken back a total of 218 questionnaires. The questionnaire is conducive to the understanding of the audience's age structure, cultural level and exhibits of interest, providing a support for the research on the future development direction.

Investigation shows:

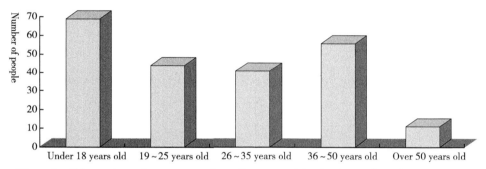

Figure 1   Column chart of age structure for audiences participating in popular science activities

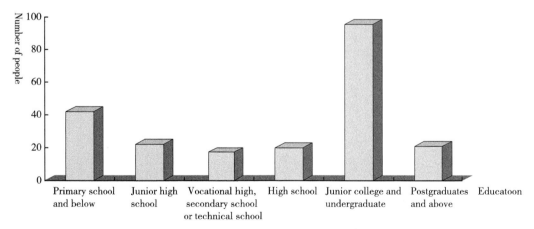

Figure 2   Column chart of Cultural Level for audiences participating in popular science activities

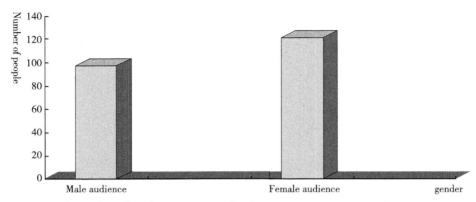

Figure 3　Column chart of gender composition of audiences participating in popular science activities

7. Conclusion

For the purpose of "promoting the scientific spirit, spreading scientific thought, advocating and popularizing scientific knowledge and scientific method, and enhancing the quality of science", with a variety of popular science activities, Yunnan Science and Technology Museum enlarges the effect of science popularization, makes good use of the two platforms of China Mobile Science and Technology Museum Yunnan Patrol, the permanent exhibitions "Scientific Exploration" and "Experience of Science", combined with the cooperation manners of museum-school, museum-community and museum-enterprise, makes all efforts for maximum participation and more experience by the children in Kunming, and moreover, brings science and technology and joy to frontiers, mountainous areas and ethnic minority areas, letting the students feel the endless charm brought by "Experience of Science".

Reference

[1] JING JIA. Planning and organization of popular science activities[M]. Huazhong University of Science and Technology Press, 2001 (05).

[2] CAO JIANPING. Discussion on the way of science popularization activities in science and technology venues[M]Scientific Chinese, 2017 (15).

[3] Baidu Encyclopedia: https://baike.baidu.com/item/%E4%BA%91%E5%8D%97/206207?fr=aladdin.

[4] China Science Communication: http://www.kepuchina.cn/.

[5] Science popularization activity resource service platform: http://www.kepu365.org/.

One Hundred Year History and Innovation Development of The Geological Museum of China

Jia Yueming[1] , Liu Shuchen, Chen Yuanyuan, Xu Cuixiang,
He Zhefeng, Ruan Jiaping

Abstract The Geological Museum of China (GMC) as a national museum was established in 1916. It is known as the first public natural science museum in China. Over the past one hundred years, groups of outstanding experts and scientists, with remarkable achievements and exploring spirit, have emerged from GMC. The museum has been dominated among the Asian geological museums and has built a high reputation for its rich collections, top scientific researches and exquisite exhibitions around the world. The GMC has adhered to the development strategies of invigorating the museum through collections, talents and technologies, cultivating the core values of erudition and fraternity. It plays an important role in supporting the work of territorial resources, carrying out community services and popular science education activities and promoting the development of national geological museums. Although facing many challenges (e.g. a single function of architecture, limited space for exhibitions, researches and collections, little international cooperation), the museum obtains the innovation and development in the new period through building a new museum building and conducting international exhibitions.

Keywords The Geological Museum of China   one hundred year history   innovation development

1. Introduction

The Geological Museum of China (GMC) as the first public natural science museum in

---

[1] Jia Yueming: director of the Geological Museum of China. Email: 522191845@qq.com.

China was established in July 1916. The museum has been dominated among the Asian geological museums and has built a high reputation for its rich collections, top scientific researches and exquisite exhibitions around the world. The GMC has adhered to the development strategies of invigorating the museum through collections, talents and technologies, cultivating the core values of erudition and fraternity. It plays an important role in supporting the work of territorial resources, carrying out community services and popular science education activities and promoting the development of the national geological museum. On July 20, 2016, President Xi Jinping sent a letter to congratulate the GMC's centennial, fully affirming its outstanding achievements in geoscience research, geoscience knowledge dissemination and its important effect on developing Chinese geological undertakings, and enhancing the scientific quality of the public. Besides, President Xi put forward specific requirements and pin high hopes on the GMC's development in the future, and pointed out the direction and goal in the letter.

2. One Hundred Year History of the GMC

2.1 One Century History

The GMC has always existed as an independent scientific organization since the beginning of 1916, lived through trials and hardships. The GMC is a witness to the occurrence of modern Chinese science, is an inheritor of Chinese geological undertakings, and is an inaugurator of the science popularization in China. Therefore, the GMC is a legend in the history of Chinese science, Chinese geoscience and Chinese museum. On July 14, 1916, the first 18 geological students trained by Chinese themselves graduated, and held their graduate exhibition in the No.3 Fengsheng Lane, which promoted the Exhibition Hall of Geology and Mineral Resources that is a rudiment of the GMC entering the historical arena. This graduate exhibition displayed a total of 917 specimens, which became the first collection of the GMC.

In 1935, the Exhibition Hall of Geology and Mineral Resources was relocated to Nanjing due to the Japanese invasion of China. The temporary Exhibition Hall was located at No. 942 Zhujiang Road, Nanjing. It opened in February 1937, with 1,500 square meters and 12 showrooms. After the outbreak of the Anti-Japanese War, the Exhibition Hall moved from Nanjing to Chongqing in 1938. The exhibition hall as an important part of western China science museums reopened and displayed mineral rock, stratigraphic paleontology, vertebrate fossils and soil. Oppositely, the Exhibition hall moved from Chongqing to Nanjing after the

the victory of the Anti-Japanese War. Finally, the Exhibition Hall returned to Beijing after the founding of new China.

Currently, the museum building is located in Xisi, Beijing. The building was approved by Prime Minister Zhou Enlai in 1956 and was built in 1958. The museum was renamed the Geological Museum under the Ministry of Geology and opened to the public in 1959. It covered 11,500 square meters with the exhibition area of more than 5,000 square meters. In 1986, the museum was renamed once again as the Geological Museum of China(GMC). At the beginning of 21st century, the building went through large scale renovations, and the renovated building reopened to the public in July 2004. After the renovations, the museum was composed of five permanent exhibitions, namely the earth, minerals and rock, gemstone, prehistoric life and land and resources (now changed to "hundred years hall"), and additionally, two temporary exhibitions. The museum was visited by nearly 500,000 people every year.

2.2 Top Scientific Researches

The GMC not only has recorded and witnessed more than hundred years' tortuous development in Chinese earth science and geological undertakings but also has accumulated abundant natural essence and intangible assets. The GMC is home to Zhang Hongzhao, Ting Wen-Chiang, Wong Wen-hao, Xie Jiarong, Hou Defeng, Huang Jiqing, Yang Zhongjian, Pei Wenzhong, Cheng Yuqi, Gao Zhenxi and Liu Dongsheng, who are the founders of modern Chinese geology.

2.3 Natural Essence

At present, the GMC has more than 200,000 geological specimens, covering all fields of geoscience. Among them, there are the world-known dinosaur fossils of *Shenzhousaurusorientalis*, *Shantungosaurusgiganteus*, *Sinosauropteryxprima*, as well as paleoanthropology fossils such as *Sinanthropuspekinensis*, *Homo erectus yuanmouensis* and *Homo sapiens sapiens*. Additionally, the GMC has a variety of mineral specimens, such as the "King of Single Crystal" donated by Chairman Mao, the world's largest calcite geode, huge druses of fluorite intergrowth with calcite, the world's largest known scapolite cat's eye, tsavorite and exquisite azurite, realgar, cinnabar, orpiment, scheelite, stibnite, beryl and pyromorphite. Many of them are precious items reputed as "National Treasures" with world-class quality.

3. Development Status and Achievements of the GMC

Along with the development of modern Chinese science, the GMC inherits hundreds of thick history and scientific spirit in Geology. In addition, the GMC plays an important role in serving the development of the land and resources industry, carrying out community services and popular science education activities and promoting the development of the national geological museum. We try to do the best with President Xi's letter "Remain true to the original ideals, and keep pace with the times".

3.1 Focusing on the Center and Serving the Development of the Land and Resources Industry

The GMC, directly under the Ministry of Land and Resources of China, has assumed the responsibility of the protection of fossils, land and resources science popularization and local chronicles compilation since 2009. Therefore, the management office of Land and Resources Science Popularization Center, the office of Paleontological Experts Committee of China and the Chronicles Office of Land and Resources were set up successively under the GMC.

Firstly, since the Land and Resources Science Popularization Center was founded, a total of 176 popular science bases have been named according to the standard naming rule. Under the effective management of the center, different kinds of popular science bases could bring their own popular science advantages into full play, thus positively promoting the development of land and resources science popularization of China. Secondly, the Office of Paleontological Experts Committee of China makes a great contribution to the protection of the fossil in China, due to its remarkable achievements in making laws and regulations, strengthening the management of collections, promoting the study of fossils and so on. Thirdly, the Chronicles Office of Land and Resources is responsible for chronicles and almanacs compilation of land and resources. The books of China Land and Resources Almanac comprehensively and objectively describe the reform and development of land and resources in China. Now the Almanacs have become the most important historical archives, providing the basis for policymakers at all levels of government.

3.2 Relying on Special Funds to Significantly Improve the Level of Collection

The basis on which our activities are founded is the collection . The collection is the important symbol of distinguishing the museum, library and exhibition hall. The rich collection is of great significance to the development of the museum. Based on the special funds of the collection, purchase and comprehensive study of geological relic specimens, the GMC has purchased nearly 3,000 geological specimens, including scheelite, stibnite, tsavorite and other exquisite items since 2011. The project not only has greatly enriched our collections but also has significantly improved the level of collection. Our scientifically priceless collections are the key to our successful research activities, international collaborations, and vibrant exhibitions.

After five years of exploration and practice, the GMC has formed a set of effective management methods and operation mechanism during the work of collection, purchase of specimens. The two "Three separation" principles (the separation of research, negotiation or decision-making, and the separation of inquiry, bargaining or pricing) and the "Three mechanism" (supervision control, expert consultation and democratic decision-making) are gradually moving towards maturity and perfection, providing some reference for us to deal with the routine work of the museum.

3.3 Relying on Science Popularization to Spread the Knowledge of Earth Science

We always regard spreading the knowledge of earth science and enhancing the scientific quality of the public as our mission. For this mission, we adhere to the science popularization idea of allowing the public to be close to nature, cherish resources, protect the environment, edify sentiment, advocate science and love the motherland. Meanwhile, we regularly hold new exhibitions, and additionally, strengthen the science popularization team, expanding the popular science education activities' outreach.

3.3.1 Regularly launching the new exhibitions to serve the public

Display plays an important role in the social function of the museum and is a unique language of the museum. The permanent exhibitions are the soul of the museum, and on the other hand, the temporary exhibitions are effective supplements. Compared with the permanent exhibitions, the temporary exhibitions are closer to the times, and more attractive for people. We have held a variety of temporary exhibitions in recent years (Table 1). For example, "One Century Achievements of the Geological Museum of China", a historical exhibition to commemorate the 100th anniversary of the GMC obtained positive social effects. By early 2017, the exhibition had been visited by nearly 900,000 people, including about 400 group visits and more than 30 special receptions. The China Central Television (CCTV) news, the newspapers of the People's Daily and the Guangming Daily all reported this exhibition and related activities. The number of reports reached 1,077 times in total until the present day. The exhibition of "One Century Achievements of the Geological Museum of China" has been the largest, the most visited, and the most attractive since the 21st century. Therefore, it was rewarded one of the ten news of the process of Chinese Paleontological Popularization in 2016, and the first prize of the excellent geology and mineral achievement of "Malachite Cup". In addition, the exhibitions of "the first Mineral Treasures of China" and "Mineral Treasures of the World (2017)" displayed hundreds of high-quality gem and mineral crystals from home and abroad. The two exhibitions presented the classic display of high-quality gem and mineral crystals, offering people a visual and spiritual feast.

Table 1　The number of temporary exhibitions by the GMC

Year	2013	2014	2015	2016	2017−now
Number	11	6	7	7	6

3.3.2　Carrying out "Coming inside" popular science education activities with novel themes and interesting forms to serve the public

The Regulations of Museum, issued by the Chinese Government in 2015, put the educational functions of the museum in the first place. In recent years, the quantity and quality of popular science education activities (Table 2) of the GMC have continuously improved. A series of popular science activities with novel themes and interesting forms, such as drawing green earth, painting fan by mineral pigments and so on, greatly aroused the enthusiasm of the public, especially arousing youngster's interest in learning the knowledge of earth science.

Table 2　The number of popular science education activities

Year	2013	2014	2015	2016	2017−now
Number	10	15	13	28	25

3.3.3　Expanding the popular science education activities' outreach to benefit more people by the method of "Going outside"

In recent years, the GMC has actively assumed social responsibility and sincerely served the general public. We positively spread the knowledge of earth science to the public outside of

Beijing, in order to satisfy the people's wide demands of the geological knowledge. For example, we gradually expanded the "GMC into Campus" popular science education activities from the center to the suburbs in Beijing from 2011. By late 2015, the "GMC into Campus" activities had covered all the 16 districts and counties of Beijing (Table 3). Secondly, we went out of Beijing and walked into Hunan, Hubei and Liaoning Province, Tibet, Inner Mongolia Autonomous Region and other parts of China to organize large-scale science popularization activities (Table 3), which attracted wide attention in the society. According to the statistics, Hundreds of thousands of people benefited from our large-scale activities.

Table 3　Describing the popular science education activities of the GMC

Themes	Locations	Years	Contents and Forms	Number of Visitors
Large-scale science popularization activities	Changsha and Chenzhou in Hunan Province	2013–2017	China Mineral and Gem Show in Changsha, Hunan	>500,000
	Beijing	2015, 2016	summit forums, lectures and exhibitions	>200,000
	Donghai City in Jiangshu Province	2015	the exhibition of "the Return of Crystal King"	>300,000
	Huangshi City in Hubei Province	2017	Geological Science Popularization Conference of Huangshi, China	>150,000
The GMC into Campus	Four schools from Mentougou, Miyun, Chaoyang and Xicheng Districts in Beijing	2011	four exhibitions and four lectures	±10,000
	Two schools from Yanqing and Changping Districts in Beijing	2012	two exhibitions and two lectures	thousands
	Five schools from Pinggu, Shunyi, Mentougou, Daxing and Chaoyang Districts in Beijing	2013	five exhibitions and five lectures	±10,000
	Six schools from Fengtai, Fangshan, Huairou Districts in Beijing	2014	six exhibitions and six lectures	thousands
	Three schools from Shijingshan, Tongzhou and Xicheng Districts in Beijing	2015	four exhibitions and three lectures	thousands
	Five schools from Shunyi, Fangshan, Yanqin, Mentougou and Xicheng Districts in Beijing	2016	five exhibitions and five lecture	thousands
	Four schools from Tongzhou District in Beijing and Hejian in Hebei Province	2017	four exhibitions and eighteen lectures	±10,000
The Train of Technology	Ganzhou City in Jiangxi Province	2014	six exhibitions, two lectures and one free consultation and identification	>10,000
	Dandong City in Liaoning Province	2015	four exhibitions and two lectures	±10,000
	Chifeng City in Inner Mongolia	2016	two exhibitions and one lecture	±10,000
	Lasha, Naqu and Rikaze regions in Tibet	2017	three exhibitions and five lectures	±10,000

In addition, our teams of science popularization experts have been continuously strengthened, from purely relying on the experts from our museum at the first to the active involvement of professional people from the Ministry of Land and Resources. We started to invite famous academicians, such as Li Tingdong, Ouyang Ziyuan, Liu Jiaqi and Zhou Zhonghe, to teach students geological knowledge, cultivating the interest of students in natural science.

3.4　Guiding Other Geological Museums' Construction and Development

As a leader of the geological museums in China, we should make full use of the leader's influence so as to provide technical guidance and business support for other geoscience museum's construction and management.

3.4.1　Establishing 12 branches, and reinforcing guidance

In order to give full play to the GMC's guidance in a collection, scientific research, science popularization and the advantage of experts team, we began to set up the branches in the whole country from 1996. So far, a total of 12 branches, such as China Dinosaur Park in Changzhou, Yantai Natural Museum, Huangguoshu Jade Hall, Jiayin Dinosaur Museum and Donghai Crystal Museum, have been built. These branches all played an important role in the local cultural and economic construction.

3.4.2　Based on the platform, enhancing the communication between museums and promoting the construction of museums

At present, the Land and Resources Specialized Committee of the Chinese Association of Natural Science Museums, the Geological Museum Specialized Committee of the Chinese Museums Association, the Science Popularization Committee of the Geological Society of China and the Protection and Research Association of Fossils are affiliated to the GMC. Based on these platforms, we integrate the power of all the geological museums from the whole country and provide positive opinion and suggestion for the other geological museums, thus greatly promoting the development and construction of the geological museums in China. In particular, on the basis of the Land and Resources Specialized Committee, we organized and participated in nearly 10 international and domestic exchanges, as well as held 8 joint exhibitions with other organizations. While promoting the development of the geological museums, we will work hard to disseminate geological knowledge and encourage people to be more scientifically minded, thereby better serving the economic and social development and promoting the construction of ecological civilization.

4. Conclusions

The GMC has made remarkable achievements in scientific research, science popularization and collection over the past one hundred years. However, along with the rapid development of society and the deepening of globalization, the desire of people for high-quality facilities is increasingly higher, and the thirst of people for scientific knowledge is getting increasingly stronger, and the requirement of people for social educational functions and service of the museum is getting more and more specific. Moreover, President Xi, as the leader of Central Committee of the Communist Party of China, put forward the strategy of invigorating the country through technology and culture, and pointed out "scientific innovation and science popularization are the two wings for realizing innovation-driven development", putting science popularization on a par with scientific innovation for the first time. Overall, the GMC will usher in new opportunities and challenges in scientific research, knowledge popularization and international cooperation.

4.1 Facing the Big Challenge of Limited Space and Starving for a New Building

The current museum building has been used for nearly 60 years, with a construction area of 11,500 square meters and the display area of 5,000 square meters. The limited conditions, such as a single function of building and limited space for exhibition, researches and collections, seriously restrict the function and long-term development of the GMC, resulting in us being difficult to adapt to the pace of social development and progress and meet the public's expectations.

Since the 21st century, a lot of experts, scholars, leaders and the general public have appealed to us to build a new museum building with first class facilities and capabilities. With the significant support of the Ministry of Land and Resources, we have strived to build a new museum building with more advanced facilities, stronger capabilities, richer collections and more magnificent appearance since 2016. It is our goal to set up a new museum building that represents the national strength and national image.

4.2 Little International Cooperation and Carrying out Joint Exhibitions Based on the Featured Collection

With the deepening of globalization and internationalization, as well as the opportunity of

The Belt and Road Initiative strategy, there is still much space for us to carry out international exchanges and cooperation. In the future, under the policy support and financial support of the government, the Geological Society of China and other organizations, we will carry out various levels and various forms of exchanges and cooperation, such as joint exhibitions, scientific research and training, together with the museums from a lot of the countries along The Belt and Road Initiative, thus broadening our international visions and reinforcing international exchanges and cooperation.

"I hope that you take 100th anniversary as a new start, remain true to the original ideals, keep pace with the times, regard enhancing the scientific quality of the public as your mission, focus on serving young people, play a better role of geological research base and science popularization palace, develop the museum more advanced and distinctive and make new contributions to build a world scientific power and to realize the 'Chinese Dream' of the great rejuvenation of Chinese nation." This is a congratulation letter from President Xi. These words stimulate us to remain true to the original ideals, keep the mission in our minds, abide by the purposes, work hard, innovate and stride ahead, moreover, make new and more contributions to build a world scientific power and to realize the "Chinese Dream" of the great rejuvenation of Chinese nation.

References

[1] JIA YUEMING. Pass-it-on for One Hundred Years, Science Spirit Being Eternal: to Celebrate the Centenary of the Geological Museum of China. *Acta Geoscientica Sinica*, 2017, 38(2):133-134.

[2] ZONG SUQING. Promoting the Use of the Collections by the Power of the Society: taking the Yangzhou Museum for Example. Appreciation 2016.

[3] Wang Hongjun. The Basis for Chinese Museology.Shanghai: Shanghai Chinese Classics Publishing House, 2001: 246.

[4] The Chinese Association of Natural Science Museums. China Science Popularization Museums Almanac in 2014. Beijing: China Science and Technology Press, 2014.

[5] The Chinese Association of Natural Science Museums. China Science Popularization Museums Almanac in 2015. Beijing: China Science and Technology Press, 2016.

[6] The Chinese Association of Natural Science Museums. China Science Popularization Museums Almanac in 2016. Beijing: China Science and Technology Press, 2017.

Museum-School Cooperation Project of China Maritime Museum Promoting Students' Navigation Education

Wu Chunxia [①]

Abstract To respond positively to the national The Belt and Road Initiative strategy, with the support of the Shanghai Municipal Education Commission, China Maritime Museum carries out "Museum-School Cooperation Project of China Maritime Museum", and together with Shanghai primary and secondary schools, tries to build a new research base of scientific education with certain influence, cultivate a group of dedicated, innovative young students good at science, and promote the professional ability of science and technology teachers. Through 5 subprojects for schools, teachers and students, with various forms and rich contents, this project has made remarkable achievements in promotion of youth maritime education, and will make further optimization and improvement, forming a replicable, propagable model of education, striving to make new achievements for China's maritime industry and construction of The Belt and Road Initiative.

Keywords Museum-school cooperation   Student   Navigation Education

1. Origin and General Situation of the China Maritime Museum

From 1,405 to 1,433, Zheng He, a famous navigator in the Ming Dynasty, led his fleet into the Western Ocean seven times, marking a milestone in our maritime cause (Fig. 81). In memory of the 600th anniversary of Zheng He's voyage, July 11, 2005, was marked as China Maritime Day and the State Council of the People's Republic of China approved the establishment

---

① Wu Chunxia, Director of Education Department of China Maritime Museum, Email: ruoxi_wcx@126.com.

of China Maritime Museum by the Ministry of Transport of P.R.C and Shanghai Municipal Government. On July 5, 2010, China Maritime Museum was open to the public and filled in the gap that there was no national maritime museum in China.

Located at the Dishui Lake of Lingang New City in Shanghai and near the East China Sea, China Maritime Museum is 80 km from the downtown. The museum is an important cultural platform to make China a sea power, carrying forward the brilliant Chinese maritime civilization, promoting international maritime exchange and increasing teenagers' love for the maritime cause.

With a floor space of 46,434 square meters, the museum has an indoor display area of 21,000 square meters and the outdoor display area of 6,000 square meters. It consists of six exhibition halls: Hall of Chinese Navigation History, Hall of Ships, Hall of Navigation and Ports, Hall of Maritime Affairs and Sea Safety, Hall of Seafarers and Hall of Navy, as well as two specific exhibition zones, namely Fishery Zone and Marine Sports & Recreation Zone. In addition, the museum also has a spherical planetarium, a 4D Theatre and a Children's Center. Since its opening, the museum has kept an annual tourist reception of about 320,000, among whom teenage students account for 50%. Besides the permanent exhibitions, each year there would be temporary exhibitions characterizing the main theme of the time and tenet of our museum. For example, they are Navy of 1911, 5th Anniversary of the Gulf of Aden Escort, Ancient Chinese Maritime Cultural Relics Exhibition, Special Exhibition of Western Maritime Cultural Relics, Exhibition of 120th Anniversary of Jiawu Battle, Routes 1,600 – Exchange between the Netherlands & China, The History and Sovereignty of Diaoyu Island, Sail to the World—Exchange between China, the UK and the Netherlands, Exhibition of Shanghai International Shipping Center Construction Fruits and Exhibition of Exported Artwork in 18th and 19th Centuries.

Against the backdrop of national maritime power strategy, since its opening, China Maritime Museum has been committed to spreading maritime knowledge and science & culture. A number of diversified education activities have been designed around the main theme of maritime. For example, there are activities like Maritime Lifestyle, Maritime Theme Day, Aircraft Carrier Style, Science Show, Drama of Zheng He's Voyage and Creative Art. The museum gives priority to the interaction with the neighbourhood and schools around, forming a Three-Kilometer Culture Service Ring. The museum places emphasis on teenager students' maritime education, especially working with Shanghai Municipal Education Commission, on the

"Museum-School Cooperation Project".

2. Museum-School Cooperation Project of China Maritime Museum

The Museum-School Cooperation Project, which aims at improving the abilities of scientific teachers and students on the platform of the museum, sets its objective as the teachers and students of Shanghai primary and secondary schools. The project consists of five sub-projects, including the Development of School-based Curriculums, Museum Teachers Seminar, Mini-topic Researchers, Scientific Explainers and Culture Service Package. Each sub-project is open to the public, focusing on the permanent exhibitions and exhibits and all the education resources inside and outside the museum.

2.1 Development of School-based Curriculums (For Teachers) Makes up For the Lack of Maritime Courses

The sub-project, which promotes the cooperation with excellent teachers from some of Shanghai primary and secondary schools, undertakes the development of school-based curriculums on maritime history & culture heritage and the spread of scientific knowledge; the collection of students courses which combine the museum with school-based curriculums. Besides, for students of different grades, there are designed research courses of different types and levels which integrate class with the museum.

Up till now, there are about 60 school-based curriculums covering maritime history, humane and scientific education. All the curriculums which involve the exhibitions and education resources in the museum will be approved by experts and published on the museum website for use, free of charge for primary and secondary schools in Shanghai.

2.2 Museum Teachers Seminar(for Teachers) Makes up For the Lack of Maritime Scientific Teachers

Members of the seminar will have access to the exhibits and the exhibitions in the museum. Through communication with experts, training and group discussions, the teachers will take good advantage of the museum and improve their teaching abilities.

Each member of the teachers' seminar has designed an appropriate plan for teenagers extracurriculum activities to make the students take a liking to the museum. In the first stage, 200 teachers from 23 schools,10 districts signed up. After censoring, there were 48 teachers attending the project. The course covered many aspects, such as museums, navigation, and so on. A total of 20 experts teach and communicate. More than 40 navigation thematic activities were planned.

2.3 Mini-topic Researchers (for Students) Will Make the Students Start to Research and DIY

The sub-project aims at instructing the secondary school students, high school students, in particular, to further explore and research. Researchers and experts instruct students to do research on a certain mini-project, taking advantage of the exhibitions, education, research and cultural relics in the museum.

To ensure the diversification of the main themes of the sub-projects serving students of different levels, the mini-topic involves both practice and research. The research topic focuses on maritime exhibition and education, with problem-studying at the core. There may be such topics as clean maritime, study on fast boat (clipper) structure, and development of ironclad techniques.

The practice topic focuses on interesting projects that require observation and experiment. There may be such topics as model ship making, ceramics making and repairing, desktop aquarium. There were over 50 students taking part in the first stage.

2.4 Scientific Explainer (for Students) Provides a Platform for Students to be Trained and Grow

There were more than 120 students enrolled in the sub-project and, after censoring, there were 87 students attending the sub-project. The project sets up a four-stage system, including knowing, learning, thinking and explaining. All-around Scientific Explainer training plans have been covering what explaining is, what to explain, how to explain well and what good idea to present in explaining. Students, after being trained, will raise their personal awareness of science, activate their thinking, and develop their personal senses of mission and social responsibility.

During the first stage of the project, there were 60 students participating in the project, and after training, the first"Little Scientific Explainer"Contest was held. The award-winners were granted the privilege to take part in maritime theme activities in advance. Some of the outstanding learners would be selected for further study. All the explainers' work would be included in educational activities in the future.

2.5 Culture Service Package (Face Schools) Makes up For the Disadvantage of the Location of China Maritime Museum

China Maritime Museum, due to its location and other disadvantages, will have to extend maritime knowledge and culture to a wide range of teenage students as possible by providing exhibitions, lectures, activities and teaching aid for various schools. Common activities for schools include more than a certain exhibition, lecture or experience. The exhibitions are divided into those of maritime science, history, art & culture, etc; the lectures cover maritime history and technology; the experiences involve some DIY maritime activities. During the last summer vacation, altogether ten Maritime Summer Camp activities were successfully held, covering six main themes. There were more than 1,500 students enrolled in the project, and after censoring, there were 447 students attending the project from 133 schools.

China Maritime Museum has accomplished its goal in the first stage of the Museum-School Cooperation Project. We have set up a series of courses for maritime research and designed a group of education and practice activities. A team of scientific teachers is being formed for teenagers' maritime education. A certain group of maritime experts is now gathering for the program. We display teenagers' maritime research & practice and work with some primary and secondary schools on the project to work out the possibility of the Museum-School Cooperation Project.

Meanwhile, the project gets the students to acquire the basic Chinese maritime history and be acquainted with some famous figures, to know what Chinese navigation techniques contribute to the world, to learn and utilize modern navigation techniques, to know how important the right of sea is to a country and its people and inspire them to learn by themselves.

Later on, the museum will perfect the second and third stages of the Museum-School Cooperation Project.

3. Conclusion

At present, the Museum-School Cooperation Project is at the exploratory stage, the practical experience is not rich. Meanwhile, because of the restrictions on the national education policy, there is no wide range of practice for a period of time. But, with the prosperity and development of culture and the demand for the comprehensive quality education of the students, the Museum-School Cooperation Project will be a trend. Museum as an important place for informal education will get more and more attention from the society. Its value and significance will also be truly reflected and upgraded.

References:

[1] SONGXIAN. Research on Museum-School Cooperation Project [M]. Shanghai Scientific and Technological Education Publishing House, 2016 (8).

[2] Development Plan in 13th Five-Year of China Maritime Museum.

[3] China Maritime Museum's Implementation Scheme of Museum-School Cooperation Project.

	1	2
	3	4
	5	

1 Aircraft Museum—Dhangadhi, far western Nepal
2 Aviation Museum—Kathmandu
3 General view of the Museum from South—West
4 Viktor Ambartsumian
5 Viktor Ambartsumian's bedroom and the place where he slept

6 A view of Nammal Gorge, Salt Range, Pakistan, Permian to Eocene succession is exposed in the Gorge

7 A view of Khewra Gorge, Salt Range, Pakistan, the Precambrian and Cambrian Sequence is well exposed

8 16 Herakleidon street building

9 in the classroom

```
    10
   ─────
    11
   ─────
    12
   ─────
    13
```

10 The entrance of the National Dinosaur Museum
11 In the garden of the National Dinosaur Museum
12 Display specimens amongst items for sale in
    the Museum shop
13 National Museum of the Republic of Kazakhstan

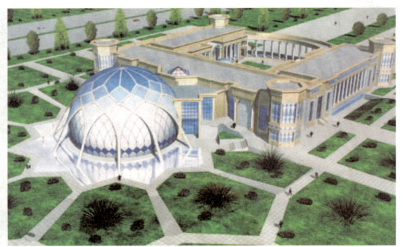

14	15
16	17
	18

14　Hall of Ancient and Medieval History

15　felt dwelling—kyiz ui

16　Golden Man

17　3d model of the completed building of Khayyam Planetarium

18　ASM public activities

19
20
21
22

19 Newly constructed 10 m dia. Planetarium located at the Ferdowsi University of Mashhad
20 Beijing auto museum
21 Collected and repaired vehicles
22 Vehicle composition, literature and model and so on

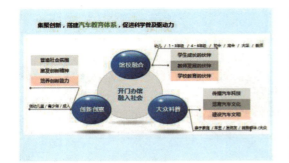

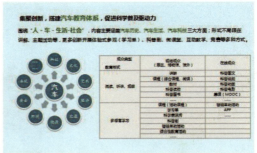

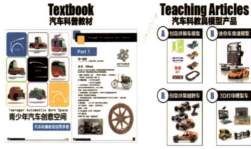

23、24、25、26、27　Educational systems

相传，早在5000年前黄帝就制造了指南车。

It is said that 5,000 years ago, the Yellow Emperor invented south-pointing chariot.

中国古代伟大的思想家、教育家孔子曾乘车周游列国，传播文化思想。

Over 2,500 years ago, the great thinker and educator Confucius made a tour of numerous kingdoms in a carriage, spreading Chinese culture and ideas.

2500 years ago

2200 years ago

秦始皇统一中国后，实行了"车同轨"制度，车辆制造进入了标准化阶段。

China's first emperor, the First Emperor of Qin, unified China under one rule for the first time, including the rule "the same rutting". He standardized certain aspects of the tracks and vehicles throughout the empire.

28	28	South-pointing Chariot
29	29	Confucius made a tour of numerous kingdoms in a carriage
30	30	the rule "the same rutting"

多元文化促进国与国、城与城、人与人间的互动交流。
Multicultural events have promoted interactive exchanges between countries, cities, and people.

一路同行世界梦！
我们期待让车载着人、人载着思想，共赴"人-车-社会"的美好与远方
We look forward to continually working together toward the World Dream! We also look forward to the
continuation of cars carrying people and people carrying ideas. All of these things will work together in
propelling our progress toward the beautiful and far-sighted goal of integrating people, cars, and society.

31

32

33

31   Multicultural

32   World Dream

33   Bird's-eye view of Cozumel, Planetarium

34
35
36

34 Screening of the fulldome show Mayan Archaeoastronomy: Observers of the Universe

35 The historic 1933 Queen's Park wing and the 2007 Crystal renovation of the Royal Ontario Museum

36 The 26-metre long skeleton of the Blue Whale salvaged from Newfoundland by the ROM in 2014 and now the centerpiece of a traveling exhibition inspiring audiences throughout the world

37 | 38
39

37   A.F. Kohts, the founder of the Darwin museum, as a university graduate

38   The State Darwin Museum, present day

39   The interactive exhibition "Walk the Path of Evolution"

40	41	42	43
44	45	46	
47			

40、41、42、43 Scientific Publications

44、45、46 Awards of the State Museum Director

47 Cretaceous trip of Liaoning Provincial Science and Technology Museum

48	
49	50
51	

48   VR RIDE
49   The Earth Ark
50   The Domino of the Extinct Animal
51   The Legend of Elk

52	53
54	55
	56

52   Deer of All Lands

53   New Chongqing Natural History Museum

54   Dinosaur Hall of Chongqing Natural History Museum

55   Dinosaur on stamp

56   Panoramic view of Shaoxing Science and Technology Museum

57
58 | 59
60
61

57　The old museum, at No. 104, Zhaojiatiao, Jiang'an District

58、59　A corner of the exhibition hall

60　New Wuhan Science and Technology Museum

61　The audience waited in line for a visit

```
      | 63
   62 |----
      | 64
   ---+---
   65 | 66
```

62、63、64 View of interior of new hall of Wuhan Science and Technology Museum
65、66 Rich and colorful science activities

67	68	69
70	71	72
73	74	75

67、68、69 Bilingual popular books for ethnic minorities (Chinese and Dai writing)
70、71、72 Bilingual science books for ethnic minorities (Chinese and Tibetan)
73、74、75 Bilingual science books for ethnic minorities (Chinese and Yi writing)

76	77
78	79
80	
81	

76、77、78、79、80 Scene of supporting science popularization

81 China Maritime Museum

"一带一路"科普场馆发展
国际研讨会论文集

目　录

"一带一路"科普场馆发展国际研讨会主旨报告

《北京宣言》

1. 背景与由来

为推进"一带一路"沿线国家科学文化的交流，加强沿线国家自然科学博物馆之间的合作，由中国自然科学博物馆协会主办、中国科学技术馆和上海科学技术馆承办的首届"一带一路"科普场馆发展国际研讨会，于2017年11月27—30日在北京召开。来自"一带一路"沿线22个国家的24家科普场馆和机构确认参会。

在与参会人员的联络过程中，有部分代表提出，期望本次研讨会成为"一带一路"沿线科学类博物馆开展深度合作的一个良好开端，重要的是形成长效合作机制，在人文交流、教育合作、科技发展等多个方面互通有无，共商合作、共建平台、共享成果，以此推动社会进步和全人类的共同发展。

部分参会代表向大会主席建议，应在研讨会正式举办之前共同探讨如何巩固和丰富研讨会成果，提议形成大会宣言，以彰显合作共识、固化研讨成果、扩大会议影响，为未来可持续发展奠定基调。此建议得到了大会主席的充分肯定与支持。

因此，来自12个"一带一路"沿线国家的涵盖自然、天文、地学和综合科技等不同类型科普场馆的参会人员组成了《北京宣言》核心起草小组。经过小组成员的多次讨论、修改形成了《北京宣言》草案。后经首届"一带一路"科普场馆发展国际研讨会圆桌会议讨论定稿，大会正式发布《北京宣言》。

2.《北京宣言》

当前，世界发展机遇与挑战并存。各国都在追求和平、发展与合作。中国在2017年5月15日成功举办"一带一路"国际合作高峰论坛圆桌峰会，为构建人类命运共同体注入了强劲动力。

出席2017"一带一路"国家科普场馆发展国际研讨会的全体代表，于2017年11月27—28日汇聚中国北京，秉承"一带一路"高峰论坛的宗旨，围绕"协同共享、场馆互惠、共建科学传播丝绸之路"主题，思考科普场馆如何在促进联合国可持续发展目标的实现方面发挥不可替代的作用，开展深入而细致的讨论，共商合作、共建平台、共享成果，最终形成以下宣言，即《北京宣言》。

我们认为，在信息爆炸、跨界融合的时代，"一带一路"（丝绸之路经济带和21世纪海上丝绸之路）沿线国家科普场馆之间的合作交流是未来博物馆领域可持续发展不可缺少的要素。"一带一路"沿线国家科普场馆之间不断加强的合作与交流，能够真正实现各国家之间的政策沟通、设施联通、文化互通、民心相通，使"一带一路"沿线各场馆的教育、展示、收藏和研究成果在不同的文化环境中相互交流，更好地推进科技发展与创新，进而促进社会经济的可持续发展和人类的共同进步。

我们认为，建立"一带一路"沿线国家科普场馆之间的信任体系，能够发挥值得信赖的纽带和极具价值的传播者的作用，鼓励宽容和批判性思维。共建互信是共享科学的基础，而共享科学是"一带一路"建设的重要内容之一。"科学让生活更美好"的理念，将通过"一带一路"沿线国家科普场馆的多元合作，形成科学传播"新丝路"，打破科学知识传播的壁垒，共同构筑科学、文化与精神传递的命运共同体。

我们认为，"一带一路"建设，不仅需要科学知识和科学文化，还需要科学方法和科学精神。科普场馆则是开展科学传播不可或缺的力量。"一带一路"沿线国家的科普场馆应寻求多种合作与交流方式，将以物为本的传统展示技术与最新科技相结合，促进公众参与科学。

为此，我们愿意：构建"一带一路"沿线国家更广泛的科学传播联盟，开展展览的巡展与交流，博物馆教育资源合作开发，人员定期互换交流与考察，定期举办科普场馆人员培训班，开展博物馆研究与藏品征集合作，合作举办科技节、青少年科学竞赛等，为推动创新与发展，促进资源的共享与场馆的互惠共赢，促进彼此的交流，提升科学传播的能力做出积极贡献。

我们应当：利用"一带一路"沿线国家科普场馆的重要资源，构建积极有效的沟通交流平台，搭建机制性合作框架，建立协作网络，促进更广泛而深入的文化与科技交流，共同为建设人类命运共同体贡献力量。

我们建议：依托"一带一路"沿线国家科普场馆交流的机会，积极推动学术性非营利组织"新丝路科普场馆学会"的成立，以此更好地促进各场馆之间互通有无、增进了解、促进创新、谋求共同进步。

我们建议：以此次"一带一路"国家科普场馆发展国际研讨会为契机，积极推动成立"一带一路"科普场馆合作发展基金会，并定期推送科普场馆的最新资讯，以此支持和促进各馆之间合作交流的持续开展和进行。本次论坛取得圆满成功的同时，产生了一系列研究成果，因此，我们建议，"一带一路"沿线国家的科普场馆论坛每两年举办一次，该活动将秉承"平等、自愿、互惠"的原则，在合作与共商、共享、共赢的基础上举行，以推动"一带一路"沿线国家的深度合作。

本着开放、包容的精神，欢迎认同《北京宣言》目标的其他国家的机构参与。

"一带一路"背景下自然科学博物馆
对《2030 年可持续发展议程》的潜在贡献

约斯兰·努尔 [①]

摘　要　"一带一路"倡议和联合国《2030 年可持续发展议程》的性质和范围各不相同,但均以可持续发展为总体目标。因此,"一带一路"倡议对《2030 年可持续发展议程》的实施具有巨大的推动作用。自然科学博物馆如何为实现《2030 年可持续发展议程》,特别是在"一带一路"的政策沟通、设施联通、贸易畅通、资金融通、民心相通五大优先领域中做出贡献? 为此讨论四个问题: 自然科学博物馆在推动"一带一路"国家之间科技创新合作方面的潜在作用; "一带一路"国家的自然科学博物馆之间实现互联互通; "一带一路"背景下自然科学博物馆的海外发展援助潜力; 自然科学博物馆在加强文化理解方面的作用。

关键词　2030 年可持续发展议程　"一带一路"倡议　自然科学博物馆

首先,我谨代表联合国教科文组织对中国自然科学博物馆协会成功举办首届"一带一路"科普场馆发展国际研讨会表示祝贺。本届研讨会以"协同共享、场馆互惠、共建科学传播丝绸之路"为主题,探讨自然科学博物馆的发展如何能够为实现联合国《2030 年可持续发展议程》(以下简称《2030 议程》) 做出贡献,呼吁自然科学博物馆为"一带一路"国家之间科技创新合作、构建命运共同体增砖添瓦。下文将围绕自然科学博物馆在《2030 议程》以及"一带一路"倡议中所扮演的角色和担当的责任展开。

---

①　约斯兰·努尔: 联合国教科文组织下属自然科学与能力建设部自然科学司方案专家,博士。E-mail: y.nur@unesco.org。本文根据作者 2017 年 11 月 27 日在"首届'一带一路'科普场馆发展国际研讨会"上所做的主旨报告进行编译。

1.《2030 年可持续发展议程》

　　《2030 议程》由联合国 193 个会员国在 2015 年 9 月举行的历史性首脑会议上一致通过。《2030 议程》是整体的、不可分割的，并兼顾了可持续发展的三个方面：经济、社会和环境，涉及发展中国家和发达国家在内的 70 多亿人。联合国前秘书长潘基文指出："这些目标述及发达国家和发展中国家人民的需求，是人类的共同愿景，也是世界各国领导人与各国人民之间达成的社会契约。既是一份造福人类和地球的行动清单，也是谋求取得成功的一幅蓝图。"《2030 议程》包括 17 个目标、169 个具体目标和230 项指标，让所有人享有人权，实现性别平等，增强所有妇女和女童的权能，并强调"不让任何一个人掉队"。

　　消除贫穷与饥饿，让所有人能够平等地享受经济社会发展的成果，依然是可持续发展目标之首。随着经济和科学技术的发展，粮食短缺问题得到较大缓解。然而，长期困扰发达国家的营养失调和营养过剩的问题在发展中国家日益凸显，糖尿病、肥胖、心血管疾病正日益严重地威胁着人类的健康和未来。由于淡水资源匮乏、分布不均、水质恶化等问题，使得全球约 1/5 的人无法获得安全的饮用水。与此同时，有些地区饱受洪水泛滥的侵扰，面临因缺乏清洁安全水源和环境卫生系统受到破坏而引发的霍乱、麻疹、疟疾等传染病以及营养缺乏的危险。解决这些问题的关键是全球高度参与，将各国政府、私营部门、民间社会、联合国系统和其他各方召集在一起，调动现有的一切资源，协助落实《2030 议程》所列出的 17 个目标。而推动可持续发展的长期实践证明，科学技术和创新是根本途径。2015 年 9 月召开的联合国发展峰会把加强科学、技术和创新作为落实《2030 议程》的重要执行手段。2016 年 6 月召开的联合国首届"科学、技术与创新促进可持续发展多利益攸关方论坛"和 7 月召开的联合国"可持续发展高级别政治论坛"上，与会各国代表一致强调科技创新支撑和引领可持续发展的重要性和不可替代性，强调科学、技术和创新应贯穿所有可持续发展目标。

　　当今世界存在的不平等、歧视、仇外心理和相关不容忍等现象，成为人类安全、和平与发展的重大威胁。弱势群体遭受着资源被剥夺的境遇，更容易受到环境恶化造成的影响。"不让任何一个人掉队"是《2030 议程》的核心承诺，确保人人被尊重，体现了议程制定对普惠性的高度重视。这个承诺促使全球重新思考弱势群体的定义、特征和成因，以及社会、经济、政治和环境等制约因素对弱势群体的影响等。同时，通过《2030 议程》增强弱势群体的权益，造福弱势群体，确保社会公正、民主、性别平等、持续性以及有包容性。

实现《2030 议程》需要从个人、社区、国家和国际社会各个层面，以相得益彰的方式推动可持续发展。一是加强制度建设、增强能力、改革政策、强化民主治理，需要国家和社区对扮演什么角色以及识别并抓住新型机遇的能力进行广泛反思。二是根据上述原则提供全球公共产品，在货币、金融、贸易技术和环境问题的全球治理方面，降低权力不对等。三是需要国家内部、国家之间加强合作。四是制订可持续发展中所需的国家开发计划、预算和商业模型，例如，减少含碳产品的使用、形成良好的低碳产业群和产业链，实现低碳经济。

各国面临的情况不同，重点需求、能力、资源也不同。因此在国家层面，一是需要根据《2030 议程》制订与可持续发展目标相关的新的战略规划、政策、计划和投资方法。二是设置新的指标，建立全面的监测评估系统，明确设定目标和具体路线图，追踪实施情况并衡量影响。三是形成实现可持续发展目标所需的一整套强有力的金融、科技、自由贸易、信息获取等实施手段。四是呼吁政府及其合作伙伴以新的模式在各部门和各领域之间开展合作，以促进全面的实施行动。

然而，有相当一部分政府官员、政策制定者、科学家和商业运行决策者对可持续发展目标认识过于片面，认为可持续发展仅与环境问题相关。因此，政策决定者、科学家、商业运行决策者和公众需要充分了解可持续发展目标，发现其中蕴含的机遇和责任，深化利益相关方的关系，紧跟政策步伐，同时减少对可持续发展议程的负面影响，并做出更多积极贡献。

在实现可持续发展目标方面，自然科学博物馆发挥着突出作用，有两个案例可佐证：其一，2016 年 11 月 10 日是第一个"国际科学中心和博物馆日"，邀请公众共同回顾他们围绕联合国可持续发展目标举办的活动，探讨其实践与理念，是科学中心与科学博物馆在国际舞台上展示自身在鼓励公众参与科学技术话题方面产生的影响。其二，2017 年 11 月 15—17 日，第二届世界科学中心峰会以"连通世界，实现可持续的未来"为主题，发布了由世界六大地区科学中心协会组织领导人签署的《东京议定书》。《东京议定书》号召科学中心应该重点考虑如何加深公众对可持续发展目标意识的重要性和紧迫性的了解，以及参与有助于实现可持续发展目标的行动。我们应该采取行动，与当地社区进行沟通和合作，以便能够贯彻可持续发展目标；我们应该担当交流平台，让科研工作者能够走向社会。

自然科学博物馆以五类人群为目标，开展有针对性的活动，促进可持续发展目标的实现。第一，政策制定者作为让科技和政策相辅相成、促进科学技术创新融合的最佳人选，通过鼓励政府制定新的科学、技术、创新战略，促进知识型社会的建设。自然科学博物馆可参与制定、评估、改革现有的科学技术创新系统，并协助制定战略和行动计划。第二，科学家是推动可持续发展目标实现的重要力量，自然科学博物馆积极与科学家合作，鼓励科学家积极参与实现可持续发展目标。第三，生产部门（包括工业和农业）利用科学知识提高生产力，提高经济效益。第四，公众是可持续发展目标的受益者，同时，公众参与也是实现可持续发展的必要保证，可持续发展的目标和行动，必须依赖社会公众和社会团体最大限度地认同、支持和参与。第五，年轻人是未来所在，获得良好的教育，掌握科学知识，才有希望在可持续发展中获得最终的成功。

2. "一带一路"倡议

2000 多年前，亚欧大陆上勤劳勇敢的人民，探索出多条连接亚欧非几大文明的贸易和人文交流通路，后人将其统称为"丝绸之路"。2013 年 9 月和 10 月，习近平主席在出访中亚和东南亚国家期间，先后提出共建"丝绸之路经济带"和"21 世纪海上丝绸之路"的倡议（简称"一带一路"倡议），得到国际社会高度关注。"一带一路"倡议旨在通过基础设施建设、电信、数字网络信息技术把不同地域联系起来。习近平主席 2017 年 5 月 14—15 日在北京举行的"'一带一路'"国际合作高峰论坛"上解释了"一带一路"倡议的核心精神。

第一，深化中国和邻国之间的政策沟通。在"一带一路"倡议下，中国和其他国家已经签署 120 多项务实合作协议。这些协议不仅涵盖了"硬件"基础，如运输、基础设施和能源；还包括"软件"基础，如电信、海关和检疫等。协议还包括经济和贸易、工业、电子商务、海洋和绿色经济合作的计划和项目。

第二，加强基础设施互联互通。中国和相关国家一道共同加速推进雅万（雅加达—万隆）高铁、中老（中国—老挝）铁路、亚吉（亚的斯亚贝巴—吉布提）铁路、匈塞（匈牙利—塞尔维亚）铁路等项目，建设瓜达尔港（巴基斯坦）、比雷埃夫斯港（希腊）等港口，规划实施一大批互联互通项目。目前，以中巴（中国—巴基斯坦）、中蒙俄（中国—蒙古—俄罗斯）、新亚欧大陆桥等经济走廊为引领，以陆海空通道和信息高速路为骨架，以铁路、港口、管网等重大工程为依托，一个复合型的基础设施网络正在形成。

第三，加强贸易互联互通。2014—2016 年，中国同"一带一路"沿线国家贸易总额超过 3 万亿美元，中国对"一带一路"沿线国家投资累计超过 500 亿美元。中国企业已经在 20 多个国家建设了 56 个经贸合作区，为有关国家创造近 11 亿美元税收和 18 万个就业岗位。

第四，强化金融互联互通。亚洲基础设施投资银行为"一带一路"沿线国家的9个项目提供了17亿美元贷款。丝路基金已投资40亿美元，中国与中东欧16+1金融控股公司正式成立。这些新的金融机制和世界银行等传统的多边金融机构各有侧重、互为补充，形成层次清晰、初具规模的"一带一路"金融合作网络。

第五，加强民间互联互通。在此背景下，中国开展了科学、教育、文化、卫生、人文等领域的合作。中国政府每年向有关国家提供1万个政府奖学金名额。中国地方政府还设立了专门的丝绸之路奖学金，以鼓励国际文化和教育交流。

如何将可持续发展目标与"一带一路"倡议连接起来？"一带一路"倡议和《2030议程》均以可持续发展为总体目标。联合国秘书长安东尼奥·古特雷斯在"'一带一路'"国际合作高峰论坛"上指出：《2030议程》与"一带一路"倡议有共同的宏观目标，旨在创造机会，带来有益全球的公共产品，并在多方面促进全球联结，为基础设施建设、贸易、金融、政策以及文化交流，带来新的市场和机会。因此，"一带一路"倡议对联合国《2030议程》的实施具有巨大的推动作用。

3. 自然科学博物馆作为可持续发展目标的传播平台

在"一带一路"倡议的背景下，自然科学博物馆如何发挥好自己的作用、扮演好角色？自然科学博物馆能够提供什么人才和智力的支持？

首先，自然科学博物馆是"一带一路"沿线国家互联互通的桥梁和纽带。本次研讨会共有来自17个国家的40名自然科学博物馆代表参加，并签署多项合作协议以加强"一带一路"国家自然科学博物馆之间的对话与交流。通过对话平台，有助于"一带一路"沿线国家自然科学博物馆之间加深了解，找准共同点，共同开发合作项目。

其次，在"一带一路"倡议背景下，自然科学博物馆挖掘自身国际援助的潜力。中国自然科学博物馆向"一带一路"沿线国家的场馆提供相应的援助。援助包括三类：第一，技术援助——通过分享中国在自然科学博物馆发展方面积累的知识和经验，实现中国与"一带一路"沿线国家的自然科学博物馆之间优势互补、良性合作、融通发展。第二，能力建设援助——通过设立培训班，培养自然科学博物馆的高层次人才及未来核心竞争力。第三，资金援助——"一带一路"沿线国家的自然科学博物馆在发展过程中，有望获得中国的资助以达到双赢的结果。联合国教科文组织非常愿意推动这些活动的开展，积极推进中国、"一带一路"沿线国家和非"一带一路"沿线国家之间的战略对接和合作。

再次，科学中心和科学博物馆应该成为可持续发展目标传播的一个平台。我们在这个平台上要做什么呢？可以参考上文提及的《东京议定书》。

最后，通过自然科学博物馆加强人与人之间的互联互通。第一，作为非正规科学教育平台，自然科学博物馆提供了与正规教育不同的学习环境。从本质上讲，学习变成了一种自愿的、自主的行动。自然科学博物馆的展览方式强调参与体验，主题涉及科学技术的发展及其对社会的影响等，科学原理蕴含在展览和展品中，通过与展品互动，将科学与公众的生活紧密相连，激发公众对科学技术的兴趣。第二，促进科学界、社会和政策制定者之间的对话。自然科学博物馆作为科技创新政策对话的平台，提高对"科学地制定政策"和"为科学发展制定相关政策"重要性的认识。由于没有足够的预算开展相应的活动，很多发展中国家缺乏为科学发展制定的相关政策，或者空有虚名。在科学沟通中最重要的目标群体是政策决定者，他们为科技创新以及科学政策提供相应的预算。自然科学博物馆有能力影响决策过程，但绝非易事。第三，在"一带一路"倡议下，自然科学博物馆是跨文化理解、促进和平的平台。科学史是自然科学博物馆一直关注和探索的主题，可以提高公众对科学和技术发展历史的认识。让公众了解科学的历史根源，为他们提供一个了解科学和技术发展的多元文化维度的机会，正如我们今天所知的那样。几千年来，伊斯兰、中国、希腊、印度等文明都通过"丝绸之路"相互交流。通过理解各国文化，我们会更加尊重文化间的差异。科学和技术可以是一种和平的交流方式，它是科学中心和科学博物馆可以为世界各地的和平建设做出贡献的一种方式。第四，自然科学博物馆本身并非技术创新者，但通过向公众展示最新的科学研究和新技术，推动技术创新，帮助公众理解创新过程——设计、建造和使用技术的技能——以及科学技术的影响。在过去10年，科学中心和科学博物馆在向公众介绍新的创新方面，特别是与当前全球性挑战有关的技术创新方面，发挥了越来越重要的作用。

博物馆：我们共有文化精神的前沿实验室

安来顺 [①]

摘 要 当代公共博物馆在经历了 250 多年的发展演变后，其发展的社会、文化、专业甚至经济环境，都在发生着重要变化。新的趋势无疑为博物馆带来新机遇、新可能，同时也伴随着新挑战。在今天博物馆需要回答的诸多问题中，有一个问题始终处于核心位置：我们的博物馆机构是否对社会很重要？围绕这个问题的相关讨论，让当代博物馆新的文化观逐渐清晰起来，让博物馆为适应新形势而进行的战略选择有了新的坐标，并最终引导出关于博物馆价值体系可能的重构以及新的运行规则。

关键词 博物馆 趋势 战略 前瞻

1. 引言

公共博物馆历经 250 多年的进化和发展，无论是认识论、类型学，还是功能学、方法学，都与 19 世纪末欧洲的经典博物馆有着巨大差异。最明显的趋势之一是越来越多的科学普及和推广机构的出现，与传统意义上的艺术史、人类学、考古学博物馆承担相同的社会责任，遵从相同的职业伦理，致力于"为社会和社会发展服务"这一相同宗旨。作为全球博物馆及其从业者的专业组织，国际博物馆协会在 1961 年的章程中，将从事科学普及的公共机构界定为极为重要的博物馆类型。国际博物馆协会 30 个国际专门委员会中，科学与技术国际委员会自 1972 年正式创建以来，一直显示出巨大活力和潜力。该委员会由各国科技界的博物馆专业人员构成，不仅服务于传统意义上的科学和技术博物馆，而且为推动科学中心等机构的发展发挥了越来越重要的作用。中国的科技博物馆界深度参与了该委员会的工作并担任着重要角色。国际博协作为重要的交流与合作平台之一，以相同的目标和责任把世界上不同学科领域的博物馆

① 安来顺：国际博物馆协会副主席，中国博物馆协会副理事长兼秘书长，研究馆员。研究方向：博物馆学。地址：北京市西城区阜成门内大街宫门口二条 19 号北京鲁迅博物馆，邮编：100034，E-mail：an_laishun@vip.163.com。本文系作者 2017 年 11 月 27 日在"首届'一带一路'科普场馆发展国际研讨会"上所做的主旨报告。

和博物馆专业工作者联合在了一起。以下拟围绕当代博物馆与社会发展之间的能动关系（即当今社会如何影响博物馆的发展？博物馆又将如何作用于当今社会），从"形势""战略"和"前瞻"三个方面展开探讨。

2. 形势：博物馆发展的社会环境经历变革

21世纪社会环境的变化对当今博物馆的发展产生了极其深远的影响。

第一，从广义的文化角度看。人类赖以生存的物质文化、人类追求生活品质的健康文化、人类共同遵守的社会政治文化、人类闲暇生活中追求美的审美文化，这四种主要文化类型的内涵和外延都在发生变化。40多年来出现的这些变化，都不是线性的和单一方向的，而是相互联系和螺旋式的。无论哪个国家、何种学科属性、多大规模的博物馆，都无法游离于这些从内容到表现形式的文化变革影响之外，因为归根结底上述四种类型的文化是博物馆保存、研究、展示和传播的主要对象。博物馆及其从业者在这些新趋势面前，无法仅站在某个单一学科的立场上卓有成效地开展工作，而跨学科甚至多学科融合成为切实的解决方案。

第二，从参与型社会的角度看。公众的深度参与是现代社会的重要特征。参与型社会的特征给博物馆传统理念带来的冲击是巨大的，影响也是深刻的。其实，真正意义的公众参与已经远超出了公共机构（如博物馆）在自身业务功能框架内吸引公众的有限介入，而是这些机构在精神和伦理上的民主化和社会化，它要求包括公共机构认同公众应当拥有的平等地位，要求将公众视为自身存在与发展的主体，而不是无足轻重的客体；要求公共机构尊重公众共享公共机构资源的权利；要求公共机构鼓励公众作为具有独立意识和权利的主体参与到机构中来；要求公共机构更加包容，在保持站在自身学科的立场上加以表达的权利的同时，也允许公众在机构中发表自己的意见。上述特征势必给博物馆的传统理念带来巨大冲击，因为在传统博物馆理念中，博物馆始终认为自己在所属学科是唯一的权威，而这种权威是公众不可以挑战的。

第三，从公共机构运行的经济环境看。世界范围经济形势不确定、不稳定因素逐渐增多，由此造成公共机构财政状况变得越来越复杂。经济形势虽然未必最直接、最迅速地传导到博物馆领域，但其间接和长远影响不容忽视，应该引起博物馆的政策制定者、博物馆管理者和运行者的高度警觉。正是由于这样的警觉或者说已经感受到的压力，一些博物馆为了证明自己的价值和吸引更多地方财政支持，将生存和发展途径转向置身于社区和教育性活动作为调整的战略之一，这种在博物馆学界被称为"第三代博物馆"甚至"第四代博物馆"现象确实值得研究。现实中的情况还远不止如此，备受关注的是，变化的经济环境已迫使博物馆进入他们原本相对陌生的经济领域。这其中确实存在着机遇，但是挑战也无处不在，甚至争论已直指诸如"博物馆如何坚守作为公共文化机构的属性""在经济活动中博物馆的职业道德标准是什么"等本质性而非简单操作性的问题。

第四，从与博物馆相对邻近的领域看。日新月异的数字技术无一例外地渗透到所有的社会细胞中，能够提供多重选择的休闲产业在与博物馆竞争公众的文化消费，博物馆必须给公众充分的理由，让他们不断走进博物馆。知识经济使得今天人们对知识和技能的掌握不再通过一次性学习取得，因此，为重新唤起公众对终身学习的热情，博物馆的"第三空间"、旅游产业和当地社会经济发展、非物质遗产、当前社会热点话题、新技术的运用、更广泛和多学科的信息提供，成为了21世纪博物馆六大新的关注点。

3. 战略：当代博物馆功能的提升与转型

当今博物馆必须回答一个根本性的问题：我们的博物馆是否对社会很重要，无论是眼下还是未来？选择功能提升与转型是博物馆回应上述问题的战略。

第一，具有当代特征的博物馆文化观逐步形成。进入21世纪以来，正在逐渐形成的博物馆文化观具有了新的价值取向：博物馆不再仅是为了保存既往的痕迹，而且是为了保留于当代社会，于国家和民族具有重要意义的文化灵魂；博物馆不再仅是受托于我们的祖先延续遗产的生命，而且是受托于当代和子孙更好地发挥它们的作用；博物馆不再是某些群体利益或学科体系的代言者，而是拥有特有的社会身份和价值观，通过其功能的实现，成为一个国家和民族良知的代表。上述价值取向的转变，无疑是当代博物馆文化的一个极其重要的进步。

第二，顺应社会变革而调整和优化的博物馆战略。社会变革的挑战是客观存在的，同时也为博物馆提供了新的机会，即保存、记录、见证和参与变革的机会。为此，博物馆开始探索符合社会预期的战略选项。这里包括博物馆"必须做"的：博物馆必须承担起文化记忆库（广义的文化记忆：物质文化、健康文化、社会政治文化、审美文化）的职责，因为记忆功能是博物馆最原始、最核心的功能之一；博物馆必须使历史文化能够为今人所理解，在公众与最直接、最真实的博物馆资源之间架起一座桥梁；博物馆必须反映社会关注的问题，至少做到应有的敏感，让自身成为呼吁为社会进步而采取行动的渠道；博物馆必须成为所在社区公共文化体系的有机组成部分，把博物馆的服务延展到所有人群，并充当现有的、不同文化的融合剂。这里还包括博物馆

"应当做"的：博物馆应当增进民族之间、国家之间更深入的理解，把一个国家的文明引向广阔世界，让各民族的知识自由流动，文化信息得以有益的交流；博物馆应当在诠释传统文化的同时诠释新的文化，以宽容的态度让人们获得信息，使其在尊重自身成就的同时，对其他文化保持好奇的心态和尊重的态度，推动文化传承和创新；博物馆应当成为新文化的催化器，通过陈列展览等方式，在博物馆与其收藏或展品结合中生成新知识、催生新文化。在不同博物馆的实践中，上述这些"必须"和"应当"分别被置于不同的优先级。

在这里要特别指出，科学技术博物馆通过战略调整，在可持续环境领域承担更重要的责任。城市化是社会发展的一个重要趋势，而且步伐还在不断加快，同时也将生物多样性等可持续发展环境问题摆在人们面前。这不再是遥远的、抽象的问题，而是当下的、具体的问题，是与人们日常生活密切相关的问题。博物馆在可持续发展环境方面的战略之一，是鼓励全社会把看似属于科学技术的问题与突出的社会问题紧密联系在一起，努力帮助人们在理解过去的同时，吸取今天的教训，致力于创造可持续的未来。

4. 前瞻：博物馆价值体系重构正成为可能

国际博物馆协会自 1946 年成立以来，主导了对博物馆定义界定的 8 次修订，现在正在进行第 9 次修订，目的是试图回应博物馆不断变化的发展形势。目前权威的博物馆定义形成于 1974 年，修订于 2007 年，即"博物馆是一个不以营利为目的的、为社会和社会发展服务的、向公众开放的永久性机构。它以教育、研究和欣赏为目的收集、保存、研究、传播和展示人类及其环境的物质的和非物质的遗产。"在国际博物馆协会的定义中，"为社会和社会发展服务"是一个极易引发讨论的表述。可以预期的是，一个新的博物馆定义将在 2019 年国际博物馆协会全体大会上正式提出，它应该是反映当今社会条件下博物馆新趋势、新条件、新使命和新可能的表述。

毫无疑问，一个成熟的博物馆行业必定有一套大家共同拥有的价值观、精神自律体系，以及行之有效的基本行为标准，这就是职业道德准则。1986 年国际博物馆协会通过了《国际博物馆协会职业道德准则》（*ICOM Code of Ethics*），得到了 150 多个国家近 37000 个博物馆机构和从业者的认同和遵守。该准则确定了涉及博物馆举办者、博物馆收藏、博物馆与公众、博物馆合法运行和专业化运行以及博物馆从业者个人行为的行为准则。随后经过四次修改，较好反映了当代博物馆发展的新形势、新要求，成为许多国家博物馆的"软法"，为此这些国家不再制定专门法律。

联合国教科文组织于 2015 年 11 月通过了由国际博物馆协会起草的关于《联合国教科文组织关于保护和加强博物馆与收藏及其多样性和社会作用的建议书》（以下简称《2015 建议书》）。对于博物馆而言，《2015 建议书》是一个强有力的指导文件。最引人关注的是，《2015 建议书》把博物馆"为社会和社会发展服务"表述置于一系列特定的现代价值体系中，如公平、自由、团结、社会包容、社会融合、可持续发展等。与国际博物馆协会现行 2007 年的定义相比，教科文组织要求博物馆在涉及人、人性等方面扮演更加积极的行动者角色，将博物馆的作用领域由收藏、历史遗迹等方面向更广阔的人文领域拓展。

　　2016 年 11 月，联合国教科文组织为落实《2015 建议书》在深圳召开了国际博物馆高级别论坛。中国国家主席习近平给论坛发来贺信，指出："博物馆是连接过去、现在、未来的桥梁，它们不仅是历史的保存者和记录者，也是当代中国人民为实现中华民族伟大复兴的中国梦而奋斗的见证者和参与者"。绝非巧合，在该论坛上，时任教科文组织总干事伊琳娜·博科娃强调了"博物馆是我们共有人文精神的前沿实验室，它保护我们的遗产，激发新的创造力，帮助我们捕捉世界的纷繁复杂。"这也许预示着，一个面向今天与未来的博物馆时代正向我们走来，相应地，博物馆新的价值体系构建会在不久的将来成为可能。

参考文献

［1］盖尔·洛德. 文化变革与博物馆［R］. 东京：2011 年亚洲博物馆论坛，2011.

［2］安曼努尔·阿林茨. 博物馆的社会角色［R］. 乔治敦：圭亚那国家博物馆的演讲，1999.

［3］张文彬，安来顺. 城市文化建设与城市博物馆［J］. 装饰，2009（3）：15-17.

［4］埃里可·多夫曼. 变革的世界，变革的博物馆［R］. 上海：国际博协第 22 届大会自然历史国际委员会会议，2012.

［5］博闻. 为面向今天与未来的世界博物馆事业激发新创造力［N］. 中国文物报，2016-11-23.

顺应大势、勇于担当，共同开辟"一带一路"科普场馆发展的光明前景

程东红 [①]

摘 要 自然科学博物馆是进行科学传播和开展科学教育的机构。通过回顾自然科学博物馆的发展历程，分析可持续发展目标、地区发展不均衡、科技与社会交互对自然科学博物馆发展产生的主要影响，目前国际社会和自然科学博物馆为解决上述问题而采取的行动，以及自然科学博物馆未来发展面临的挑战，本文指出"一带一路"倡议为自然科学博物馆应对上述挑战提供了一系列解决方案。据此，提出促进"一带一路"沿线国家自然科学博物馆之间互惠共享的多项对策和建议：搭建交流平台、开展合作、建立长效机制、加强能力建设，为"一带一路"沿线国家之间开展更深层次的科技、教育和文化合作交流注入活力，最重要的是为构建"一带一路"沿线国家自然科学博物馆命运共同体而共同协力。

关键词 自然科学博物馆 "一带一路"倡议 科学传播 互惠共享

在全球范围、国家领域、城市层面直至乡镇基层，自然科学博物馆都在为科学文化的传播发挥着重要的平台作用。哪怕是一个小型的科技馆，也能在科普工作中发挥出极大的作用，将科学和应用技术惠及当地百姓，特别是少年儿童，以丰富其日常生活，厚实其未来人生。"一带一路"倡议为自然科学博物馆的发展提供了新机遇和新途径。

1. 自然科学博物馆的历史沿革与发展历程

随着社会的不断发展变化，博物馆的定义也在不断地与时俱进。国际博物馆协会自 1946 年成立以来，章程中对博物馆的定义已经修订了八次，其中，2007 年修订的第八版定义指出："博物馆是为社会及其发展服务的、向公众开放的非营利常设机构，

① 程东红：中国自然科学博物馆协会理事长，博士。E-mail：dhcheng@cast.org.cn。本文系作者 2017 年 11 月 27 日在"首届'一带一路'科普场馆发展国际研讨会"上所做的主旨报告。

为着教育、研究、鉴赏之目的而去征集、保护、研究、传播并展示人类及人类环境之物质和非物质遗产"。据国际博物馆协会副主席安来顺博士称，博物馆定义的第九版目前正在讨论阶段。

加拿大魁北克大学传播系教授、中国科学技术馆新馆内容建设国际顾问委员会主席伯纳德·希尔博士（Bernard Schiele）长年专注自然科学博物馆的研究，他认为，自然科学博物馆是"社会参与的、并向公众传播最新科学发现和技术应用的发展机构"。自然科学博物馆的发展经历了四个阶段，在每个发展阶段中，"一带一路"沿线国家的自然科学博物馆均不乏突出表现。

第一个发展阶段为 1683—1929 年。在这个发展阶段中，自然科学博物馆致力于充实并展示其馆藏，同时呈现自然、科学和技术发展的历史。始建于 1802 年的匈牙利自然历史博物馆，在参与本次国际研讨会的众多博物馆中，堪称历史最为悠久的博物馆。除此之外，国家达尔文博物馆（俄罗斯，1907 年）、皇家安大略博物馆（加拿大，1914 年）以及国家地质博物馆（乌兹别克斯坦，1926 年）都是在这一发展阶段先后建立起来的。

第二个发展阶段为 1930—1959 年。这个发展阶段以展示现代科技和强调知识本身为特点。著名的亚美尼亚地质博物馆建立于 1937 年，展品数由建馆之初的 1410 件增加到 14000 件。同在这个阶段建成开放的，还有贝尔格莱德的尼古拉·特斯拉博物馆（塞尔维亚，1952 年）。

第三个发展阶段为 1960—1975 年。这个发展阶段对自然科学博物馆的发展产生了重大影响，其特点是促进科学大众化，并加速知识的扩散。在这一特定历史时期，谢尔盖·科洛列夫航天博物馆（乌克兰，1970 年）建成开放，备受公众欢迎。

第四个发展阶段为 1976 年至今。对于这个最新的发展阶段，研究人员虽众说纷纭、莫衷一是，然咸认为这是一个科技创新并与社会互动的阶段。在这个发展阶段中，自然科学博物馆如雨后春笋，在"一带一路"沿线国家纷纷建立起来，举其荦荦大者，有于 1976 年建立的巴基斯坦自然历史博物馆，于 1988 年建立的澳大利亚国家科技中心，于 1993 年建立的澳大利亚国家恐龙博物馆，于 1998 年建立的亚美尼亚维克多·安巴尔楚缅天文台，2004 年建立的希腊赫拉克莱冬博物馆，2009 年建立的埃及亚历山大天文馆暨科学中心，2010 年建立的波兰哥白尼科学中心，2012 年建立的菲律宾思维博物馆，2014 年建立的哈萨克斯坦共和国国家博物馆，2015 年建立的肯尼亚科学中心，2016 年建立的马来西亚槟榔屿技术中心，2017 年建立的尼泊尔航空博物馆，以及计划在 2018 年建成的伊朗伽亚谟天文馆暨科学中心。准此以观，"一带一路"沿线国家的自然科学博物馆不仅独具历史特色，而且伴随着科技的发展而不断革故、趣时以取新，同时更与世界各国的科普场馆保持着密切的沟通与交流。

回顾自然科学博物馆事业在中国的发展，其兴衰起落，主要受两个因素的影响：一是社会经济发展状况，这是任何一个自然科学博物馆的发展环境和经济基础；二是国家、地方政府部门对博物馆领域的投入程度。2016年，适逢中国地质博物馆建馆100周年，国家主席习近平亲发贺信，以志纪念，而策来兹。此举足证中国政府和社会层面均极重视自然科学博物馆领域的发展，具有重大的象征意义，使全国自然科学博物馆界同仁备受鼓舞。

1949年中华人民共和国成立之初，整个中国大陆仅有自然科学博物馆22座，其中既有公立的，也有私立的；中国人经营的有之，外国人经营的也有之。在政府和社会的大力支持下，中国自然科学博物馆人筚路蓝缕，以启山林，实现了自然科学博物馆几何式的增长。截至2015年，全国自然科学博物馆的数量已增至969座。

深入探究中国自然科学博物馆的发展历程，有三个场馆的发展极具代表性。一是北京天文馆，它是中国第一座天文馆，1957年建馆，当时规模极小。进入21世纪后，该馆在原址基础上进行扩建，形成A、B两馆，2004年建成开放。二是中国科学技术馆，今天的观众，置身于其雄伟的建筑之内，游目于其先进的设施设备，无不叹为观止，而鲜知其当初的艰难缔造。由于当时资金拮据，1988年，中国科学技术馆一期工程竣工之时，展厅面积相当局促狭小。2000年世纪之交，中国科学技术馆二期新展厅建成，并与一期展厅同时开放，国家科技馆初具规模，服务观众的能力得到了显著的提升，但仍与社会公众对科学文化的需求相去甚远。2006年，中国科学技术馆异地重建新馆，并于2009年（中华人民共和国成立60周年）国庆前夕在北京奥林匹克公园中心区卒底于成、敬启管钥，自此常年门庭若市，一直保持人气长盛不衰，实体馆年观众量连年稳中有升，2017年参观人数近400万人次。三是上海科学技术馆，该馆承亚洲文会（1933年）和上海自然博物馆（1956年）之旧，并在其基础上发扬畜皇、终成其大。2001年，上海科学技术馆建成开放；2015年，上海自然博物馆新馆开放，成为上海科学技术馆自然分馆。目前在建并计划于2020年建成开放的上海天文馆，将成为上海科学技术馆的天文分馆。

上述三个案例足以证明，自然科学博物馆事业在中国这样一个发展中国家，其发展经历了萌芽、发展、壮大的过程，也曾经历过缺少资金、研究不足和专业人员匮乏的阶段。正如"罗马不是一天建成的"，今天中国的近千座自然科学博物馆也绝不是反掌折枝、唾手可得的，是经过中国的博物馆人数十年如一日，赴之以毅力，持之以恒心，才成就了今天中国自然科学博物馆的辉煌。应邀参与本次研讨会的"一带一路"沿线国家自然科学博物馆中，有些面积也不大，我认为"大、小"并不重要，重要的是因陋就简、量力而行，先把场馆建成并运行起来，而不是急功近利，盲目求大求全，经营兴作非崇其堂庑、重檐藻棁不可。有时候，小反而意味着更多的灵活性和更大的提升空间，希腊赫拉克莱冬博物馆就是一个典型的例子。建设自然科学博物馆，本来就是一个由小到大、聚沙成塔的过程，从来不是、将来也不可能是一蹴而就的。

今天我们能够聚首北京，共商"一带一路"沿线国家科普场馆共同发展大计，中国自然科学博物馆协会可谓功不可没。自 1980 年成立以来，中国自然科学博物馆协会以学术立会，以科普立身，在交流信息、研究理论、提升能力、推动创新等方面，均发挥了重要的作用。截至 2016 年年底，协会拥有团体会员 658 家，个人会员 1264 人。

2. 自然科学博物馆面临的新问题、新挑战

自然科学博物馆不是自成小天地的象牙塔，而是提供公共科学传播服务的基础设施，是社会公众参与科学的重要平台。科学技术（如信息科学、生物工程、纳米材料和技术、新能源和可再生能源、海洋技术、空间技术）的发展及其应用，一方面给我们带来了巨大的福祉和便利，另一方面也带来了诸如人口膨胀、环境污染、食品安全等一系列问题。公众现在已经开始反思，甚至质疑我们为此付出的代价是否值得，由此产生了所谓的"具有科学背景的社会问题"和"具有社会意义的科学问题"，这些问题引发了自然科学博物馆领域的思考和改变。

本报告将从可持续性、发展不平衡、科技创新及其与社会的互动三个方面所面临的新挑战、新做法和新问题进行探讨。

2.1 可持续性

根据《世界观察研究所 2015 年度报告》，人类的各种需求正在不断增长，而根据测算，各种自然资源已无法支撑这种无节制的需求增长，有些资源已近枯竭。一个健康的社会、环境及经济体必须刻不容缓地采取行动，并提出创新性的解决方法。为了实现千年发展目标，自 2000 年以来，联合国和其他国际组织已经行动起来，发起了一系列倡议。2015 年 9 月，联合国可持续发展峰会通过了雄心勃勃的《2030 议程》，提出了未来 15 年需要实现的 17 个可持续发展目标和 169 个具体指标，在全球和国家政策推进层面，实现了从千年发展目标向 2030 年可持续发展目标的顺利过渡。

与此同时，联合国教科文组织认识到要科学创造可持续未来。通过科学创造知识，促进理解，使我们能够找到应对当今严峻的经济、社会与环境挑战的措施，以及实现

可持续发展与构建绿色社会的方法。^① 基于此，自然科学博物馆一直致力于公众的技能和能力建设，不遗余力地培养公众的好奇心，为公众理解科学提供对话的平台，促进公众在校外学习科学，推动全社会的创新创业和应对各种挑战的能力。通过这些行动，自然科学博物馆就科学之于可持续发展的重要性这一问题，向全社会发出清晰而强烈的信号。^②

自然科学博物馆通过不断创新展览和教育活动的内容和形式，及时回应国家、社会和公众日益增长的需求，借以促进可持续发展理念在全社会的普及与实践。以埃及亚历山大天文馆暨科学中心为例，该馆每天在其"公共科学传播世界公园"内举办"七大洲面临的环境问题"主题教育活动，每场活动时长 15 分钟，共有 20 名学生参与其中。该活动将地球的七大洲设置为七个不同站点，每个站点所传达的科学内容主要针对该大洲各国目前所面临的极度恶化的环境问题。^③

可持续发展是一个全球性的、跨国界的问题，不仅十分复杂，而且令人困扰。目前，各国各行其是、零敲碎打的做法，固不足为训。同时，自然科学博物馆面对可持续发展问题，开展的实践多为立足当地和本馆的孤立行动，回应仍然不够，内容较为局限，模式比较单一，必须加以改进。在今天这个高度相互依存的世界，正如任何一个国家都不可能独力解决其所面临的许多问题一样，各国自然科学博物馆也必须联合起来、一致行动、综合发力、全面施策，方能于事有济，带来改变。

自然科学博物馆怎样联合起来？如何行动？这是非常值得我们深入思考和探讨的问题。可持续发展促进了各国及其民众对人类共同命运的关注，全球化实现了全人类跨文化的对话，这为自然科学博物馆带来了新机遇，提高了自然科学博物馆的可及性，打破了国家和地域界限，也为自然科学博物馆有效应对全球的社会和环境关切，并通过贡献各自的技能和能力提供实际的帮助，赋予了重要的社会责任。当务之急，是用好用足信息技术这一联结自然科学博物馆和全球公众的强固纽带，努力拓展国际视野，切实提升全球意识，大力开展跨区域和跨文化的实时交流，借以消弭公民和科学技术之间的鸿沟。

2.2 发展不平衡

在过去数十年间，世界实现了巨大的经济增长，但由此带来的财富与繁荣的分配竟然如此不均，以至于人们将其视为世界许多地区每况愈下的社会问题、政治动乱和

---

① 详细内容参看：https://en.unesco.org/themes/science-sustainable-future。

② 详细内容参看：http://www.un.org/sustainabledevelopment/blog/2016/11/unesco-science-museums-vitally-important-for-sustainable-develop ment。

③ 详细内容参看：https://www.iucn.org/content/egypts-planetarium-science-centre-activities-rio20。

教育资源不均衡现状的根源。目前，全球仍有 1/3 的人生活在较低人类发展水平状态。即便是在中国这样的人类发展指数处于高水平的国家，依然面临着经济社会发展不协调的问题。直言之，在联合国可持续发展目标（确保包容和公平的优质教育，让全民终身享有学习机会）与客观现实之间，横亘着一道巨大的鸿沟。

为缓解经济增长不平衡的问题，联合国开发计划署每年斥资 10 亿美元，用于实施经济转型、自然资源管理和早日康复计划等旨在改善民生的倡议。至今，这些倡议已经帮助 98 个国家创造了 200 多万个新的就业机会（其中 36% 是为妇女创造的），94 个国家实施了低排放和适应气候变化的发展措施，全球有 119 个国家共计 2470 万人（其中 51% 是妇女）从中受益。

为应对教育发展不平衡的问题，2015 年 11 月，联合国教科文组织发布了《教育 2030 行动框架》，为各国政府和合作伙伴提供了如何将承诺变为行动的指南。《教育 2030 议程》旨在确保优质教育成果为全民所共享，凸显了教育领域的包容性、公平性和性别平等，并特别强调在一个多元的、相互依存的世界里，公民通过终身接受教育和培训去获得工作技能的重要性。[1]

为应对博物馆发展的不平衡，国际博物馆协会发布了《国际博物馆协会 2016—2022 战略发展计划》，以期通过扩大参与、改进服务、增强沟通和能力建设提升会员的价值。[2]

2017 年 11 月 15—17 日在日本东京召开的第二届世界科学中心峰会，是国际科学中心和科学技术馆界三年一度的盛会。作为国际社会的利益相关者，来自各大洲的近千名自然科学博物馆、政府部门、科研机构、公司企业的与会代表，围绕峰会"联通世界，实现可持续的未来"主题，就当前人类社会共同面临的全球性问题展开了广泛而深入的讨论，以寻求具有长效影响的战略对策。峰会还发布了由全球六大地区科学中心和科技馆网络组织负责人共同签署的《东京议定书》。如何有效实现科普资源的均衡分布和科普服务的公平普惠，也是峰会热议的问题之一。

在这方面，澳大利亚国家科技中心的"科学马戏团"项目尤为引人注目。所谓的"马戏团"，实际上就是一个"迷你"的流动科技馆，由一辆卡车和 25 件便携式互动展品构成，辅以若干现场科学表演，主要以学生、教师和社区居民为服务对象。自 1985 年启动以来，该项目已经覆盖了澳大利亚 500 多个社区（其中包括 90 个原住民社区），

---

① 详细内容参看：http://unesdoc.unesco.org/images/0024/002456/245656E.pdf。

② 详细内容参看：http://icom.museum/fileadmin/user_upload/pdf/Strategic_Plan/ICOM_STRATEGIC_PLAN_2016-2022_ENG.pdf。

在学校举办了15000多场科学秀。参与该项目的教师中，有85%认为这个项目帮助提高了学生对科学的兴趣。与"科学马戏团"项目异曲同工的是泰国国家科学技术馆的"科普大篷车"项目。该项目于2005年启动，每年运行200天，巡回40个府（泰国全国共76个府，不包括曼谷）。

这仅是一个开始。当前一个不容回避的事实是，欠发达地区的自然科学博物馆所能提供的科普资源捉襟见肘。更棘手的是，其自主开发资源的能力非常薄弱，并且缺乏获取可共享资源的渠道和机制。而形成鲜明对比的是，经济发达地区和中心城市的大中型场馆则拥有相对优质的资源。在这种情况下，后者有责任和义务通过开展跨区域跨场馆资源的整合、共享和流动，为前者提供资源和服务。这不是一种居高临下的慈善行为，而是一种历史使命与时代担当。

2.3 科技创新及其与社会的互动

层出不穷的科学新发现、新成果，持续地推动着世界的转变，不断地刷新人们理解科学和认识世界的方法。关于科技创新及其与社会的互动问题，可以分解成三个层面：一是突破传统自然科学博物馆以已有的、成熟的科学发现和技术成果作为主要展陈内容的窠臼，在展览中及时反映最新的科技进展；二是采用先进的技术作为工具和手段，有效提升展览和教育活动的质量；三是促进具有科学背景的社会问题和具有社会意义的科学问题进入自然科学博物馆，通过科技专家与公众之间的互动，搭建沟通平台。希腊塞萨洛尼基科学中心暨技术博物馆"将课堂设在太空"的活动就是一个典型的例子，该活动是欧洲航天局的教育活动之一，通过将欧洲哥伦布实验室设在国际空间站内，来自欧洲各国的教育工作者受邀可以围绕在其中开展的实验项目进行切磋，提出自己的想法。

科技进步固然不可逆转，但是在取得巨大成就的同时，更给我们带来了诸多挑战。为此，自然科学博物馆必须做到：第一，将最新的科技成果融入其现有的展陈内容以及教育活动当中，以避免出现滞后性；第二，充分利用信息技术等高新技术去寻求展览教育、观众服务、用户体验等方面的创新，同时将其服务拓展至观众参观之前和参观之后，并及于不能到达现场参观的公众；第三，搭建新的科技交流平台，充分发挥自身特性和优势，努力增进公众对科学与社会以及科学与伦理等问题的理解和探讨。

总之，自然科学博物馆必须通过其展陈内容和教育活动的与时俱进，去及时回应全球对可持续发展的关切，去阐述具有科学背景的社会问题和具有社会意义的科学问题，去满足公众对理解和参与科学的新需求和新期待。

3. 解决之道：合作与共享

面对上述挑战，我在这里提出的对策是：携手合作，资源共享。

3.1 互惠共享的中国探索

前面所述的三个问题——可持续性、发展不平衡、科技创新及其与社会的互动——中国莫不如是。以言可持续发展，则空气污染、水资源匮乏、城市发展无序……凡此种种，不一而足，多见诸于媒体；以言发展不均衡，则不仅体现在经济社会发展方面，同时也体现在自然科学博物馆的分布方面，至今仍是我们的痛点。

近年来，中国自然科学博物馆界在应对上述问题的过程中，不断探索实现公共科普资源和服务公平普惠、互惠共享的路径和做法，取得了显著成效，案例如下。

案例一：科普大篷车。科普大篷车是面向基层科普需求而推出的多功能科教专用车，以车载小型化科普展品（20 ～ 30 件，因车型而异）和展板的形式，为县以下的城镇社区、学校、农村，特别是贫困边远地区提供科普服务，因其具备机动灵活的特点，素有"科普轻骑兵"之称。科普大篷车项目于 2000 年由中国科学技术协会启动，2012 年正式移交中国科学技术馆运行，不仅解决了科普服务"最后一公里"的问题，而且弥补了科学技术馆建设在区域和城乡间不均衡的问题。

案例二：流动科技馆。中国流动科技馆项目以"体验科学"为主题，于 2010 年 6 月正式启动。顾名思义，流动科技馆具备小型科技馆的基本展教功能，主要服务对象为当地没有实体科技馆的县级区域的公众。其由 50 余件互动展品组成 3 个主题展区和 10 个分主题展区，并与科学表演、科学实验、科普影视等相结合，展览面积 800 平方米，可在当地体育馆或会议中心等公共建筑内进行展出，展出周期通常为 3 个月。

案例三：农村中学科技馆。为了解决农村地区公共科学教育服务不足的问题，自 2012 年以来，中国科技馆发展基金会向社会筹集资金 2000 万元，实施"农村中学科技馆公益项目"。这些微型科技馆以 20 件（套）科普展品、4 ～ 5 台电脑、1 台 3D 打印机、1 台多媒体投影仪和 1000 本科普读物为内容，深入到偏远乡村中学，并直接在教室内"安家落户"，使当地的青少年在成长过程中，能和城市里的孩子一样拥有参观科技馆的美妙体验。

案例四：东西科普场馆携手、实现资源优势互补。青海省德令哈天文科普馆是西部地区第一座以天文为主题的科普场馆。其位于天高云淡、人迹罕至的青藏高原，拥有非常有利的天象观测条件和一流的设备。不久前，该馆与北京天文馆签订了战略合作框架协议，以期实现优势互补、互惠互利、长期合作。根据这一框架协议，北京天文馆对德令哈天文科普馆提供场馆管理、人员培训、群众活动的组织以及相关业务指导等方面的帮助；德令哈天文科普馆则承接北京天文馆组织的天象观测活动。通过这种协作关系，不仅激活了彼此的资源优势，而且进一步提升了天文场馆的服务水平和科普能力。

案例五：科普影片联合采购、联合研发。 在中国大陆地区，多数自然科学博物馆都设有 IMAX 影院、4D 影院等特效影院，由于自身影片制作能力不足，片源主要依赖进口，但影片租金高，使影院承担负担较重。为此，中国自然科学博物馆协会成立了特效影院专业委员会，搭建影片联合采购平台，48 家成员单位集体谈判、统一订片、轮换放映，打破国际特效影片发行商的垄断独登。与此同时，通过建立影片联合开发平台，场馆与企业共同投资开发特效影片，在降低制作成本的同时，提高影片制作的数量、质量和水平。此外，通过举办学术论坛、展示交流会、放映技术培训班等活动，解决成员单位面临的设备升级、数字化转型和电脑控制等技术挑战，共同提升设备管理和业务技术水平。

案例六：中国数字科技馆。 如何将前沿科学知识和高新技术引进展厅并融入教育活动当中？如何根据学校、社区、公众多元化的需求开发出丰富多彩的科普资源？对于任何一个自然科学博物馆来说，这些都是极具挑战的难题。2005 年 12 月，中国科学技术协会联合中华人民共和国教育部、中国科学研究院共建"中国数字科技馆"（https://www.cdstm.cn），将展览展品、教育活动、科普视频等组织、开发、整合成为数字化科普资源，提供给科普场馆、公益组织和社会公众免费获取和使用。会上，中国科学技术协会书记处第一书记怀进鹏代表中国科学技术协会郑重宣布，愿意通过"科普中国"和"中国数字科技馆"现有的网络平台，将这一科普资源库无偿向"一带一路"沿线国家科普场馆开放使用，使既优既渥，既需既足。

以上探索，成效几何？让我来数说兼图说一二。

2000—2017 年，科普大篷车共开发出 4 种车型，面向全国累计配发 1445 辆，开展活动 184000 次，惠及公众约 2.06 亿人次。

2011—2017 年，流动科技馆累计配发展览 288 套，巡展 2261 站，服务公众约 8332 万人次，并已正式启动第二轮全国巡展工作。

2012—2016 年，农村中学科技馆已覆盖 29 个省/自治区，受益青少年达 137 万人次。

2006—2017 年，中国数字科技馆官方总资源量 9.79 万亿字节，日均浏览量超过 293 万次。

位于青藏高原的德令哈已经成为中国第一个将天文课列为全体学生义务教育必修

课的城市。

以科技馆等科技类场馆为例，近期调查的数据显明：由于缺乏资源而从未参观过科技类场馆的人群已由 57.7% 降至 22.6%，而参观过科技类场馆的人群则由 7.9% 增至 22.7%，如下图所示。由此可见，公民通过科技类场馆获取科学知识和科技信息的机会增多，对场馆的利用率提高了。

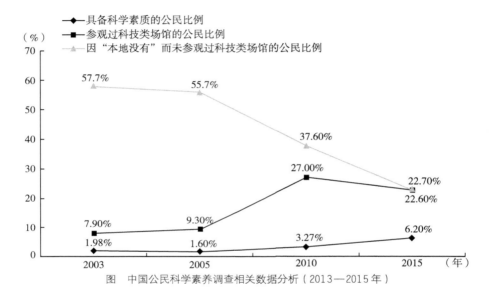

图　中国公民科学素养调查相关数据分析（2013—2015 年）

因此，通过自然科学博物馆的建设发展，极大地弥补了中国科普资源城乡分布不均衡的问题，提高了科普资源的利用率，使得公共科普服务能够覆盖全国各地区、各阶层人群，推动了中国科普服务的公平普惠与效能提升，取得了良好的社会效益。

3.2 "一带一路"倡议为自然科学博物馆提供发展机遇

"一带一路"倡议是由中国国家主席习近平提出的，旨在加强欧亚国家之间的合作，在世界多极化和经济全球化两大趋势下，推动经济全球化深入发展，促进区域可持续发展及均衡化，拓展社会、经济、文化发展空间与交流互鉴。"一带一路"倡议秉持"和平合作、开放包容、互学互鉴、互利共赢"和"共商、共建、共享"的原则。概括起来，"一带一路"倡议的主要内容包括"一个核心理念"（和平、合作、发展、共赢）、"五个合作重点"（政策沟通、设施联通、贸易畅通、资金融通、民心相通，统称"五通"）和"三个共同体"（利益共同体、命运共同体、责任共同体）。我则更关注这个倡议的人文内涵以及搭建人文平台和纽带，因之为沿线各国人民之间，同时也是为自然科学博物馆人之间，提供了一个相互提升、彼此协调和共赢合作的机制。

"一带一路"沿线国家在历史上创造出了形态不同、风格各异的文明，并通过交流形成了多彩绚烂的丝路文化。直言之，交流互鉴，是推动人类文明进步和世界和平发展的重要动力；而沿线国家科普场馆之间面对相似的挑战和机遇，不断加强合作与交流，能够真正实现各国家之间的政策沟通、设施联通、文化互通、民心相通，使"一带一路"沿线各场馆的教育、展示、收藏和研究成果在不同的文化环境中进行相互交流，更好地推进科技发展与创新，进而促进社会经济的可持续发展和人类的共同进步。

3.3 "一带一路"倡议下自然科学博物馆互惠共享的思考与建议

互惠共享，是我们回应当代新问题、提供新服务的解决方案。受"一带一路"倡议所倡导的"构建开放、包容、均衡、普惠的区域合作架构"原则的启发，愿与各国代表，共同探索构建"一带一路"沿线国家自然科学博物馆之间互惠共享机制的可行性。为此，谨提出以下思考和建议。

3.3.1 加强交流沟通

建议本着平等、自愿和互惠的原则，每两年组织一次"一带一路"科普场馆发展国际研讨会，以进一步促进各馆之间的了解和信任，联络领域内同行间感情，建立合作关系，开展资源共享，交流彼此经验，在此基础上建立"一带一路"沿线国家场馆、研究机构等长效交流机构。中国自然科学博物馆协会愿为此进行筹维。

3.3.2 开展合作项目

项目是合作干实事的载体。我欣喜地看到今天上午中国和"一带一路"沿线国家的科普场馆之间签署了如此多的合作协议。它们堪称开展项目合作的先锋，值得我们激赏不已，并期待这些项目硕果盈枝。此外，还有许多中国的自然科学博物馆也表达了与"一带一路"沿线国家场馆分享其优质科普资源的意愿。我乐见研讨会期间以及研讨会之后将有更多的项目合作协议达成。

构建"一带一路"沿线国家更广泛的科学传播联盟，目前可谓万事俱备，我们理应抓住机遇、乘势而上，开展展览巡展与交流，合作开发博物馆教育资源，人员定期互换交流与考察，定期举办科普场馆人员培训班，开展博物馆研究与藏品征集合作，联袂举办科技节、青少年科学竞赛等，为推动创新与发展、促进资源的共享和场馆间的互惠共赢，为提升科学传播的能力做出积极贡献。

3.3.3 建立长效合作框架和机制

"行百里者半九十"。每两年举办一次研讨会，并在一些场馆之间签订合作协议，这些尚不足以支撑我们为建立"一带一路"沿线国家自然科学博物馆命运共同体而勠力合作的强烈需求。为避免持坚无术、末路蹉跎，需要建立起长效的合作框架和机制。

上海科学技术馆王小明馆长等国内外参会代表历经数月，以工作小组的形式共同酝酿起草了《北京宣言》，其中就开展长期互利合作方面，有许多建设性的意见和建议。

作为对这些意见和建议的回应，中国自然科学博物馆协会将以编发电子工作通信的形式，最充分地利用各馆的资源，定期推送"一带一路"沿线国家科普场馆的最新信息，借以建立起积极有效的沟通平台，为各馆开展更广泛更深入的文化和科技交流合作服务，为构建沿线国家科普场馆命运共同体，提供长期、持续的组织保障和有效连接。为此，我希冀明天论坛结束之际，《北京宣言》得以一致通过。

3.3.4 能力建设：人才是我们的核心竞争力

人才，尤其是青年才俊，乃是自然科学博物馆的核心竞争力和未来希望之所在，此诚可谓经验之谈。通过组织开展展品设计、教育活动开发、观众研究分析、收藏与研究等方面的在职培训，我们可以加强场馆青年员工的能力建设和知识更新，提升员工职业能力和专业水平，最终为场馆运营、服务提供有效支撑，为自然科学博物馆事业可持续发展造就一大批业务骨干和领军人才。为此，我们需要努力筹措一笔经费，建立起专项资金，借以扶持新人，奖掖后进，支持"一带一路"沿线国家科普场馆员工特别是青年员工开展能力建设。

4. 结语

"一带一路"倡议源自中国，更属于世界；根植于历史，更面向未来；重点面向亚欧非大陆，更向所有伙伴开放。"一带一路"沿线各国都有自己独特的历史渊源，更有文化互鉴、交流融合的传统。在科技引领未来的新时代背景下，"一带一路"沿线国家自然科学博物馆联合起来，构建命运共同体，将会是世界自然科学博物馆领域的一道独特风景。让我们共同努力，为世界各国公民科学素质提升服务，为全球和平发展贡献力量！

参考文献

MASSIMIANO BUCCHI, BRIAN TRENCH. Routledge Handbook of Public Communication of Science and Technology [M]. 2nd ed. 2014.

"一带一路"科普场馆发展论坛入选论文

尼泊尔航空博物馆简介

Laxmi Basnyat [①]

摘要 这是尼泊尔创新之举，它包括两个独立的展览，即：一个设在首都加德满都，另一个则设在尼泊尔遥远西部的丹加迪。位于加德满都的飞机博物馆于 2017 年 11 月 5 日对外开放，设在丹加迪的飞机博物馆则于 2014 年 9 月 7 日对外开放。两家博物馆均由 Bed Uprety 信托公司资助。建设两座博物馆的主要目的是：

· 教育和鼓励年轻的尼泊尔人投身航空行业。

· 提供有关航空历史的资料。

· 让民众意识到航空行业中的各种挑战与机遇。

· 打造一个旅游目的地。

· 帮助癌症患者。

1. 丹加迪飞机博物馆

商业飞行员兼前尼泊尔军队步兵上尉 Bed Uprety 先生创建了尼泊尔的飞机博物馆，将旅游业以及民众对航空行业的意识带到尼泊尔偏远的西部地区。在博物馆投运的第一年，就成功地给所有前往飞机博物馆参观的人们留下了深刻印象。

博物馆设在一架已弃用的尼泊尔考斯米克航空（Cosmic Air）提供的 100 座飞机机身中，参观者进入博物馆的入口时会看到要求脱下鞋子的提示牌。参观者按要求登机后，飞机舱门打开，一位身穿空中服务人员制服的男人向所有人致意。游览便在飞机内部蜿蜒地展开。在飞机驾驶舱停留期间，导游会讲述许多关于驾驶和操纵飞行的相关技术知识（彩页图 1）。

整个参观过程中，游客可观赏到在飞机舱壁上排成一列的 200 多架商业飞机和喷气式战斗机模型，它们均由当地儿童建造。博物馆的这些微缩模型也是 Bed Uprety 社

---

① Laxmi Basnyat：尼泊尔飞机博物馆负责人，尼泊尔加德满都市泰国国际航空公司副经理。邮箱：lmibasnyat@gmail.com。

区外展活动的一部分，旨在增强年轻人对航空的兴趣，飞机博物馆本身推动了新游客的参观热情，他们不断地涌入该地区一睹博物馆的风采（彩页图2）。

2. 加德满都飞机博物馆

加德满都航空博物馆是由空客300—330飞机改装而成，非常富有创意。如前所述，这是Bed Uprety上尉的独创构想。他的灵感来自美国前总统肯尼迪，这位总统曾经说过："不要问你的国家能为你做什么，而是你能为你的国家做什么。"纽约的肯尼迪机场就是以他的名字命名的。上尉Bed Uprety曾驾驶过多种机型，穿梭于尼泊尔、印度、印度尼西亚、新加坡、马来西亚、泰国、阿曼、塞浦路斯、意大利、法国和美国等80多个国家。

博物馆可以成为世界各地游客的关注焦点。一些目的地之所以能够吸引数百万的游客，是因为那里有著名的博物馆。这架飞机的飞行历史仅持续了8个月，2015年在加德满都机场遭遇了跑道偏离事件。将这架偏离跑道的巨大飞机归正是困难的，为此机场被关闭了4天之久。Bed Uprety上尉以前每次驾驶飞机看到停在加德满都机场上的飞机时，就梦想着有一天能够得到这架飞机，在加德满都建立一个博物馆，这样不仅可以将其作为旅游景点，而且可以教育学生了解航空知识。尼泊尔和国际技术人员用了4个月的时间把飞机拆成若干小块，晚上用卡车运输。又过了9个月，工程技术人员在Sinamangal（离唯一的国际机场很近）完成了对该飞机的重新组装。

Bed Uprety信托公司是一家非营利组织，它已在丹巴迪采用一架废弃的佛克尔100型飞机建立了一个航空博物馆。该博物馆的收入被用于帮助癌症患者。加德满都航空博物馆是丹巴迪博物馆的续篇。信托公司与尼泊尔民航管理局（CAAN）合作建立了这座博物馆。加德满都飞机博物馆的筹建资金来自Bed Uprety上尉的个人积蓄以及尼泊尔珠峰有限银行2700万卢比 [①] 的贷款。这个项目总计注资7000万卢比。只要出具学校的证明，全尼泊尔9～12年级的学生都可以免费参观博物馆内350多架微缩飞机模型（机型包括从莱特兄弟的第一架飞机到第一次世界大战和第二次世界大战期间的战斗机）。其他学生可凭身份证享受50%的折扣。航空博物馆的部分收入将被用于协助Bed Uprety信托公司为尼泊尔西部的癌症患者提供治疗资金。部分收入还被用于资助丹巴迪飞机博物馆以及正在尼泊尔西部Mahendranagar建造的Siddhanath Baba（印度神）雕像。

加德满都航空博物馆由这架巨型飞机改装而成，驾驶舱和空中交通管制录音系统一应俱全，博物馆中的展品包括从怀特兄弟时代的飞行器到最新的现代化飞机，以及尼泊尔的航拍照片集，创始人为Bed Uprety上尉。

---

① 卢比：尼泊尔所使用的货币名称。

传承科学精神的维克多·阿姆巴楚米扬故居博物馆

格里高尔·布娄提安 [①]

摘 要 维克多·阿姆巴楚米扬（Viktor Ambartsumian，1908—1996）是 20 世纪伟大的科学家之一，是世界著名的天体物理学家、数学家和理论物理学家。他的工作成果改变了人类对宇宙的认识。位于布拉堪的维克多·阿姆巴楚米扬故居博物馆再现了这位科学家的生活和科学活动情况。博物馆向观众介绍了从古代到 20 世纪末期人类对宇宙认识的演变，并将维克多·阿姆巴楚米扬的经历作为阐释一个人影响科学发展的生动范例。

关键词 维克多·阿姆巴楚米扬 故居博物馆 科学研究 科学精神

1. 博物馆历史

维克多·阿姆巴楚米扬故居博物馆筹建于 1998 年（即维克多·阿姆巴楚米扬去世两年后），设在阿姆巴楚米扬位于布拉堪的故居（1950 年由他本人修建）原址（彩页图 3）。该博物馆曾是亚美尼亚科学院主席团下属的一个机构，自 2014 年起，它已经成为布拉堪天文台下属的机构。布拉堪距亚美尼亚共和国埃里温市 35 千米，是阿拉加措特恩区下辖的一个村庄，它位于阿拉拉特山南坡，海拔约 1400 米。

2. 维克多·阿姆巴楚米扬生平

维克多·阿姆巴楚米扬是 20 世纪伟大的科学家之一（彩页图 4）。尽管维克多·阿姆巴楚米扬所从事的科学研究涉及现代科学的三个不同领域：数学、理论物理学和天体物理学，但他在天体物理学领域的成就最为著名。他的著作彻底改变了人类对宇宙及其演变过程的理解。由于阿姆巴楚米扬的科学成就，现代天体物理学已经不再是描

---

① 格里高尔·布娄提安：亚美尼亚布拉堪天文台博士，维克多·阿姆巴楚米扬故居博物馆馆员。E-mail：gbroutian@gmail.com。

述天体及其辐射和运动的物理本质的科学，已经演化为探究天体及宇宙演变的科学。他是几乎所有国际科学院的成员。他还是唯一当选为国际天文学联合会主席（1961年）和科学联盟国际理事会的苏联科学家。维克多·阿姆巴楚米扬的两个最重要的发现——恒星共生体（又称为星协）和星系核的活动（分别发表于 1947 年和 1958 年）表明：我们所处的星系的恒星形成尚未完成，这一过程仍在进行中。宇宙中存在不同年龄的恒星，诞生仅十几万年的恒星与非常古老的恒星同时存在。同时，星系的形成过程也在持续，由于星系核的活动，新的星系正在不断形成。

3. 博物馆开放情况

博物馆的展品安置在故居的各个房间内，总面积约 170 平方米，这里曾是维克多·阿姆巴楚米扬及其家人的居所，所以没有专门的展览室。故居周围还有一座占地 6400 平方米的花园，观众可以看到阿姆巴楚米扬生活的实际环境。博物馆每年接待观众 5000 ～ 6000 人次，旺季为春秋两季。

故居博物馆一楼有 4 个房间陈列展品。首先是门厅，设有讲述阿姆巴楚米扬家族起源的小型展览。此处展示了从 17 世纪末起直至他孙辈的家谱。此外，还简要介绍了维克多·阿姆巴楚米扬曾获得的各种奖项，其中的一些证书陈列于此。观众可以从门厅参观至休息室（或称娱乐室），这里没有安排专题展览，而是保留了这位科学家人生最后几年的生活原貌。

起居室（餐厅）是博物馆内最宽敞的房间，这里安排了内容最丰富的展览。在这里，观众可以看到阿姆巴楚米扬用亚美尼亚语、俄语和英语（他通晓 7 门语言）书写的手稿。另一个小型展览介绍了阿姆巴楚米扬出版的著作。维克多·阿姆巴楚米扬生前出版的著作有 1000 多部，此展览收录了其著作中最重要的部分。观众可以在这里看到维克多·阿姆巴楚米扬的著作涉及现代科学的多个门类：物理学、数学、天体物理学和哲学。维克多·阿姆巴楚米扬也是亚美尼亚科学院的主要组织者之一。他执掌亚美尼亚科学院 50 年之久，其中担任副院长 4 年，担任院长的时间长达 46 年。一个专题展览向观众介绍了维克多·阿姆巴楚米扬在这个领域内的活动。

这个房间中的一个微型展览向观众介绍了维克多·阿姆巴楚米扬之父阿马扎斯普·阿姆巴楚米扬的生平。他是 20 世纪初亚美尼亚文化的杰出代表之一，是杰出的诗人、语言学家、哲学家，而且是希腊和拉丁作家荷马、埃斯库罗斯、欧里庇得斯、索福克勒斯和维吉尔杰所著经典作品的杰出翻译家。一楼的最后一个房间是维克多·阿姆巴楚米扬的办公室，他的科学图书馆也完好地保留于此。位于一楼的卧室中没有专题展览，只是保留了原貌，观众得以在此感受他简朴的生活（彩页图5）。

博物馆每周向观众开放6天（每周一闭馆）。迄今，博物馆坚持向观众免费提供解说。

博物馆观众中有一半以上是在校学生。学校及公共机构时常邀请馆方进行有关维克多·阿姆巴楚米扬及其科学成就方面的演讲。

博物馆每年会在9月18日举行"一日科学会议"，以纪念维克托·阿姆巴楚米扬诞辰。会议演讲在博物馆的花园中进行，临时搭建的观众席可容纳40～50名听众。科学会议的主题涵盖非传统领域中最新的、往往是革命性的现代科学成果。

博物馆安排讲解及说明时认为：亲身示范是最有效的教育方式，人人都可以理解。因此，我们介绍维克多·阿姆巴楚米扬的生活和科研活动时常常采用这种方式。

博物馆优先向学生提供讲解。通常的讲解活动安排为：博物馆工作人员向观众介绍维克多·阿姆巴楚米扬的生平事迹；潜心科研事业并影响整个国家的生活，甚至是世界大战，如何获得至高赞誉。我们希望告诉观众的是：一个科学家能够为祖国做出怎样的贡献，如何让世界变得更美好。我们希望观众了解的还有：科学，尤其是基础科学具有强大的力量，能够帮助人们改变这个世界。我们向观众展示维克多·阿姆巴楚米扬的生活以及科研活动，为年轻人认识现代科学领域中科研工作的重要性提供了生动事例，鼓励他们不要畏惧困难，要在科学工作中勇于探索。

故居博物馆自身也面临着各种困难和问题。困扰博物馆工作的首要问题是空间有限。博物馆的参观人数每年都在增加，17年间已增至每年6000人次。我们需要接待大量观众，特别是在夏秋季节以及周末。鉴于博物馆空间有限，必须限制观众在博物馆参观的时间，这给博物馆达到自身的开放目标造成了很大困难。此外，大量的观众会对博物馆展品的存放条件产生影响，导致展品的寿命缩短。因此，我们另辟蹊径，以应对博物馆遇到的此类问题。博物馆已经决定建造一座新的建筑物，以便开展所有的讲解和教育活动。将老建筑作为故居纪念馆，使观众在听过演讲了解相关内容后能够看到原物及环境。博物馆的新建筑将建在故居老建筑的附近。新建筑将配备大型演讲厅或会议厅、多个展厅以及其他辅助用房。此外，出于教育目的，屋顶上还会装设两部小型天文望远镜。我们正在寻找赞助以便启动这个项目。老建筑——维克多·阿姆巴楚米扬故居也需要大修，因为这幢建筑的历史已经超过60年，目前破损严重。

此外，我们还计划在亚美尼亚筹建一座新的天文博物馆。因为在亚美尼亚共和国境内，从新石器时代到著名的布拉堪天文台，天文学有着悠久的历史。亚美尼亚各山区中有着众多的遗迹，显然具有天文和历法的意义。亚美尼亚的各个考古遗址也出土了一些文物，同样有着天文和历法方面的解释。此外，还有许多中世纪亚美尼亚科学家手稿，包含关于天文以及历法方面的内容。亚美尼亚还有许多中世纪天文仪器可以制作复制品。当然，博物馆中最丰富的藏品来自布拉堪天文台，以此来讲述维克多·阿姆巴楚米扬及其学生的成就。所以，筹建新博物馆已经有了很好的基础。我们对这项宏伟工程充满期待。

为肯尼亚青年谋求 STEM 教育的平等机会

肯尼斯·孟杰罗　查理·特劳特曼　格拉汉姆·沃克尔 ①

摘　要　肯尼亚人口中 18 岁以下的青少年占 50%，他们是肯尼亚社会中一股重要而活跃的力量，大力发展青少年教育是促进发展的有效方式。肯尼亚科学中心是一所筹建中的 STEM 教育机构，旨在为所有青少年提供平等的教育机会，并接受 STEM 教育方面的启迪。本文介绍了建设和发展肯尼亚科学中心的各种工作方式，期望能够借鉴国际上其他科学中心的经验，以达成独具特色的发展目标。

关键词　肯尼亚　科学中心　平等的教育机会　公众参与　能力建设

1. 科学中心的作用

科学中心是一类校外（又称为"非正式"）学习机构，在某些方面，它类似于博物馆。但与传统博物馆相比，科学中心里的绝大多数展品都是为了接触和试验设计，这一切便是完全意义上的动手实践。在这样的场所，年轻学员及其父母、学生以及公众通过各种科学、技术、工程和数学（STEM）方面的展览、节目和活动，在互动与亲手实践的氛围中得到参与科学、接受培训和启迪的机会。本文所提及的许多研究都涉及并围绕一系列非正式学习环境展开。例如，科学中心、艺术博物馆和其他类型的组织机构，它们都利用了交互式展览、节目、活动和与公众互动的方式。在大多数情况下，一个非正式互动学习环境得到的结果类似于其他非正式的学习环境。一家大型国际研究机构声称：由肯尼亚科学中心（SCK）提供的非正式教育经验类型能够促进学生在校内外长期保持好奇心并养成终生学习的习惯。具体而言，有研究表明：互动式科学中心可增加参观者的科学知识以及对科学的了解，并可提供令人难忘的学习经历，会对体验者的态度和行为产生持久的影响。科

①　肯尼斯·孟杰罗：肯尼亚科学中心主任。Email：kentrizakari@gmail.com。
查理·特劳特曼：纽约伊萨卡科学中心名誉理事。
格拉汉姆·沃克尔：澳大利亚国立大学国家公众科学认知中心职员。

学中心具有广泛的个人和社会影响，并可促进不同年龄段的学习者相互学习。科学中心可促进公众和科学界之间的信任和理解，并为他们所在社区的发展产生积极的影响。

2. 公众参与肯尼亚首家科学中心的规划

为了更好地了解公众对作为肯尼亚教育机构的科学中心的认识，调查人员就第三届国家科学、技术和创新周（2014 年 5 月 19—23 日于肯尼亚内罗毕举行）所提供的公共科学展览和项目，对参观者进行了初步调研。调研旨在了解观众对科学中心的看法、理解和期望。调研人员对 56 位参观者进行了问卷调查，他们在体验了各种展品以及演示节目后提供了反馈。71% 的受访者为男性，29% 为女性。70% 的受访者从未听说过科学中心，30% 的受访者曾经听说过科学中心，但他们都称从未在学校中听说过。没有任何人曾经真正参观过科学中心。除一个人外（98%）都表示从刚体验过的展品或演示中学到了一个新的科学概念。近一半（42%）受访者称：最有吸引力的活动是从胡萝卜中提取 DNA，这表明参观者对作为一种活动的科学过程和内容感兴趣。超过 60% 的人表示：自己所在的县和国家政府未来应为科学中心提供资金。所有受访者（100%）均支持设立肯尼亚科学中心。研究得出的结论是：尽管约 2/3 的受访者从未听说过科学中心，但一旦他们体验后，都能欣然接受并赞同科学中心的理念和演示。同时，这些理念和演示对所有人都具有普遍的吸引力。总之，这项研究为建设发展肯尼亚青少年科学中心提供了有力的支持。

3. 肯尼亚科学中心的筹建与试运行

肯尼亚农业和畜牧业研究组织（KALRO）在 2014 年公众调查的基础上，同意为筹建科学中心划拨用地空间，并为其初步规划和项目开发提供各种资源。下一步是规划和建设一座新的科学中心的关键，包括：制定目标、任务、架构和初始资金需求方面的方案陈述。肯尼亚科学中心工作人员于 2015—2016 年与国际知名科学中心的专业人员进行了磋商。通过这些努力，肯尼亚科学中心制定了自己的愿景和目标，创造性地规划了一系列试运行教育活动项目，获得了一批科学展品，并依据肯尼亚农业和畜牧组织划拨的用地空间为科学中心设计了平面图。

4. 能力建设

为了在试运行和未来规划工作方面做好准备，肯尼亚科学中心通过参观全球各地的科学中心和其他博物馆，并参加南非、美国、加拿大和澳大利亚有关建设开发科学中心的国际会议和培训计划，目前已经具备了一定的工作能力。这些能力建设活动的主要目标是：一是了解世界各地的其他科学中心如何开展（尤其是针对青少年）科学、技术、工程和数学方面的公众教育活动；二是如何在创建肯尼亚科学中心的过程中借鉴其他国家的经验，更好地满足肯尼亚青少年的特殊需求。

截至2017年年底，工作人员已经走访了25家博物馆，其中非洲5家、亚洲1家、澳大利亚4家、加拿大6家、美国9家。这些博物馆大多数是科学中心。访问其他类型博物馆的主要目的是了解更多关于博物馆建筑、商业实践、展品开发过程及游客服务方面的信息。

自2014年以来，工作人员为了筹备肯尼亚科学中心曾经参加过11次科学中心会议、博物馆培训课程及其他能力建设项目。通过这些会议和培训课程，肯尼亚科学中心工作人员了解到了有关启动和运营科学中心的各个方面内容，包括如何为场馆发展筹集资金和其他资源，如何进行商业实践，要采取多元化手段为青少年提供服务，要建立国际合作网。而且，这些由国际会议构建起来的遍布全球的合作网络同样有助于搜集展览等各类资源。

例如，沃尔顿科学博物馆可持续发展计划旨在为发展中国家的科学中心提供培训，帮助其应对有关可持续发展的全球挑战。该方案使得来自不同国家科学中心的从业人员齐聚一堂，学习成功的教育活动方案，并且帮助科学中心构建起国际网络，以便相互支持、资源共享以及交流经验。这一计划可提供内容丰富的成套资料和其他资源。科学中心因其对青少年的教育而被全世界公认为青少年的信息与灵感之源，肯尼亚科学中心可有效利用这些资源，促进青少年对科学教育的兴趣，并将科学应用到解决肯尼亚所面临的一些重大挑战中，如食品生产、水资源供应以及不断向社会输送接受过STEM教育的公民。

5. 方案与展品

在试运行阶段，肯尼亚科学中心已经尝试了一系列的项目和展品，促进多样化的青少年群体参与到STEM教育中来。例如，肯尼亚科学中心建造了一个带有科学元素障碍的户外科学迷你高尔夫球场，旨在鼓励青少年参与到科学和可持续发展中来。在另一个项目中，肯尼亚科学中心帮助孩子创建、种植和维护菜园。对于很多参与其中的孩子来说，这是他们人生中第一次得到种植作物并可时常观察其生长的机会。肯尼亚科学中心正是利用这样的机会，将青少年吸引到农业、食品生产和食品科学领域中的。

肯尼亚科学中心开发的活动还包括：非洲科学巡展——通过在学校和公共场所进行科学演示，激发非洲青少年们对科学的好奇心；非洲青少年呵护地球家园计划——与澳大利亚"青少年呵护地球家园计划"合作开展的一项活动，旨在促使学生致力于实现可持续发展目标；合成生物学冒险——与加拿大康科迪亚大学合作开展的一项青少年计划；青年学员健康教育——与青少年医学院（美国）合作开展的一项健康计划；家庭学校计划——适用于在家受教育的孩子的 STEM 教育。

这些及其他的活动方案正为肯尼亚青少年提供越来越多接受 STEM 教育的机会，并让他们看到自己在 STEM 经济中拥有的未来。公众对这些项目的热情始终有增无减，2014—2017 年，肯尼亚科学中心的观众人数增长了 350%，目前每年的观众总数已经超过 6000 人。

6. 科学中心的空间设计

能力建设之旅已经表明：科学中心的建筑类型各不相同。有些科学中心利用现有的建筑物，在其基础上进行翻新；而有些科学中心则花重资修建了标志性建筑，其成了当地旅游景点。许多成功的科学中心开始的建筑规模并不大，全部建设工程持续多年，而不是一次建成一座大型建筑物。在某些情况下，一次建成的大型建筑物存在着难以解决的各种问题，例如，很多空间得不到有效利用，在作为公共空间使用或进行 STEM 教育时存在某些设计缺陷。一些情况下，大型标志性建筑的建设存在问题，增加了建筑成本，导致展品预算减少，并且在开馆后无法达到公众参观的预期。

我们看到的另一个现象是：最为成功的科学中心拥有许多不同的收入来源，而非仅依赖一两个来源，如政府或单一的基金会及捐助者。就肯尼亚科学中心而言，肯尼亚农业和畜牧组织在人员配备和建筑物使用方面的支持至关重要，而基金会和国际科学中心协会也曾帮助我们并为肯尼亚科学中心提供了诸多支持，从而使中心的工作达到了目前的水平。肯尼亚科学中心仍需要更广泛的支持，以实现下一阶段的发展。

7. 肯尼亚科学中心未来的发展愿景

肯尼亚科学中心现已成功地证明，公众对中心提供的多种 STEM 展览和活动有着很高的需求。为此，中心领导正在制订下一阶段的发展计划。肯尼亚科学中心的发展愿景"通过 STEM 教育，为肯尼亚青少年创造更美好的未来"指导着科学中心发展规划的各个方面，其主要内容如下。

空间：扩大现有空间，使之能够容纳包含 50 件交互式 STEM 展品的展览，使得针对学校团体的教育方案能更好地实施；建设公共卫生间、大厅、员工办公室、纪念品商店等。

展品：通过收购或建造 50 件以 STEM 为主题的展品，对现有展览进行补充。

人员：额外增聘 15 名工作人员，从事展品设计、教育方案编写、公关、宣传和维护科学中心等工作。

8. 结论与建议

在肯尼亚开展的一项调查表明：受访者对科学中心的认知程度较低，但对肯尼亚科学中心提供的各类科学展品和活动表现出强烈的兴趣。所有受访的肯尼亚人不分性别和受教育程度，都非常支持建立科学中心。

通过走访 25 家科学中心和其他类型的博物馆，肯尼亚科学中心建立起一个国际行业合作网络。这些国际业内人士愿意协助肯尼亚科学中心，并能提供建议、展品和相关资源。这项能力建设工作也为科学中心在保证正常运行的基础上不断发展提供了极大的帮助。科学中心工作人员所参与的国际培训项目和会议，为中心的发展、教育资源的积累以及业内咨询工作打下了坚实的基础。

作为功能完善的公共教育中心，肯尼亚科学中心为了从成功的试运行阶段转向正常运营，将要采取三项措施：设计和建造永久的科学中心大楼；开发 50 件互动式科学展品；增聘 15 名员工。

确保这些工作的完成将是肯尼亚科学中心下一阶段的工作重点。

参考文献

[1] BONNEY R，BALLARD H，JORDAN R，et al. Public Participation in Scientific Research：Defining

the Field and Assessing Its Potential for Informal Science Education. A CAISE Inquiry Group Report. [J].
Online Submission, 2009: 58.

[2] DIERKING L D, BURTNYK M S, BÜCHNER K S, et al. Visitor learning in zoos and aquariums: A
literature review [D]. Annapolis, MD: Institute for Learning Innovation, 2002.

[3] LEONHARDT G, GREG M. Burning Buses, Burning Crosses: Student teachers seecivil rights. [J].
2002.

[4] WINTERBOTHAM N. Museums and schools: developing services in three English counties 1988—2004
[J]. University of Nottingham, 2005.

[5] TRAVERS T, GLISTER S. impact and innovation among national museums. [J]. Proceedings of the
National Museums Directors' Conference, 2004.

[6] MONJERO K. Building communities through science-Kenya Agricultural Research Institute [J].
Proceedings of the 6th Science Centre World Congress, 2011.

[7] MONJERO K. Developing Science Centres in Africa for Equality of Opportunity and Global Sustainability,
the case of science centre Kenya [J]. Proceedings of Science Centre World Summit, 2017.

[8] MONJERO K. Do I Belong?Identity and Diversity Issues in Science Centres, serving new audiences,
especially those from indigenous and underserved backgrounds inKenya [J]. Proceedings of Science
Centre World Summit, 2017.

[9] TRAUTMANN C H. "The Business of Science Centers," ASTC Dimensions, Association of Science-
Technology Centers [M]. Washington, DC: May-June.

[10] MONJERO K, MITEI D, KARIUKI E, et al. Science awareness: the case for a science centre in
Kenya [J]. Proceedings of the 2nd National Science, Technology and Innovation Week. Nairobi
Kenya, 2013.

[11] MONJERO K, MITEI D, KARIUKI E, et al. Status and Progress of the Science Centre Movement
in Kenya and the Region [J]. Proceedings of the 15th Conference Southern African Association of
Science and Technology Centres, 2013.

亚美尼亚共和国国家科学院地质科学研究所 H.KARAPETYAN 地质博物馆

盖娅尼·格瑞高瑞安 [1]

1. 简介

亚美尼亚地质博物馆成立于 1937 年 6 月，拥有著名地质学家 Hovhannes Karapetyan 教授丰富而多样的藏品，曾隶属于苏联科学院亚美尼亚分院地质科学研究所。建馆之初有 1410 件展品，目前已有 14000 多件展品。展览的组织源自第 17 届国际地质大会（在莫斯科举行）与会者提出希望了解亚美尼亚的地质情况。随后他们领略了亚美尼亚展品的风采并参加亚美尼亚地质游览。博物馆的建立在当时是亚美尼亚共和国地质科学的一件大事，一个整合展品材料集中、加工、展示和存储的中心便应运而生。随后，博物馆将研究考察所收集到的样本纳入藏品库中。

就地质学而言，亚美尼亚是高加索地区最令人感兴趣的地区之一。在这里，从上元古界变质片岩开始直至厚第四纪溢流岩及现代沉积，几乎所有地质时代的地层序列均得以展现。

历经 80 年的建设，博物馆得到了一定的发展，其馆藏也变得愈加丰富。博物馆下设的部门包括：矿物学、古生物学、岩类学、自然资源、火山学、亚美尼亚的矿泉水及亚美尼亚的地质科学天然遗迹。

博物馆致力为不同年龄的人开设各种户外课堂，使参与者有机会熟悉独特的天然遗迹、举办博物馆之夜活动、国际博物馆日、知名地质学家纪念活动、会议、研讨会，前往设有流动博物馆的村庄进行参观、开展野外调研，以及与其他博物馆开展合作。

博物馆占地 600 平方米，累计参观者达到 20120 人次，受众包括老年人、学生、儿童、游客等。参观人数逐年增加。我们开展与学校课程相对应的教育课程。与教师组织圆桌会议，商讨如何为在博物馆教授的课程提供支持。组织各种科学会议，在

---

① 盖娅尼·格瑞高瑞安：亚美尼亚共和国国家地质博物馆馆长。邮箱：gayane347@gmail.com。

"世界水日""地球星球日""国际山岳日"等场所讨论并交流有关现代生态问题的最佳体验。

博物馆需要的各项辅助设施包括：礼品店、使用新技术及工艺的显微镜、电台指南、将档案文件保存在相应标题下的特殊用品。博物馆可以与其他博物馆合作组织如下活动。

亚美尼亚有两个自然博物馆。

亚美尼亚国家自然博物馆

亚美尼亚国家自然博物馆于 1952 年建成于埃里温的一个蓝色清真寺院子，当时曾被称为共和国自然科学博物馆。1960 年该博物馆更名为亚美尼亚国家自然博物馆。

亚美尼亚动物研究生 NAA 动物学博物馆

动物学和水生生态学中心的动物学博物馆始建于 1920 年，它是在沙皇时期亚美尼亚将军谢尔科夫尼科夫的努力下建立起来的。

2. 结论

（1）针对儿童、学生及老人举办为期 7～14 天的教育实践活动，亚美尼亚以其地质构造而闻名，吸引着专家和游客的造访。

（2）组织各种科学会议，在"世界水日""地球星球日""国际山岳日"等场所讨论并交流有关现代生态问题的最佳体验。

（3）交换矿物样本，以充实博物馆的资金。

（4）创造全球资金，支持自然博物馆的发展和现代化。我相信此次会议将有助于与会的各博物馆取得更大进步并提高声望。

未来的巴基斯坦地质公园——盐岭

撒肯德·阿里·拜格　默罕默德·瓦恰斯　缅恩·哈森·艾哈迈德 ①

摘　要 盐岭包括六个峡谷，具有出露充分的地质学和古生物学特征，并且交通便利。这些特点吸引了研究者对这个独特地区的地质时代测定方面的浓厚兴趣，开始对寒武纪三叶虫、二叠纪腕足类、三叠系下统菊石、下第三系大型有孔虫和上第三系及第四纪脊椎动物产生的古地理进行了研究调查。在地质学家眼中，这些盐岭峡谷充满着如此美妙的地质遗址，堪称世界上独特的野外自然历史博物馆，同时也可能被建设为巴基斯坦的一个地质公园。

关键词 盐岭　化石产地　地质公园

1. 简介

盐岭是一条东西走向的波谷，东边界是杰勒姆河，西边界为印度河。在印度河的更远处，形成了一个 U 形急弯，发展成了南北向。盐岭是喜马拉雅山活跃的正面逆冲带，从旁遮普平原北部崛起，并在北部含油气的波特瓦盆地内轻微下沉。盐岭的名称来源于其具有世界上最厚的岩盐层的产状，埋藏在盐岭组的前寒武纪亮红色泥灰岩中。

盐岭堪称一部开卷的地质史学书籍。盐岭所拥有的出露充分和交通方便的特点吸引了来自世界各地的地质学家和古生物学家。如纳玛尔峡谷（彩页图 6）、克乌拉峡谷（彩页图 7）等路边的地质现象，还有各种各样的地质特征以及富含化石的成层岩。由于缺乏植被，寒武纪地层、含有腕足动物的二叠系碳酸盐岩层系、二叠 – 三叠系界线、早三叠世菊石层、下第三纪海相地层（含具有定年意义的有孔虫）得以充分出露，从而为地质学各领域的详细研究提供了绝佳机会。因此，盐岭在科学和教育研究方面具有突出的普遍意义，需要被认定为地质公园，从而保护其所包含的地质历史。

---

① 缅恩·哈森·艾哈迈德：巴基斯坦地质调查局主任。E-mail：mianhassanahmed@gmail.com。

2. 先前的研究

格里首次发表了比例尺为 1 : 50000 的盐岭地质图。戴维斯和品菲尔德记录了盐岭的下第三纪较大有孔虫类。瓦根研究了二叠纪腕足类。格兰特描述了盐岭的二叠纪三叶虫。沙朗汇编了巴基斯坦的地层情况。瑞曼公开发表了巴基斯坦的首个地层规范。卡曼尔和泰科特描述了二叠纪和三叠纪岩石的详细地层情况。哈格描述并记录了盐岭讷默尔峡谷中第三纪较小有孔虫类的分布。萨米尼在其博士论文中确立了盐岭始新世演替的蜂槽虫状生物地层学。萨米尼和布特、萨米尼和霍廷格曾讨论过盐岭上新世和下第三系岩的微化石生物地层情况。

3. 地层学和古生物学

3.1　前寒武纪

盐岭组是暴露于东部盐岭中唯一的前寒武纪时期岩石地层单位。它包括厚层蒸发相，由盐、石膏和泥灰岩组成。被称为"克乌拉陷阱"的高度风化的火成岩体也是该组的一部分，包括高度分解的辐射状浅色矿物（可能是辉石）。该组缺乏化石。它一致地被含有寒武系动物群的碎屑层序所覆盖，因此盐岭组的时代被认为是前寒武纪。

3.2　古生代

盐岭的古生代沉积层厚度达数百米，但具有强烈的不整合性，表明奥陶纪和石炭系的缺失。寒武系岩石以碎屑层序为代表，上部含有较厚的碳酸盐相。

二叠系层序由混合（旋回）碳酸盐岩、硅质碎屑岩相以及底部的冰碛岩相组成。这些化石包括大量的苔藓虫、腕足类、双壳类、腹足类、鹦鹉螺类、菊石类、三叶虫类、蜓类、牙形石和海百合。二叠纪岩石因存在长身贝属而著名。化石还包含冈瓦纳舌羊齿属和恒河羊齿属的孢粉化石。卡曼尔和泰科特详细讨论了二叠系和三叠系之间的似整合关系。

3.3　中生代

中生代岩石的特点是在三叠系底部，三叠系和侏罗系之间，侏罗系和白垩系之间以及白垩系顶部存在着一些重要的缺失。沉积物大部分是由页岩、砂岩、石灰岩和白云石组成的浅水海洋和陆源物质。三叠系主要由石灰岩、白云岩、砂岩和页岩组成。

页岩单位含有丰富的化石，并含有保存良好的齿菊石属。侏罗纪主要由石灰岩、页岩、砂岩（带有附属的白云铁矿床）组成。白垩系岩石主要为砂质和泥质。泥质岩沉积于还原环境中，且含有黄铁矿结核和箭石。

3.4 新生代

古新世 – 早始新世地层大部分暴露于盐岭的中西部。这些岩石轻微地向北倾斜浸入含油的波特瓦盆地，并被新第三纪和第四纪磨砾层覆盖。中新世至渐新世数量的不整合被认为是由区域隆起和特提斯海的闭合造成。特提斯起源的下第三系 – 下始新统地层中含有丰富的具有定年意义的货币虫和榄核虫科等大型底栖有孔虫化石。新第三系和第四系磨砾层记录了喜马拉雅隆起的历史，并以产出哺乳动物、鸟类和爬行动物的化石而闻名。

4. 结论和建议

盐岭富含几个独特的地质学和古生物学遗址，均具有极高的学术研究价值，应考虑予以保护和保存。盐岭适合建成巴基斯坦的一个地质公园，而这需要古生物学者、政府机构共同努力，培养当地居民的保护意识，从而有效保护这些地质遗迹。

参考文献

［1］PASCOE E H. The early history of the Indus，Brahmaputra，and Ganges［J］. Quarterly Journal of the Geological Society, 1919, 75：138–157.

［2］SAMEENI S J. PaleoParks–The protection and conservation of fossil sites worldwide；The Salt Range：Pakistan's unique field museum of geology and paleontology［J］. Carnets de Géologie/Notebooks on Geology，Brest，2009.

［3］COLBERT E H. Siwalik mammals in the American Museum of Natural History［J］. Transactions of the American Philosophical Society. New，1935，26：i–x，1–40.

［4］GEE E R. The saline series of north–western India［J］. Current Science，1935，II：460–463.

［5］GEE E R. The age of saline series of the Punjab and Kohat［J］. In–Proceedings of the National Academy of Sciences of India，Calcutta，（Section B），1945，14：269–312.

［6］SHAH S M I. Stratigraphy of Pakistan［J］. Geological Survey of Pakistan，Memoirs，Quetta，2009，21：381.

［7］WAAGEN W. Salt Range fossils；Productuslimestone fossils［J］. Pal–ontologiaIndica，I（4）：329–

770.

[8] GRANT R E. Late Permian trilobites from the Salt Range, West Pakistan [J]. In-Palaeontology, Oxford, 1966, 9 (1): 64-73.

[9] SAMEENI S J, HOTTINGER L. Elongate and larger alveolinids from Choregali Formation, Bhadrar area, Salt Range, Pakistan [J]. Pakistan Journal of Environmental Science, 2003, 3: 16-23.

[10] SHAH S M I. Stratigraphy of Pakistan [J]. Geological Survey of Pakistan, 1977, 12: 138.

[11] REHMAN H. Stratigraphic Code of Pakistan [J]. Geol. Surv. Pakistan, 1962, 4: 8.

[12] KUMMEL B, TEICHERT, C. Relations between the Permian and Triassic formations in the Salt Range and Trans-Indus ranges, West Pakistan [J]. NeuesJahrbuchfür Geologie und Palontologie, Abhandlungen, 1966, 125: 297-333.

[13] KUMMEL B, TEICHERT C. Stratigraphy and paleontology of the Permian-Triassic boundary beds, Salt range and trans-Indus ranges, West Pakistan. [J]. University of Kansas, Special Publication, 1970, 4: 1-110.

[14] HAQUE A F M. The smaller foraminifera of the Ranikot and Laki of the Nammal Gorge, Salt Range. Pakistan [J]. 1956, 1: 300.

[15] SAMEENI S J. Biostratigraphy of the Eocene succession of the Salt Range, Northern Pakistan [D]. Punjab University, Lahore, 1997.

[16] SAMEENI S J, BUTT A A. Alveolinid biostratigraphy of the Salt Range succession, Northern Pakistan [J]. Revue de Paléobiologie, Genève, 2004, 23 (2): 505-527.

匈牙利自然历史博物馆的历史、设施和活动

伽伯·西萨巴 [①]

摘要 匈牙利自然历史博物馆是世界上建成较早的自然历史博物馆之一，具有独特的传统，并在社会中发挥着决定性的文化作用。馆内收藏有丰富的植物学、动物学、人类学、古生物学和地质学的藏品、原生型标本以及藏书，为世界各地研究人员的研究活动提供收藏品和设施。该馆每年举办的展览吸引观众约 30 万人次，组织近 500 场教育活动，受众达到了 44000 人。该馆意识到博物馆当前所面临的挑战与机遇，积极应对诸如气候变化、生物多样性危机、新发传染病、生物入侵等全球性问题，在探索与相关政府部门以及其他科学组织和机构合作方式过程中发挥主导作用。

关键词 匈牙利自然历史博物馆 沿革 馆藏 教育活动 挑战与机遇

1. 简介

匈牙利自然历史博物馆建于 1802 年，是世界上建成最早的自然历史博物馆之一。请永远不要忘记匈牙利自然历史博物馆的独特传统以及在社会中所发挥的决定性文化作用，我们的使命是提高并保持对自然多样性的认识和接受度，并确保公众对保护自然环境的承诺。匈牙利自然历史博物馆下辖两个附属的乡村博物馆（位于匈牙利珍珠镇的马特劳博物馆和位于齐尔茨镇的包科尼山博物馆）。作为匈牙利的科学和文化机构之一，全馆雇用了 230 多名员工。

博物馆收藏了 1000 多万件涉及植物学、动物学、人类学、古生物学和地质学的藏品，以及 65000 多种原生型标本，博物馆的图书馆藏有 40 万册书籍和期刊。这些馆藏品以及来自世界其他地方的极具价值的物品，不仅代表了匈牙利和喀尔巴阡山脉的自然历史，也使该馆成为匈牙利和中欧生物和地理多样性的终极标志。每年有 350 名来自世界各地的研究人员使用匈牙利自然历史博物馆的收藏品和其他科学设施。各种研

---

① 伽伯·西萨巴：匈牙利自然历史博物馆副馆长。E-mail：csorba.gabor@nhmus.hu。

究活动涉及分类学、系统学、生态学、进化生物学、流行病学、保护生物学、地质学、古生物学和体质人类学等领域。博物馆举办的展览每年吸引约 30 万人次（约占全国总人口的 3%）。

匈牙利自然历史博物馆成功地参与了欧盟的几个项目，包括数据库管理、数字化计划、研究基础设施行动、科学家交流以及生命＋保护生物学项目。该馆是欧洲分类学设施联盟（CETAF）、国际博物馆协会（ICOM）、欧洲科学中心协会（ECSITE）以及全球基因组生物多样性网络（GGBN）的成员。

2. 历史沿革

该馆的起源可追溯到 1802 年，费伦茨·塞切尼伯爵设立了匈牙利国家博物馆，他为国家捐赠了 1700 册藏书、手稿和钱币藏品。同年，他的妻子捐赠了匈牙利首个自然历史藏品，其中包括来自匈牙利的有价值的矿物和腊叶标本。1811 年，首个古生物学和动物学收藏品被送到博物馆。后来，在匈牙利的贵族化时代，伴随 19 世纪改革期间的爱国主义情绪高涨，由于捐赠和采购数量的增加，馆藏品开始迅速飙升。

1869 年是首次记录有关观众人数数据的年份，当年有近 65000 名观众参观了该馆。由于馆藏资料的种类不断扩大，自然历史的收藏品不得不从国家博物馆独立出来。1870 年建立了独立的动物学、矿物学、古生物学和植物学部门。由于独立性越来越高，越来越多的专家加入到这个部门，成长速度愈加迅猛。到 19 世纪末，自然历史标本的数量超过 100 万个。

1956 年匈牙利革命期间，博物馆遭受了历史上最大的损失。由于苏军炮击引起的火灾，导致博物馆著名的非洲陈列品以及大部分的矿物学和古生物学藏品损毁。容纳于另一栋建筑中的动物学分部展厅也遭到了燃烧弹的袭击，成千上万的标本被焚毁。在接下来的几十年里，博物馆在收藏品方面的策略是把重点放在收购上，以取代其丢失的藏品和陈列品。博物馆组织了几次前往非洲以及朝鲜、越南、蒙古等地区的收集之旅。因此，从这些地区收集到的资料是全球同类中最丰富的。从 20 世纪 80 年代开始，亚洲、非洲、南美、澳大利亚和大洋洲的更多国家成为收集藏品目的地。

20 世纪末，匈牙利自然历史博物馆仍然缺乏可被公众识别的永久性建筑。直到 1994 年匈牙利政府才决定为博物馆提供一处永久地点，并为其常设展览和某些藏品指定了当前位置。根据当时的计划，到 2002 年之前，所有馆藏品都将被安置在一个地点。1996 年，我们在这座新改建的大楼里开设了"匈牙利人与自然"的历史与生态常设展览。1999 年，一座全新的建筑向公众开放，展出人类学分部连同动物学分部的两个分支部门（哺乳动物和鸟类）的藏品。2010 年，我们现在的常设展览"生命的多样性"向公众开放。

2013 年，两个乡村自然历史博物馆（地处匈牙利东北部的珍珠镇的马特劳博物馆和地处匈牙利西部齐尔茨镇的包科尼山博物馆）成为布达佩斯匈牙利自然历史博物馆的成员机构。

遗憾的是，由于有关博物馆的政治决议发生了一些变化，其他馆藏品无法被移到这个地点，而是留在了它们单独的建筑中。为所有馆藏品和陈列品找到一个永久性的放置地点，是该馆现在面临的最大挑战。

3. 馆藏品亮点

动物学分部是欧洲较大的动物学馆藏之一，拥有近 800 万件藏品。它收录了来自世界各地的重要资料，收藏了来自巴尔干半岛、亚洲内陆、东南亚以及远东的藏品，而来自新几内亚和东非的历史资料则在世界范围内独一无二。

植物学分部包括 200 万件标本，其中大部分是蜡叶标本。博物馆藏品包括可追溯到 18 世纪中期不寻常的历史资料，其中包括林奈标本、中欧蜡叶标本和化石植物遗迹。

来自喀尔巴阡盆地的数十万件单独登记的化石则被安置在古生物学分部，它们包括来自匈牙利上白垩统的第一批恐龙遗骸。

矿物学和岩石学部门中放置了约 8 万件标本，它们大部分来自匈牙利，但也容纳了来自世界各地的相当数量的样本。博物馆收藏的陨石和月球岩石样品被认为具有重要价值。

人类学分部包含代表喀尔巴阡盆地整个后更新世人类的约 5 万具人类遗骸。匈牙利尼安德特人遗骸以及位于匈牙利瓦茨镇、在自然状态保存的木乃伊是其中最值得一提的对象。

4. 教育活动

我们的展览分部每年组织近 500 场教育活动，受众达到了 44000 人，教育部为4～18 岁的学童提供近 800 次的一小时课堂课程，在正规的教育课程中对课程的主题进行了界定，且与我们馆藏品的科学范畴有关。在马特劳博物馆中，欧盟资助的项目于 2018 年 1 月启动，通过导入有针对性的教育方法，利用为教育目的特定开发的信息和通信技术与弱势群体展开互动。

匈牙利自然历史博物馆在参与高等教育方面有悠久的历史，这主要通过两个途径，其一是开展针对其工作人员的教学活动，其二是为各大专院校学生的硕士和博士项目提供安置及指导服务。目前，匈牙利自然历史博物馆是布达佩斯兽医大学的辅助大学科系，主要负责动物分类学、动物系统学和动物学课程。除此之外，匈牙利自然历史博物馆还在每学年举办近 20 场有关馆藏品的研讨会。

5. 挑战与机遇

在若干社会经济因素中，新文化预期的兴起能够显著地影响文化消费。各博物馆之间不仅会争夺消费者，还会与其他文化供应商以及数字世界展开竞争。根据 2013 年发布的欧盟文化接触与参与情况民意调查报告，2007—2013 年的文化参与度呈总体下滑的趋势。在匈牙利，那些从未参加过文化活动的人数比例在这段时间内上升至 54%。20 世纪 80 年代和 90 年代信息技术的蓬勃发展给人们支配空闲时间的方式带来了变化，并且创造了所谓的 Y 世代和后来的 Z 世代，这些人离开电脑和互联网就无法生活下去，而他们支配闲暇时间的方式就是玩电脑游戏。2017 年，匈牙利自然历史博物馆已经认识到游戏化应用所起的作用，这是在匈牙利所谓的逃生室中被成功运用的技术。为了贴近青年，该馆基于馆藏品资料，在博物馆的公共区域内设立了两个逃生室。通过涵盖各种展示方式、传播知识并向公众提供接触馆藏资料的途径，这一全新的服务在满足青少年需求的同时，也完成了文化价值传递的使命。

在自然历史收藏品的科研、策展和其他工作中，大量的数字数据得以生成和存储。数字数据是进行现代研究与收藏必不可少的工具和产品。它们是对实体对象物的补充，也是收藏品的一部分。它们所具有的社会优势是毋庸置疑的，能够提高对自然历史和生物多样性课题的认识，为教育（更新）提供信息验证，向公众提供数据，增加公众对自然科学和博物馆的兴趣，最终促进公众对科学的理解。然而，我们的主要任务是要寻找财力和人力资源，以完成有关数字化存储，以及向公众提供这些服务的各项任务。博物馆将拥有不同类型的数据库，因此它必须在以下方面采取措施：进入的权利（有限或免费）、隐私、道德（涉及软件许可）、数据复制、所有权与权利、授权、权限与信用。

如今，分类学（涉及自然历史对象物的识别和分类）的重要性被低估。但是只有在了解并掌握生物多样性专业知识的情况下，才能解决如生物多样性丧失以及人类活动对生态系统影响等各种全球性挑战。作为匈牙利唯一拥有这种专业知识和材料收集能力的机构，我们的使命是让公众和决策者相信分类学和系统学对于经济、健康和日常生活的重要性。作为一个自然历史博物馆，我们必须应对如气候变化、生物多样性危机、新发传染病、生物入侵等全球性问题。作为一个拥有大量馆藏品和专业分类学家的知名机构，匈牙利自然历史博物馆应该在确定特殊与普遍性问题，以及在探索与相关政府部门和其他科学组织机构合作方式中发挥主导作用。

生物多样性及其保护与社会变革密切相关，只有通过大力投资环境教育才能得以实现。当下需要提供正确的培训和能力建设，以提高教师在生物多样性教育领域的能力，有效地提高教师和学生的生物多样性素养水平，激励他们了解生物多样性，参与保护欧洲的生物多样性，点燃他们对科学的热情。

科学、艺术与数学的教育计划

保罗·费罗斯 [①]

摘　要　赫拉克莱冬是一个小国家的小型博物馆，它试图在一个文化产业并不总是受欢迎的环境中努力前行。在过去的 10 年中，该馆设法将艺术、科学和数学整合到一个独特的教育计划中，旨在使数学学习变得更容易、使学生热爱艺术、使科学更易于理解。为此，该馆正在与教育专家、数学家、美术教师在内的各路专家一起工作，使该馆开展的工作能适应并满足学生的各种需求。

关键词　博物馆　科学　艺术　数学　教育计划

我曾经看过一些描绘古代丝绸之路中多条路线的地图，令我惊讶的是：最西端的目的地确实是希腊。2500 年后，我们再次汇聚于此，重新肩负起振兴祖先所发起事业的使命。但不同的是，我们所进行的是一项不同但更为重要的交流，它不限于交换商品，而侧重文化和思想的交流。

1. 博物馆

希腊人口约为 1000 万，可能还不及今天到会的规模在前五名的博物馆年参观人数的总和。然而，我们每年吸引超过 3000 万的参观者前来雅典参观卫城、博物馆、国家考古博物馆以及所有构成古希腊城邦的遗址。

令人吃惊的是，希腊没有私营博物馆的传统。当我和妻子决定建立赫拉克莱冬博物馆时，我们惊讶地发现，只有少数类似的机构存在，而且最重要的是没有现成的立法可将我们的博物馆归类为真正的博物馆。因此，我们便开放了自己的博物馆，作为一家致力于文化传播的非营利组织，这是一种用来描述世界上任一家博物馆很好的方式（彩页图 8）。我多么希望当时能够遇到类似于中国自然科学博物馆协会（BRISMIS）

---

① 　保罗·费罗斯：希腊雅典娜赫拉克莱冬博物馆创始人。E-mail：hyperpol@gmail.com。

这样的组织来帮助我们克服前行道路上遇到的种种障碍。

我们今天响应"一带一路"倡议所做的事情就是寻求博物馆之间高效率、富有成效的合作，而这一点也正是我们当初创建自己的博物馆时不得不亲自去寻找的东西。我们很快得知中国将召开首届"一带一路"科普场馆发展研讨会的消息并向同行们伸出了求援之手，是 BRISMIS 给我们提供了帮助。

2. 教育

我曾邀请坐拥世界上参观人数最多的博物馆之一巴黎卢浮宫博物馆馆长来雅典参观我们的博物馆。我对他提出的两点评论感到惊讶。首先，赫拉克莱冬博物馆馆长的办公室比他的大！第二个评论，也是更为重要的：他希望卢浮宫的参观人数能再少一些，以便他可以投入更多的资源实现博物馆的各项目标，同时在人群控制方面花费更少的精力。

换言之，无论博物馆的规模大小，都要注重其目标——教育。我们对前来参观博物馆的成年人进行教育，但最重要的是，我们正在教育那些为了学习而前来参观的少年儿童。

教育是赫拉克莱冬博物馆的主要目标，我们对于能够在学生与其他观众保持 1∶1 的比例而感到非常自豪。我们的辅导教师都是大学毕业生，每天同时为四个班级举办教育活动，教材始终是当期展览。

十多年来，我们一直坚守着数学和艺术的界限。

数学和艺术的界限早在 2500 年前柏拉图所著的《理想国》中就得到了界定，在此书中，他以为数学与推理（逻辑）部分存在关联，而艺术与下等（情绪）部分相联系。

在研究了柏拉图对待艺术的观点后，我确信，如果柏拉图能够起死回生，并且关注埃舍尔的著作，他就会在后者的脸上认出这是热切追求理想状态的艺术家之一。寻找那些能够聪明地找到慈善之本的创造者，而他拒绝那些"只会机械地把镜头举起来对着自然拍照"的模仿者，他把这种做法归于只能表现一些表面的、外在的特征。柏拉图在 2500 年前讲的这些话对艺术代表寄予了厚望，他拒绝把艺术看作对现实的一种现实－奴隶式的描绘。在他的另一段对话——"菲勒布斯"中，他对"绝对美丽"的描述留给了那种采用几何形状作为基本组成部分的绘画作品。许多人把这作为 20 世纪现代艺术出现的一个重要标志，拿保罗·克利的话来说就是"无法再现可见的东西，而是起码使之可见"。

十年来，我们始终致力于数学和艺术课程，努力将像数学这种令人生畏的东西转化为有趣、充满艺术的东西，我们正在成功地做到这一点。我们的老师正在使用埃舍尔、瓦萨雷利、卡罗尔·瓦克斯等艺术家的作品，或者参考在雅典帕台农神庙附近发现的纪念碑，与学生们讨论黄金比例、分形、透视、音乐、韵律、和谐及对称等。我们的藏品中能够找到埃舍尔和其他人的著作，帕台农神庙就位于博物馆外的几百米处，艺术蕴含着隐藏的或明显的数学，在建筑的任何地方都可以找到黄金比例。这就是为什么我们总是致力于找寻一种能够将正在进行的展览与科学／数学相结合，使学习变得更容易，使数学更受人喜爱。

3. 教育计划

通过播放视频投影，让学生平行地巡游于艺术史和数学史两个层面，旨在引导学生探索人类思维和行为这两个方面产生交集并展开互动的节点。我们将重点放在了希腊几何艺术、古典艺术（帕台农神庙比例－黄金比例）、对线性视角（文艺复兴）的分析、现代艺术的几何学（立体主义、建构主义、包豪斯）以及现代分形学的"数学艺术"。同时，通过涉猎埃舍尔和瓦萨雷利的选集，向学生介绍自然界及一些重要数学概念中所蕴含的哲学原理。有针对性地巡回展览可将数学观念或科学概念与具体的展品联系起来。

我们的教育活动结合当期展览，根据国家数学教学方案设计，曾为学生开展过中文折纸、龙的几何形状、算盘等中国计算工具之类的教育活动（彩页图9）。一般来说，我们的教育活动如下所述。

第一部分：在课堂（60分钟）

中学一年级：寻找合适的关于艺术（埃舍尔、瓦萨雷利、卡罗尔·瓦克斯等）与几何的背景。邀请学生探索所选定版画的构造，并在可能的情况下，借助铅笔、尺子和指南针将它们再现。在这种情况下，学生们显然必须从数学上解构这些图像，从而进入数学概念的领域中。

中学二年级：介绍基本的几何变换（反射、中心对称、传输、旋转）。

中学三年级：将学生在学校里学到的"存在且似有"的欧里庇得斯悲剧《海伦》与哲学上和数学上"存在且似有"之间的区别联系起来。这种联系旨在通过利用适当的绘画（其中"直觉"可能带有误导性）来引导学生们认识到使用推理和证明的必要性。

中学四年级：①使用适当的绘画（实际上是视错觉）会导致对决定平行四边形的原则进行基于经验及其后理论上的提取。②埃舍尔所创的名为"Verbum（语言、标志）"的版画是特定于比例数学概念及其哲学色彩的一次头脑风暴会议（即为集体讨论会）。③介绍非欧几里得几何。

中学五年级：①研究网格分级变更中基本曲线的几何不变量。②通过埃舍尔的曲面细分，我们能够研究平面的正则和半正则划分（分别通过正则和半正则多边形）。③构成芝诺悖论的跳马，旨在对极限概念和自相似概念进行介绍。

中学六年级：①研究网格分级变更中基本曲线的几何不变量。②正在利用恰当的绘画，目的在于探讨极限、无穷大和无穷小、不连续、连续及其哲学色彩。③自然数集的"可数性"、有理数集的密度，以及实数集的"过计数性"。④函数和反函数的概念。⑤数学结构的概念。

第二部分：在展厅（50分钟）

为学生提供展览的导览服务，不仅可以促进学生与艺术家作品或文物的互动，而且还能够激励他们发表自己的言论。

第三部分：在教室（10分钟）

回到教室后，要求学生们完成评估反馈表，他们可以采取匿名方式记录对刚刚参与过的计划进行评论，包括对整体活动的简短评估，学到了什么新东西，以及有哪些困惑。

4. 我们的未来

"一带一路"倡议无疑将把我们大家聚集在一起，并且能够促进博物馆之间展开更大规模的合作。博物馆同行们可在共享信息、交流营销知识和技术诀窍方面获益匪浅。但以我个人的经验来看，这可能还远远不够。两个博物馆并不总是有可供交换的共同东西，有时需要3个或更多的博物馆才能实现"交易"馆藏品。我们通过创建一个独立的实体来解决这个问题，其目标是促进这种交流。我们委托了一家经纪公司，将4件馆藏品在欧洲国家及美国等进行巡展。道理很简单，博物馆可以确定一个巡展的馆藏品，并规定愿意出借给其他博物馆的各项条款和条件。然后，我们通过国际博物馆提供的联络方式，找到愿意参与这项交流的机构。我们欢迎在您所规定的条款和条件下，将您的馆藏品添加到我们的展览清单中，其余的事情就交给我们来办。博物馆之间的这种合作可使我们的业务充满活力。博物馆之间的合作至关重要，同时是我们对所服务社区的道德义务所在。

参考文献

［1］Allentuck M E. Herbert Read, The Philosophy of Modern Art［J］. Art Journal, 2015（3）: 302-303.

［2］M C ESCHER. Regular Division of the Plane, 1958.

［3］DORIS SCHATTSCHNEIDER. The Polya-Escher Connection［J］. Mathematics Magazine, 1987,60（5）: 293-298.

［4］SCHATTSCHNEIDER D. The Plane Symmetry Groups: Their Recognition and Notation［J］. American Mathematical Monthly, 1978, 85（6）: 439-450.

［5］H S M COXETER. The Non-Euclidean Symmetry of Escher's Picture'Circle Limit III'［J］. Leonardo, 1979, 12（1）: 19-25.

澳大利亚堪培拉国家恐龙博物馆的互动教育

托马斯·卡皮坦尼 [①]

摘 要 位于堪培拉的国家恐龙博物馆是一座寓教于乐的互动式博物馆，供成人、儿童和学生游乐。尽管博物馆规模相对较小，但通过提供最大化互动、参与和教育机会，博物馆已能够得到较高的参观人数，并通过使用脸书或微信等社交媒体推广业务，使游览变得愉快而有趣。专注有效的零售已使国家恐龙博物馆能够产生稳定的盈利，这些盈利继而被用于增加馆藏、展示和储备。

关键词 博物馆 社交媒体 互动 教育

国家恐龙博物馆（彩页图 10）坐落于澳大利亚首都堪培拉，它是该地区最重要的旅游景点之一，于 1993 年建立，目前已发展成为澳大利亚最大的恐龙和史前化石材料永久展示场所之一。博物馆侧重于寓教育、互动和娱乐于一体，旨在为学校和公众提供各式各样的活动。展品包括栩栩如生的电子恐龙以及博物馆内和周围场的大量逼真的复制品。

国家恐龙博物馆拥有约 10000 平方米的公众展示区以及大型零售店，零售店的收入可为博物馆购置足够的新标本、展品及存货，以及定期开展大规模修建和改进提供资金。商店的零售额每年可达到约 100 万美元，商店面积占地约 2000 平方米。此外，博物馆还有一个大型户外恐龙园（彩页图 11），里面充满了栩栩如生的恐龙复制品，免费向公众开放。博物馆每周向公众开放 7 天，晚上通过预约，可供学生和私人使用。

1. 互动

国家恐龙博物馆成功的关键在于其侧重提供高度的公共互动，许多标本分设于恐龙园和馆内。儿童和成年人可与恐龙模型进行互动，他们还可以触摸恐龙、岩石、陨

---

① 托马斯·卡皮坦尼：国家恐龙博物馆常务董事。E-mail：tomk@crystal-world.com。

石、矿物和化石。甚至诸如特殊造型的垃圾箱之类的简单物品都可用来让公众拍照合影。

为了让孩子们参与进来，博物馆定期在学校放假期间提供各种免费活动，以便进一步让孩子们参与和互动，如自己挖掘化石。

除为参观者提供优质的娱乐服务外，博物馆还借助脸书、微信等社交媒体，有效地对博物馆开展免费的营销推广。

媒体和电视节目在做有关恐龙、化石和古代地质学的新闻报道时，经常会讲到国家恐龙博物馆。我们鼓励这样做，因为它可为博物馆提供免费的广告宣传，与社交媒体起到同样的作用。

2. 教育

博物馆对教育非常重视，为此，会定期举办公众教育之旅和校园参观，每天可安排多达 1000 名学生。这些参观旅游在时间安排上包括下班后和晚上两种选择。博物馆还会带上"恐龙"前往新南威尔士州和堪培拉的社区和学校，向公众宣传恐龙和博物馆。

博物馆的许多员工是古生物学或地质学专业的大学生。通过聘用充满激情、知识渊博和有魅力的主持人，博物馆希望能够起到激励未来科学家的目的。

3. 零售

许多其他小型博物馆没有大量的零售店铺，与它们的情况不同，国家恐龙博物馆将其 1/5 的建筑空间用于零售。我们发现这样做能使我们有资金提高馆藏品质，同时雇用更多的员工。

为确保博物馆展示空间的连续性，许多展示标本被陈列于商店内。博物馆商店中，既有只供观赏的陈列标本，也有供出售的商品标本（包括模型）。所以，博物馆商店看起来也成了博物馆展厅和博物馆教育的一部分，而不是彼此分开的业务（彩页图 12）。

国家恐龙博物馆通过直接从中国供应商处购买大量标本和非标本类商品，从而设立整体规模较大的零售部门，已经实现了很高的库存周转量并获得了良好的利润率。如通过从中国浙江义乌采购了许多非标本类商品，能够把毛利率做到 500% ～ 1000%。通过直接向供应商（而非本地批发商）采购，使毛利率达到 100%，这样不仅可以确保得到更高的利润率，而且仍可保持合适的定价以吸引观众购买。凭借如此巨大的利润，博物馆可以购买更多的馆藏标本。

4. 结论

触摸、玩耍、互动、乐趣、分享，这五个要素构成了国家恐龙博物馆的核心。

尽管国家恐龙博物馆面积不大，但却能够凭借其提供的娱乐、动手实践活动、教育互动方式和展示吸引大众的关注，赢得了极大的成功。我们极为重视社交媒体的影响力，因为它可为我们提供免费的广告和促销宣传（很大一部分包括博物馆参观者直接在社交媒体上分享体验）。

由于公众分享体验的缘故，参与和互动获得了免费宣传和增强社交媒体宣传的效果。免费的社交媒体广告反过来会带来更多的参观人数，并增加门票收入。有趣的是，从每1澳元的入场费，可以拿到1.1澳元的零售额。这种卓越的零售收入得益于我们直接从中国供应商采购大量的玩具、游戏和有收藏价值的物品。庞大的零售空间是扩大馆藏、陈列品以及进行人员配置的绝佳收入来源。

总之，一个小博物馆完全可以通过优化其自身业务、让人玩得开心而从中获利！如果博物馆的业务让参观者兴趣盎然、玩得开心，那么他们就会在社交媒体上分享体验。动手实践活动、互动和引人入胜的展示和拍摄照片等机会使博物馆非常受欢迎。当人们玩得开心时，就会倾向产生更多消费，从而给博物馆带来成功和活力。

哈萨克斯坦共和国国家博物馆

阿里波娃·塞娅 [①]

摘 要 哈萨克斯坦共和国国家博物馆是中亚地区最年轻、规模最大的博物馆。本文介绍了该馆的基本情况，并详细介绍了该馆的古代与中世纪历史展厅、历史展厅、民族志展厅、黄金大厅等重点展厅展示的历史文化遗产以及创新的展示技术等。

关键词 博物馆 文化遗产 与历史对话

哈萨克斯坦共和国国家博物馆是中亚地区最年轻、规模最大的博物馆。该博物馆是在哈萨克斯坦共和国总统努尔苏丹·纳扎尔巴耶夫所倡导的"文化遗产"国家计划框架下建造的。

2013 年 7 月 2 日，哈萨克斯坦共和国政府第 675 号法令颁布，宣布成立隶属于哈萨克斯坦共和国文化部的哈萨克斯坦共和国国家博物馆。

博物馆位于国家主广场的独立广场，与"世纪哈萨克国"纪念碑、独立宫、和平与和谐宫、"Hazret Sultan"清真寺以及国立艺术大学和谐地融为一体。许多在"文化遗产"国家计划中被确定有价值的文物，组成了哈萨克斯坦国家博物馆的无价财富。

博物馆的建筑外观非同寻常、引人注目。这个规模宏大、独特的博物馆建筑面积74000 平方米，由七栋楼组成，最高的楼高达九层。展览空间占地 14 间，总面积超过14000 平方米（彩页图 13）。

博物馆装备有国际标准的各种设备，采用了各种现代展览技术：带有特殊内容的独特弧形屏幕、媒体楼层、现代阿斯塔纳中心部分的动态布局、众多媒体渠道、全息图、LED 技术、触敏式信息亭以及可采用三种语言提供信息的多媒体导览器。

哈萨克斯坦国家博物馆设有以下展厅：阿斯塔纳大厅、哈萨克斯坦独立大厅、黄金大厅、古代与中世纪历史展厅、历史展厅、民族志展厅、现代艺术馆。负责研究国家遗产的机构是博物馆的研究所。此外博物馆还设有儿童博物馆、儿童艺术中心、两

[①] 阿里波娃·塞娅：哈萨克斯坦共和国国家博物馆研究部主任。E-mail：muzei_nauka@mail.ru。

个展室、修复工场、实验室、专业储藏室、带阅览室的科学图书馆、会议厅和纪念品摊位。

1. 古代与中世纪历史馆

地处欧亚大陆中心的哈萨克斯坦是一个有着丰富历史和文化积淀的国家。哈萨克斯坦全国范围内的科学家开展的长期调查帮助收集到丰富而独特的资料，能够完美地说明哈萨克斯坦古代和中世纪的历史，参观者可在哈萨克斯坦共和国国家博物馆的二楼展厅中看到这些资料（彩页图 14）。供观众参观用的 500 件展品仅是 5000 件藏品中的一部分，其余的仍然存于馆中，并由研究人员逐步加以补充。

由科学家收集的博物馆考古藏品向参观者介绍了从古代到中世纪居住在哈萨克斯坦境内的各部落在历史、经济、宗教、工艺和建筑方面的情况。

展览由石器时代、青铜时代、早期铁器时代和中世纪四个部分组成。艺术品则按主题和时间顺序排列。

马是哈萨克斯坦境内最早被人类驯化的动物。在哈萨克斯坦北部的 Botaisk 定居点，考古学家发现了大量与马驯化有关的文物。这一事实为证明居住在哈萨克斯坦的人们在整个欧亚大陆文化相互关系和进一步发展方面发挥了至关重要的作用。

在青铜器时代，哈萨克斯坦自公元前 1000 年就成了有色金属冶炼的主要中心之一。

游牧民族的原始经济文化类型在铁器时代早期便开始形成，它取代了青铜时代的农牧定居生活方式。

中世纪时，哈萨克斯坦领土并入了西突厥汗国。作为土耳其人的中世纪游牧民族，为他们的后代留下了许多石雕，它们默默地看着我们，缄守着过去的秘密。

自公元 6 世纪开始，哈萨克斯坦的领土就被伟大的丝绸之路穿越。展厅内展示了在哈萨克斯坦中世纪城镇中发掘的来自其他国家的物品。

14 世纪中叶，埃米尔帖木尔在中亚建立了一个强大的国家，它囊括了哈萨克斯坦的南部。当时，城镇和陵墓的修建以空前的规模展开。

展览结束于经修复的中世纪街道和哈萨克斯坦的标志性建筑，这里的墙壁上展现了考迦阿赫迈德雅萨维的宏伟建筑群。陵墓是按照帖木尔的命令修建的。它遵循了传播伊斯兰教的宗旨，加强了他在游牧民族中的地位和权力。参观者接触到这部分展览时，可以到达所谓的中世纪市场，沉浸在那个时代的氛围之中。触敏式展台介绍了哈萨克斯坦的古代及中世纪建筑，以及丝绸之路沿线的城镇。

2. 历史展厅

历史展厅展示了哈萨克族人的数百年历史，展现了重要的历史时刻及其关键结点。该展览开始于 13 世纪的金帐汗国（又名钦察汗国、克普恰克汗国），塞米雷奇和土耳其斯坦出现在哈萨克斯坦国之时。下一个突出的话题就是哈萨克斯坦与准噶尔部族人的百年战争主题，它在国民心中留下了不可磨灭的印记。哈萨克斯坦历史上一个特殊地方就是伟大的丝绸之路，它有助于加强游牧民族和定居民族之间的经济、文化、社会和政治联系。

3. 民族志展厅

位于国家博物馆三楼的民族志博物馆致力于展示哈萨克族传统文化，它占据展览空间的中心位置。

展览的设计着眼于让参观者能够全面了解哈萨克人传统的物质文化和精神文化及其经济秩序，游牧业与定居农业的结合尤其显示出重要性，其中渔猎对象的展示则体现了哈萨克人多层面的文化生活方式。

游牧的哈萨克族人传统使用的便携式毡质居所（彩页图 15）被称为 "kyiz ui"（哈萨克斯坦人对 "毡房" 的称呼），堪称文化标志之一，它涉及了社会生活中社会、文化、法律及监管形式等各个层面。哈萨克族人的精神文化通过音乐艺术的对象、家庭及礼仪文化、宗教信仰和习俗特性而得以表现。

民族志馆配备了一种创新技术，可以显示哈萨克人的物质文化和精神文化。在位于展厅中央的 kyiz ui 顶部的天花板通过投影产生天空效果，展厅的环形墙面则显示出草原和传统村庄的图像视频。这些视听技术可为参观者营造出自然景观环境下传统哈萨克斯坦乡村的真实视觉效果，让您体验到哈萨克民族对数百年文化的归属感。

在展厅里，一台壁挂式投影仪播放纪录片的照片和视频，展示出哈萨克族人民的文化和传统。参观者也可以通过投影到玻璃上的视频获得有关传统毡房设计、建造顺序及安装的完整信息。展厅内的一个特别之处是设有一张全息图，可向观众展示由哈萨克斯坦大师创作的独特艺术作品。

4. 黄金大厅

黄金大厅收藏了哈萨克斯坦境内发现的各种金器和独特的高价值手工艺品文物。展厅每年都会得到"文化遗产"国家计划中发现的、由黄金和贵重金属制成的独特物品来充实藏品。

Scythian-Saki 文化的黄金首饰在全世界享有盛誉。参观者会震惊于萨基首领皇家坟墓中的丰厚墓葬。各地的研究者都认可 Scythian-Saki 文化的独创性，其标志是"动物风格"。黄金展厅旨在向国际社会促进和展示哈萨克斯坦的黄金物品杰作，它们在制作过程中均采用了 Scythian-Saki 的动物风格。

展出的黄金物品涵盖青铜时代到金帐汗国时代。主要部分包括萨基考古发现的各种金器。

黄金展厅可分为两部分。萨基时代早期游牧民族的最显著文化表象就是他们对周围世界的审美方式，即"斯基泰三合一青铜器"（兵器、马具、动物纹饰艺术品），这三个要素构成了一种标准模式。在一些著名的萨基墓地发现的物品中都深刻地彰显了这些文化要素。

展厅中有一个特殊区域专门用于摆放墓葬结构中真实人类墓葬的复原模型。国王的手推车和著名的伊塞克人葬礼就是典型的例子。

遵循历史主义的原则以及时间顺序，黄金大厅展出了从东哈萨克斯坦出土的金耳环，从卡拉干达州（公元前 617 年被称为 Taldi-2）出土的 200 多件"金人"服饰，其中包括 1800 多件鳞片和鱼形的黄金饰品，以及两万多枚小型金饰。

以凯末尔·阿基舍夫为首的考古学家团队将伊塞克·巴洛发掘期间发现的"金人"作为展厅中的主要展品，是萨奇艺术"动物风格"引人注目的样本（彩页图16）。它可以追溯到公元前 4—5 世纪，包括 4000 多种采用锻造、冲压、雕刻、粒化等工艺制成的黄金制品。

科学家将古代游牧民族采用金属处理制作黄金饰品的工艺等看作世界杰作。如今，伊塞克的"金人"已经在全世界闻名遐迩，形成了轰动效应，并成了独立的哈萨克斯坦人的象征。

在哈萨克斯坦西部的 Taksai-1 丘陵地区发现了最新的"金人"——一位萨尔马提亚妇女，这是考古学家惊人的发现之一，体现了古代斯基泰人的财富和权力。

5. 结论

哈萨克斯坦国家博物馆一直致力于开展各种各样的游览项目，包括采用互动形式的、主题的、哲学的、特殊的项目，以及游戏游览项目。

国家博物馆的目标是成为一个现代化的知识文化机构，旨在对有关哈萨克斯坦历史文化遗产的陈述进行分析、比较、反思、讨论和评估。当代博物馆总是致力于与参观者进行开放式的对话。博物馆已尽其所能让参观者积极地参与到和历史的对话之中。

参考文献

［1］Minbay D. National Museum of the Republic of Kazakhstan：first steps［R］. Astana：Foliant. 2015.

［2］Minbay D. Annual report on the results of the National Museum of the Republic of Kazakhstan［R］. Astana，2016.

［3］Alipova S. Actual problems of museum activity.［J］. Cultural Heritage，2017，2（71）：78–80.

非政府组织在天文学科普方面的作用

穆罕默德·哈迪·塔巴塔巴伊·雅兹迪

萨拉姆·阿巴斯　伊曼·阿赫迪·阿克拉吉 ①

摘　要　古代伊朗科学家曾对科学，特别是天文学做出过重大贡献。当代的伊朗科学家继承了这一传统，并致力于向公众传播科学。21 世纪初，由一批科学家、热心科学传播的社会人士和私营企业组成的非政府组织积极筹建马什哈德天文馆、科学中心等科普场馆。本文通过这一案例，讨论了非政府组织在科普场馆建设和科学传播中的作用、问题和可行性，并介绍了马什哈德天文学会为此采取的相应对策。

关键词　科普场馆建设　非政府组织　作用　问题　可行性

1. 伊朗古代的天文学成就

在许多古代文明中，对天空的观察一直激励着人们思考未知的事物。天文学作为最古老的自然科学之一，甚至是其他科学之母。天文学教会了人们通过运用观察能力，将简单的思想转化为辉煌的发现。

古代伊朗文明诞生了天文学史上许多重要的发现。追溯至帕提亚时期（公元前 100 年前后），有证据表明巴比伦数学与天文学研究活动主要在伊朗东部进行，接下来几个世纪的研究活动证据也被不断发现。这些古代天文学成就在 9 世纪中叶之前一直发挥着重要作用。在那个时代最重要、贡献最卓著的天文学家包括：艾哈迈德·本·穆罕默德·本·卡希尔·法尔加尼和穆罕默德·本·穆萨·赫瓦兹米。在接下来的 10—13 世纪里，《天文学大成》在中东地区被广泛传授。虽然本书中涉及的领域已有许多进展，但基本认识并未受到挑战。阿卜杜勒·拉赫曼·苏菲被称为最早识别大麦哲伦星云的人，他于公元 964 年（10 世纪）最早记录了仙女座星系；阿尔·毕鲁尼则估算出地球

---

①　穆罕默德·哈迪·塔巴塔巴伊·雅兹迪：伊朗恩格雷物理有限公司董事长。E-mail：haditaba@yahoo.com。
　　萨拉姆·阿巴斯：马什哈德菲尔多西大学物理系教授，内沙布尔天文馆科学顾问。E-mail：abbassi@um.ac.ir。
　　伊曼·阿赫迪·阿克拉吉：赛义德技术大学电气工程系讲师。E-mail：laakhlaghi@sadjad.ac.ir。

半径，并描述了地球的运动。这个时代另一位著名的诗人、数学家及天文学家当数欧玛尔·海亚姆，他的代数著作享誉世界，他也是当时天文台的负责人。正是在那个时代，最准确的贾拉里历法问世并得到改革。贾拉里历正是伊朗官方历法的基础。

21世纪，针对青少年的出版物日益增多，在公众中传播科学思想，激发他们对科学之美的认识，影响着人们的生活，也帮助他们更好地理解和尊重宇宙。

2. 伊朗当代天文科普

在欧玛尔·海亚姆的诞生地内沙布尔市，21世纪的第一年（继海亚姆之后1000多年），海亚姆文化和科学学会（非政府组织）动工兴建一项非常特殊的工程，它名为"海亚姆天文馆和科学中心"。其中有一座装备了蔡司投影设备及一台光电机械式投影仪的20米高4千分辨率的球幕影院，拥有超过300个座位，整个博物馆面积约5000平方米。对于伊朗的科学界而言，这个梦想正在成为现实。大家都相信这项工程将在自然科学教育中掀起一场革命（彩页图17）。

内沙布尔市人口约20万，临近伊朗第二大城市马什哈德。作为宗教圣地的马什哈德每年都能吸引数百万游客，这为在马什哈德市及其周边地区（内沙布尔）开展科普活动，以及建设天文馆和科学中心等机构，提供了绝佳的条件。

但遗憾的是，这项计划的进展不如预期的那样顺利。原定于2005年开放的海亚姆大型天文馆，因为伊朗受到经济制裁的影响，很多设备无法按期到位，因此该项目长期以来陷入了停滞状态。

3. 马什哈德天文学会的建立与工作

科学工作仍在继续，并且在各大学的引领下持续蓬勃发展。2003年，穆罕默德·塔吉·伊达拉蒂教授及其学生发起成立了马什哈德天文学会（ASM），其目的是在马什哈德建立一个致力于促进科学普及的非政府组织。伊达拉蒂教授也是海亚姆天文馆的董事会成员，他希望对此工程进行适当的调整，以展示他的试点项目。他还计划在海亚姆大型天文馆开幕之前聘请研究生对珍贵的资源及展示内容进行审校，并请他们为天文馆的运行做好准备。伊达拉蒂教授于2005年去世后，他的追随者继续他所创办的非政府组织的工作，致力于天文科学的发展。以全新面貌重生的马什哈德天文学会每月举办许多俱乐部和其他活动，每年的参与人数达数千人（彩页图18）。正是由于伊达拉蒂教授及其继任者们组织开展的各项活动，使得马什哈德天文学会被评为活跃的业余天文学会。2007 年国际天文日，马什哈德天文学会获得了"伊朗最活跃的业余天文学会"这项荣誉。

这些科普活动很快就吸引了很多人的参与，同时引起了投资者和政府教育部门的关注。马什哈德天文学会组织各项活动的目的之一便是鼓励政府和有政府背景的组织与他们合作。一些私人企业在马什哈德天文学会及其合伙组织（如物理教师协会）的促进下，开始投资可能带来效益的计划，如科技博物馆或科学中心。马什哈德市内已经开设了多家文化中心，包括马什哈德天文馆和库桑吉公园科学中心等。

4. 伊朗科学普及面临的困难与问题

最早源自天文学领域的创举被逐渐推广到物理和医学等其他学科，学生们开始运用强大的多媒体来展示各种复杂的概念。他们能够让具有不同知识和背景的观众相信：科学并不是普通民众很难理解的东西，科学也可以成为一种休闲娱乐的方式。在 2007 年春季的两周时间里，马什哈德天文学会组织的互动表演活动在市内最大的公园举行，参与活动的总人数超过了 10 万。

由于马什哈德天文学会的推动，许多类似的项目得到了政府、非政府组织和私营企业的大力支持。这些项目包括：巴兹玛教授科学中心、中心公园永久性科学博览中心、拉扎维博物馆、柯拉桑·拉扎维大博物馆等。

此类项目的大幅增加伴随着出乎预料的弊端，引发了科学界的极大担忧。与许多其他国家不同，伊朗第一代科学中心的部分工程开工建设之后，便由建筑单位单独负责，并未得到大学教授和专家们不可或缺的专业指导和监督。未经授权和缺乏准确性的口头或书面材料可能会误导观众，很可能使他们获得错误的知识和信息。

为了避免偏离建设项目的初衷，马什哈德天文学会于 2015 年决定与一家私营科技企业（恩格雷物理公司）合作，在马什哈德的大学校园内合办一座中型天文馆。其主导思想是：由政府相关部门和来自伊朗一流高校的专家代表投资人和公众组成指导小组，直接对项目建设进行规范和监督。这个目标近期即将实现。所有的准备工作，包括复杂的展示说明以及场馆内高科技仪器设备的布置和安装基本就绪，整个场馆将在不久的将来向公众开放（彩页图 19）。不容忽视的问题是，由于存在许多偏见，这项工程无法从政府方面获得足够的财政支持，而获得的私人捐助也很有限。更多的人宁愿去直接帮助穷人、建造寺院或是在边远村庄中建设学校。

该项目欢迎来自个人和机构各种形式的捐助。我们希望能够通过这种方式，回报各方为科学发展、为青少年、为这个国家和全世界人类的下一代做出的贡献。

尼古拉·特斯拉的遗产

布拉尼米尔·约万诺维奇 ①

摘　要　位于塞尔维亚贝尔格莱德的尼古拉·特斯拉博物馆保存了特斯拉一生使用过的大量个人物品。博物馆的无上至宝包括特斯拉的文件、手稿、科学笔记、计算、图表、图纸和信件等个人档案，共计 163911 件。2003 年，为表彰尼古拉·特斯拉及其发明的普世意义，联合国教科文组织将特斯拉的档案作为可移动人类文献遗产的一部分，加入了"世界记忆遗产名录"这一文化遗产保护的最高形式。博物馆全部馆藏已经实现了数字化并制作成缩影胶片，此举可确保将相关资料永久地流传给后代。

关键词　博物馆　尼古拉·特斯拉　文化遗产　可持续发展文化

1. 尼古拉·特斯拉博物馆的历史

尼古拉·特斯拉是塞尔维亚裔的美国发明家和工程师，他以设计现代交流电供电系统的贡献闻名于世。1899 年发现地球共振频率之后，特斯拉看到了渗透于人类生活各个领域的一次革命，而这次革命在全球范围内都极具开创性。然而在美国金融家拒绝进一步投资他的那些似乎过于雄心勃勃和无法实现的项目之后，特斯拉开始批判性地思考当时世界发展的趋势。他早在 20 世纪 20—30 年代时就谈到需要保护能源资源，寻找新的可再生能源和替代能源，以更少的消费谋求更为温和的发展，渴望追求一种符合人类普遍价值观和具备真正教育目的的文化。特斯拉作为一位全球主义早期思想家的价值，并不为他所生活年代的广大公众所知，甚至也未得到专家们的发现和认同。

1943 年 1 月，尼古拉·特斯拉在纽约去世后，美国法庭将其财产的保管权判给了萨瓦·科萨诺维奇，他是特斯拉最小妹妹马里卡的儿子。萨瓦·科萨诺维奇是塞尔维亚政治家，当时是皇家南斯拉夫流亡政府成员旅居纽约的外交官兼公关人员。在尼古拉·特斯拉去世后，他的所有财产被打包封存并移交给美国外侨资产管理局。他的物

---

①　布拉尼米尔·约万诺维奇：塞尔维亚尼古拉·特斯拉博物馆馆长，博士。E-mail：branimir.jovanovic@tesla-museum.org。

品被从纽约人酒店转移到曼哈顿仓储公司，那里已经存放了一些特斯拉的财产。在萨瓦·科萨诺维奇的倡议下，尼古拉·特斯拉的所有个人财产和著作都被运至贝尔格莱德，萨瓦·科萨诺维奇随后将这些财产和著作交给了贝尔格莱德。

1951 年 9 月，尼古拉·特斯拉的遗产被装在 60 个包、手提箱、金属箱和桶中运送至塞尔维亚的里耶卡港。然后通过火车将这些资料运至贝尔格莱德，存放于贝尔格莱德大学电气工程学院。1952 年 6 月，这些资料被从学院转移到博物馆的所在地。

1953 年的法规对尼古拉·特斯拉博物馆的责任和目标做了如下规定：保存尼古拉·特斯拉的科学及个人遗产，继续收集和保存与尼古拉·特斯拉生活及工作相关的文件和个人物品；通过常设展览展出馆藏资料，组织并促进对资料开展研究，出版尼古拉·特斯拉的作品和著述，作为版权所有者授权出版、转载和翻译这些作品；与当地或国际科学教育机构及个人进行合作；鼓励并支持对尼古拉·特斯拉曾经从事的技术科学展开科学探索和研究，并发表相关研究报告。

1953—1955 年，博物馆整理了档案资料并设计制作了常设展览。1955 年 10 月 20 日，位于贝尔格莱德的尼古拉·特斯拉博物馆向公众正式开放。这是南斯拉夫首家技术博物馆。开幕式上的常设展览让观众看到了按特斯拉图纸精确建造的各种模型，其中最为著名的是旋转磁场。博物馆向贝尔格莱德公众展示了在 1893 年芝加哥世博会期间让观众无比惊叹的哥伦布蛋。博物馆还展出了特斯拉的第一台感应电动机、一个水力发电厂模型、一个遥控模型船、多相输电系统、各种发电机和变压器等，目前最受观众欢迎的是带天线的特斯拉线圈，它构成了荧光灯的基础。来自世界各地科学文化界的知名人士参加了开幕式，嘉宾包括当时的美国驻贝尔格莱德大使詹姆士·瑞德博格。

博物馆的下一个重要仪式在 1957 年下半年举行，特斯拉姐姐安吉丽娜的女儿米莉卡·特罗博维奇及继承人萨瓦·科萨诺维奇护送特斯拉的骨灰到博物馆永久保存。当时南斯拉夫驻美国大使馆将载有骨灰的骨灰盒移交给夏洛特·穆扎尔，他于 1957 年 7 月 13 日搭乘特里维拉夫商船抵达里耶卡，四天后抵达贝尔格莱德。在博物馆一楼举行了骨灰盒移交仪式，特斯拉骨灰盒在当时作为常设展览展出。尼古拉·特斯拉博物馆开展的博物馆学工作始于 1957 年。

自 1969 年 10 月 9 日以来，博物馆一直是贝尔格莱德市的财产，根据当时南斯拉夫政府公报中公布的协议，从联邦政府转移而来。

2. 特斯拉的遗产

无论按照什么标准来衡量，尼古拉·特斯拉博物馆在当今是塞尔维亚乃至全世界独一无二的科学和文化机构，它是保存尼古拉·特斯拉原始资料和个人遗产的唯一博物馆。其财产包括以下特别有价值的收藏品：超过 160000 份原始文件；2000 多册书刊；1200 多个历史和技术展品；超过 1500 张原始技术项目、仪器和设备的照片及照片底板；1000 多张设计图和图纸。

2.1　档案

博物馆保留了特斯拉生前使用过的大量个人物品。如今，这些物品被分为博物馆馆藏的若干部分。然而，博物馆最大的珍宝是特斯拉的个人档案，包括文件、手稿、科学笔记、计算结果、图表、图纸和信件等，共计 163911 件。这些长期（1856—1943 年）积累的资料，包括从内容到形式都极其丰富多样的文件。其中有手写铅笔笔记、印制的名片（带或不带评注）、盖戳邮票、记录在最便宜纸张上的账目、带手写内容的支票和印刷品、带副本的打字文本、在羊皮纸上用彩色墨水书写的文件、印有凹版印章的章程、描图纸上的印度墨汁图纸、计划的蓝图副本以及许多其他物品。就这些资料的年代及多样性而言，它们中的大部分都得到了相对完好的保存。整个档案保存于 548 个箱子中（这是资料处理和存储的基本单位），并按照文件主题编入七个单元。

2.2　图书馆

尼古拉·特斯拉的个人收藏还包含许多松散装订的新闻剪报专集和期刊。鉴于它们的特殊性和重要性，特将其视为这部分馆藏中的特殊部门（特斯拉的剪报资料）。

尼古拉·特斯拉的个人图书馆共有 975 种、1172 册图书和 347 种、2435 份报刊。

2.3　剪报资料

作为特斯拉遗产的一部分，有来自美国并且堪称图书馆馆藏中最有价值的部分——剪报资料。尼古拉·特斯拉所拥有的各种书籍和杂志，不但可以让我们评估他在科学领域的兴趣以及他的文学品味，这些剪辑资料也让我们深入了解相关社会背景，以及特斯拉从公众和专业界得到的回应。

从尼古拉·特斯拉个人遗产中收集到的剪报资料和报纸，包括未装订的报纸剪报、书页，标注了剪取的文章、未加注释的书页，以及所有已发行的报纸。其中已装订的剪报被收藏于 57 张个人系列专辑中，估计有两万多张剪报。这些剪报的原始记录已按主题被保存在卡片索引中。特斯拉指示其合伙人和秘书从期刊中剪辑文章，其内容大部分涉及他在欧洲和美国从事的工作和活动。这些文章主要按题目（能量、电化学、远程自动控制、X 射线等）进行分类，并按年月顺序进行排列。每篇文章还依据剪辑资源加以手写注释。对这种材料进行的详细检查已经证实：尼古拉·特斯拉聘请了专门的剪报信息服务人员。

除有关尼古拉·特斯拉的文章外，还可能发现有关在相同科技领域开展研究工作的其他科学家和发明家的研究。对这些文章进行有条不紊的分类，见证了特斯拉对收集和利用信息投入的极大专注和技巧。根据其他收集和交换信息的来源，我们可以清楚地发现：尼古拉·特斯拉创建了今天被称为互联网的模型。也就是说，他有自己的组织化信息网络，在其整个信息来源系统中被加以索引、描述并安放在适当的位置。

2.4 馆藏品

博物馆涉及如下九大馆藏内容：机械工程、电气工程、美术和应用艺术、小技术项目、化学工艺、奖章、纺织和皮革、纪念物品、特斯拉的个人物品。

其中，有超过 1200 件物品得到正式注册。在这些藏品中，保留了机械和电气工程领域的各种原创技术产品，以及尼古拉·特斯拉的服装、奖牌、精美实用的艺术品等。

3. 结论

对于科学史学家来说，尼古拉·特斯拉博物馆的收藏品是 19 世纪末和 20 世纪初电子工程早期发展的重要信息来源。特斯拉的遗产除了有关特斯拉生活和工作的独特信息，还反映了特斯拉科研生涯所处的社会和科学环境。特斯拉如今在广大公众尤其是年轻人中越来越被认可，这本身即是重要的教育资源。尼古拉·特斯拉博物馆的使命就是助力研究和教育。

以车为媒，促进国与国多元文化交流
——以中法两国文化交流为例

杨　蕊 ①　　窦立敏　刘玉洁　熊　伟

摘　要　北京汽车博物馆于 2011 年建成开放，是中国国家公益性汽车主题博物馆，集博物馆、展览馆、科技馆功能于一体，建筑造型似一只明亮的"眼睛"，寓意放眼世界，面向未来。按照"科学－技术－社会"选题方式，设立三馆一区，打破国家与品牌的界限，展现世界汽车百年发展的历史，以及中国汽车工业的起步、发展与壮大，揭示汽车工业对人类文明和社会产生的巨大影响。探索文化、科技、教育、旅游等方面的融合发展，担负起中国研究汽车文化和科普教育的责任并成为引领者，填补中国汽车文化研究领域空白，意义尤为重要，也责无旁贷。以车为载体，从博物馆承载的多元文化融合角度，讲述中国故事，分享中国博物馆的思想理念与管理方法，在世界舞台上传播中国声音。在"一带一路"美好倡议下，以中法文化交流为示范，促进国与国、城与城、人与人的对话和交流。

关键词　北京汽车博物馆　多元文化　国际交流

　　车，跨越千载，不仅可以探究古代车马文化，还可以超越国度，了解世界各国不同文化背景下的汽车文化，更可以穿越时间，畅想未来汽车将如何改变生活。北京汽车博物馆，以车为媒，促进国与国之间多元文化交流，以中法文化交流为案例来论述。以中国古代车马文化，作为历史上华夏贡献和所承载的中国传统文化精神，通过汽车重走丝绸之路的案例，讲述中法两国的创造与友谊，通过中法博物馆的对话与交流，讲述中法两国汽车文化的融合，讲述以博物馆承载的多元文化活动，以及未来在"一带一路"美好倡议下，北京汽车博物馆如何促进国与国、城与城、人与人的对话和交流。

---

　　①　杨蕊：北京汽车博物馆馆长，邮箱：56185826@qq.com。。

1."人－车－社会"综合命题下的北京汽车博物馆

1.1 "文化、科技、教育、旅游"等方面的融合发展

北京汽车博物馆（彩页图 20）填补了国家专题博物馆的门类空白，是集"博物馆、科技馆、展览馆"三位一体汽车类专题博物馆，位于丰台区，建筑面积为 5 万平方米。于 2011 年建成开放，连续实现全年满载开放，累计服务社会人群 500 万人次。践行"文化、科技、教育、旅游"等方面的融合发展，先后获得国家 AAAA 级旅游景区、中国汽车文化推广基地、全国博物馆行业首家国家级服务业标准化示范单位、连续 4 年获得优秀全国科普教育基地、爱国主义教育基地、北京市先进基层党组织等荣誉。以"开门办馆，融入社会"为宗旨，倡导"人－车－社会"和谐发展。

1.2 以"科学－技术－社会"选题方式站在中国看世界

展览按照"科学－技术－社会"选题方式，设立三馆一区，打破国家与品牌的界限，展现世界汽车百年发展的历史，以及中国汽车工业的起步、发展与壮大，揭示汽车工业对人类文明和社会产生的巨大影响。展览展示打破传统陈列方式，运用新媒体、机电一体等高科技手段，引导观众通过亲身体验认识汽车文化，学习汽车知识，采用声、光、图像以及互动机械等多元化的科技表现手段和展览展示方法来满足观众的需要，实现"科技馆中的博物馆，博物馆中的科技馆"，将历史价值、科技价值、文化价值转换为社会价值。"展览展示系统技术开发及应用"项目荣获代表我国汽车工业发展的最高水平的中国汽车工业科学技术三等奖。以基本展陈为源头，结合重要时事节点和社会热点策划临时展览，推广中国自主品牌和新技术、推广中国传统文化和核心价值观、发挥以"车"为媒的国际文化交流作用，开展中法、中美、中俄、中德等文化交流。（彩页图 21 ）

1.3 馆藏填补领域空白

藏品有万余件（套），包括中国、德国、美国等国家藏品车（彩页图 22）。拓展到车辆内外的构成、文献、模型等 6 大类 21 小类（彩页图 23 ～图 27），囊括汽车发展历程中经典车型和汽车社会进程中有重要影响的品牌和车辆。成立国内第一个以汽车为藏品的评审定级机构，提出"修旧如故"修复理念，完成 1976 年上海 SH760A 为代表的承载式车身、以 1968 年红旗 CA773 为代表的非承载式车身和以 1987 年北京 BJ212L 为代表的越野车型的修复，实现让"文物活起来"。制定国内首个《车辆类藏品修复工艺、技术及验收标准》，填补国内车辆类藏品修复领域空白。

1.4 多元汽车文化置于社会广泛的联系中

博物馆的文化交流和教育活动长年不断，发挥行业优势，通过与社会的融合，充分地置于社会的广泛联系之中。2017年继续创新理念，策划专题展览、与教育机构共同开发科普项目、与媒体共同策划文化活动、与政府共同推进公共服务事项。传播"讲好中国故事"，在里昂参加第二届中法文化论坛"一带一路：中法文化汇流"，策划组织"博物馆在地区吸引力中的角色"分论坛，并以此为契机，推进丰台区和法国里昂签订友好城市备忘录；传播社会正能量，策划"雷锋——一个汽车兵的故事"，"雷锋宣讲志愿服务"项目荣获第三届中国青年志愿服务项目大赛金奖；倡导"人–车–生活–社会"和谐与美好，建设北京市新能源汽车展示体验基地，推出"绿色北京守护蓝天"系列活动，策划"车@城@人"系列专题展，以"车"的视角看历史、科技、人文、城市变迁、社会进步，让"人"爱上"这座城"；聆听博物馆的声音，聆听博物馆之美，在博物馆之夜，用声音唤起共鸣，唤起祝福。以朗诵会的方式邀请陈铎等老一代艺术家带来一场声音的盛宴。在声波中去感受历史的声声回荡，去感召未来的朗朗之声。用跨界对话的方式将博物馆的声音传播得更远。

1.5 突破传统教育理念，强调"动手"和"体验"

汽博馆50多个可视、可听、可触摸、可参与的互动展项实现了形式与内容、技术与艺术的有机统一和结合，鼓励观众参与，"让科学很好玩"。北京汽车博物馆规划一套完整的汽博馆的教育体系，以分层次培养未来汽车社会人才。规划时考虑八个层面内容：重点和目标、可用资源、主要观众、提供的内容、时间范围、营销推广、活动评估、培训需求。注重展教结合，进行多维度的展品"二次开发"，以学习目标为导向，结合实现认知与观众认知，与日常生活和社会关注点结合。面向中、小学教育层面服务素质教育，立足学校教育，开展课程化开发与博物馆资源利用的研究和实践，探索建立涵盖"主题参观""博物馆教学""校本课程开发""创新人才培养"四个层级的博物馆–学校教育合作模式。面向汽车专业教育层面服务产业发展，在普及性科普活动的基础上，搭建创新与实践平台，培养理论加实干的综合性汽车人才，为中国汽车人才的长期培养和建设，提供服务与支持。面向大众教育层面服务汽车社会，中国已步入汽车社会，与此同时，汽车文明还有待提高，汽车与社会与人的矛盾越发明显，交通拥堵、环境污染、危险驾驶日益严重。作为汽车专题博物馆，责无旁贷地要参与到汽车文明建设这个时代大课题中。进行爱国主义教育层面传承中国精神，汽车不但是集科技大成者，也是集人文精神于一体，蕴含着凝聚人心、凝聚民族的情节在其中。2015 年举国上下纪念抗战胜利 70 周年之际,我们以军车为视角,策划"金戈铁马话军车"专题展,以抗战精神为源,传播爱国主义为本,通过自主品牌军车的发展体现我国的综合实力,提升民族自豪感。

2. 以车为媒,促进国与国多元文化交流

2.1 车,从古至今是人类移动梦想,汽车博物馆承载着历史与传承

中国是世界上最早发明和使用轮子的国家之一。相传早在 5000 年前黄帝就制造了车(彩页图 28),所以黄帝又称"轩辕黄帝",其中"轩"是古代一种有帷幕而前顶较高的车,"辕"是车的纵向构件,指车前驾牲畜的两根直木。在黄帝大战蚩尤时,黄帝的军队迷失了方向,他们利用黄帝制造的指南车来分辨方向,正由于其导航功能发挥的重要作用,才赢得了胜利。后来出现了指南车的姊妹车——记里鼓车,车上有两个木人,行一里路击一次鼓,用来计算道路里程。指南车与记里鼓车是古代皇帝出行时的仪仗车,也被古人用于绘制地图、丈量土地。

2500 多年前,中国古代伟大的思想家、教育家孔子曾乘车周游列国,传播文化思想(彩页图 29)。孔子曾说:"道千乘之国,敬事而信,节用而爱人,使民以时。"也就是说治理一个拥有一千辆兵车的国家,就要严谨认真地办理国家大事且恪守信用,诚实无欺,节约财政开支而又爱护官吏,役使百姓而不误农时。这讲述的是孔子对中等诸侯国的治国理政思想,时至今日仍然有着重要的借鉴意义。可见古代车马在日常生活、社会经济、军事战争中都起到了至关重要的作用。

2200 多年前,秦始皇统一中国后,实行了"车同轨"制度,车辆制造进入了标准化阶段(彩页图 30)。该制度规定车轮距离相同、车轮大小相同、道路宽度相同,车行和人行道路分开,城市间路路可以相通,形成四通八达的交通网络。这项措施极大地便利了往来交通,促进了经济的发展。"车同轨、书同文、行同伦"开创了中国有组织、有秩序的规模化生产的先河,也是古代中国标准化的生动实践。总之,车作为华夏贡献的重要见证,一路记载着人类的移动文明。

2.2 车,从东方到西方架起一座桥梁,汽车博物馆承载创造与友谊

北京汽车博物馆不但记录着中华文明的伟大创造,也记录着世界文明的伟大贡献。在创造馆展览流线上重点展示着人类历史上第一辆机动车模型,是 1769 年由法国炮兵工程师尼古拉斯·古诺发明制造的。古诺蒸汽车向观众讲述人类是从何时开启无马之车的新时代。每当法国人到访中国看到中国博物馆里的本国发明都非常的兴奋,感慨

中国能如此尊重他国的发明和创造并在博物馆传播，这让他们非常惊喜和自豪。

汽博馆还记录并展示了这样一段故事，早在 1907 年，世界上首次跨越欧亚大陆的拉力赛，就是从北京到巴黎，这是中法之间创造的欧亚传奇。在我馆用微缩模型重现了赛程路线，这段历史已感动了上百万观众，据统计，近 90% 的观众并不了解这次伟大的赛事。2007 年，正是北京－巴黎汽车拉力赛 100 年，为了纪念人类历史上首次用汽车把东方与西方连接起来的中法汽车拉力赛，以及宣传抗击阿尔茨海默病（又称老年性痴呆），来自法国公路与人协会的伊丽莎白·佩特和法比安·哈姆驾驶 1925 年生产的雪铁龙 5HP，沿着 1907 年探险先驱者的足迹，重走当年路，历时一个半月，途经九个国家，完成了长达 16000 千米的艰辛路程。伊丽莎白·佩特因其奶奶患有阿尔茨海默病，在行车途中还记录了此病 100 年来各个国家类似患者的发病历史及治疗情况，在人与人、人与车、人与社会的互助下完成此次壮举。最终这辆车由法国标致——雪铁龙集团回购并捐赠给北京汽车博物馆作为永久收藏，这也是北京汽车博物馆接受捐赠的第一辆国外藏品车，现陈列在五层创造馆展出，它也成了中法两国人民友谊的见证。

2.3　车，从中国到法国打开一扇窗，汽车博物馆承载两国文明与追求

以车为媒，促进国际文化交流，汽博馆架起中国与世界进行汽车文化交流的桥梁。在这一使命的感召下，中法汽车文化交流打开了北京与法国巴黎的不解之缘。2014 年是中法建交 50 周年，在这个值得中法两国人民纪念的日子，汽车文化交流又开创了区域间的合作。中国人民迎来了法国米卢斯国家汽车博物馆，并与法国米卢斯国家汽车博物馆策划了系列中法汽车文化交流活动，引入"1891—1968 年法国车身造型：艺术、技术和专业成果图片展"，促进两馆进行古董车修复技艺的学术交流。这个欧洲最负盛名的汽车博物馆，让我们看到法国人民对艺术的无限热爱，对技术的完美追求，对文明成果保护的责任和使命。

两馆围绕"艺术、技术和专业成果"开启了一系列文化交流活动，同年，策划"从 1949 走来，中国汽车红旗的故事"专题展览，将中国汽车文化首次送出国门，送到汽车的发源地——欧洲。中国国车红旗入藏法国米卢斯国家汽车博物馆，这是该馆首次收藏亚洲藏品车，也是中国汽车文化第一次走进欧洲，走进汽车工业的发源地。通过讲述中国国车"红旗"的发展历程，寻迹红旗车上的中国文化元素，让观众获知中国汽车艰苦创业的故事。

2016 年，在中法建交纪念日期间策划了"北京－巴黎不解之缘"中法汽车文化专题展，回顾了中法两国在汽车文化方面的交流，讲述了车与车之间，人与人之间，城

与城之间，国与国之间的不解之缘。前驻法大使吴建民表示，"交流就是要建立在共同的语径之下才能打动人，汽车成功塑造了对话的载体。"这是打开问题思索的一把钥匙，更能佐证汽博馆从车与车促进人与人、城与城以至国与国的交流与沟通的可能性。

2.4 车，将你和我联系在一起，汽车博物馆承载两国文化与融合

博物馆被誉为"一本有力量打开人类精神世界的书"，博物馆以史为鉴，更重要的是启迪心智。北京汽车博物馆于 2011 年建成开放，连续 5 年实现全年满载开放，累计服务社会人群 500 万人次，累计开展各类文化和科教活动 500 多场。一个博物馆的价值，不在于它有多宏伟、有多现代、有多少藏品，更重要的在于它的思想、它的视野、它的价值观，以及给观众的启迪和所传递的精神和文化。

汽车文化的多样性，使得博物馆呈载的文化形态更加多元。文化交流是中法合作的基石，以文化交流促民心相连，春风化雨，润物无声。以车为媒还可以实现以花为媒，2017 年 4 月，在法国铃兰花节前，北京汽车博物馆做了一场《艺术与香氛》的主题演讲，讲述法国娇兰香水的文化与历史。这期间，正是中国传统的花朝节，通过以车为媒，促成以花为媒，在北京丰台花乡，这个已有 800 年历史"花卉之乡"美誉的地方，也正是北京市汽车博物馆的馆址所在地，将区域文化和博物馆文化交融在一起。现场嘉宾同花艺大师一起演绎东方插花艺术，听娇兰总裁博佑先生讲述西方香氛艺术与生活，品中国唐代煎茶的清香与芬芳，尝中华老字号——宫颐府的花糕，体验"花朝节"传统民俗。用这样的方式东西相遇，中法对话，感受中法花文化碰撞的魅力。博物馆需要多元文化，"国相近、民相亲"的文化更可以增进国与国、城与城、人与人间的理解和友谊。

中法之间的文化交流建立在互信的基础上，如今中法两国通过分享同一个梦想和追求，去共同创造质量无与伦比、又充满诗意般想象力的汽车作品与产品，而这种充满诗意的想象力正是中法两国共有的民族特性。作为中法友谊的见证者和中法文化的传承者，推动两国文化活动的合作与交流，汽博馆以车为媒，积极促进中法汽车文化的交流，实现"城与城"的对话，促成北京丰台与斯特拉斯堡、里昂进行对话，2017 年在中法第二届文化论坛期间，丰台和里昂签订友好城市备忘录，为两个区域间的友好交流搭建起沟通的桥梁，促进"城与城"在文化、教育、旅游、青少年等领域展开交流与合作（彩页图 31）。

2.5 一路同行中法梦、世界梦，多元文化促进国与国、城与城、人与人的互动交流

艺术 8 创始人、法国艺术哲学家、法国巴黎罗丹美术馆理事佳玥女士，长期致力

于中法文化艺术交流。艺术8坐落于北京原中法大学，佳玥女士称这座建筑"它是一个关于梦想的故事，一个独特的地方让梦想最终变成现实。从孩提时起，我就梦想有一座房子，它热情开放，让身处其中的艺术家和形形色色的人们有着宾至如归的感受。这是一座因创作创新和传承交流而存在的建筑。"她曾经说过这样一句话，"如果要在中法两国间搭建桥梁，那么我们自身就要行走在这座桥梁上，我在中国所做的一切完全符合我的预期。我已经实现了我的中国梦，艺术8就是我的中国梦。"在中国"一带一路"的美好倡议下，共筑人类命运共同体，我们每个人的梦想，就是中国梦与世界梦的最好呈现，期待着一路同行中法梦，一路同行世界梦！我们期待让车载着人，人载着思想，共赴"人－车－社会"的美好与远方（彩页图32）！

人有国界，车无国界。这为汽车文化开展国际间交流、互通有无提供了广阔空间。除了在中法两国间开展文化交流，2016年国际博物馆日和中国旅游日期间，借助中美旅游年的契机，策划了"车轮上的生活——中美汽车文化摄影展"，通过5位国际摄影艺术家的150幅珍藏摄影作品，观察美国这一"车轮上的国家"，回顾汽车文化在美国各时期的发展轨迹。也通过中国"车轮上的生活"，诠释全球最大的汽车市场的汽车文化。此外，探索汽车博物馆之夜项目，利用闭馆时间和广场空间为观众提供服务，整合社会资源，将城市科学节引入北京汽车博物馆，邀请来自美国、英国科学家进行科学秀表演，场场爆满。汽车与科学密不可分，众多科学原理都应用在了汽车的发明创造上，科学更是一种国际交流语言，这为汽车博物馆以车为媒，开展国际间的文化交流开阔了思路和视野，从不同层面促进国际间的文化交流。以车为媒，还可以策划中德、中意、中英、中韩，甚至更多建交年的汽车文化交流，让中国与世界可以用汽车文化作为交流语言开展对话。

3. 科普互惠共享资源在解决"人民日益增长的美好生活需要和不平衡不充分的发展之间的矛盾"时应勇于实践

为进一步促进北京汽车博物馆展示教育、收藏研究、休闲旅游、国际交流等能力的全面提升，拓展文化传播和科学普及的覆盖面和影响力，编制了《北京汽车博物馆"十三五"发展规划》，以全面落实丰富人民群众精神文化生活的根本任务，构建"人－车－社会"文化体系，推进当代博物馆教育体系建设，提升博物馆运行管理水平，为观众提供优质服务，为可持续发展提供文化新势力，办好人民群众满意的文化事业。以"开门办馆、融入社会"为宗旨确定了发展理念和目标，致力于成为行业服务标准化的示范平台，成为彰显城市品质的公共服务平台，成为科教、文化、旅游的融合展示平台，成为传播汽车科学文明的科普教育平台，成为世界汽车文化的交流传播平台。以担负起展示汽车文化的责任，代表中国与世界交流为使命，成为汽车文化和科普的引领者，促进"人 – 车 – 生活 – 社会"和谐与发展。将"传承、创新、合作、卓越、责任"发展理念，致力于成为服务一流、效率一流、展示一流的世界先进水平的中国汽车博物馆。结合博物馆的工作，基于为解决"人民日益增长的美好生活需要和不平衡不充分的发展之间的矛盾"能做哪些实践进行思考，策略方向可以有三方面的尝试：一是基于展陈、展品的载体运用的"多元化开发"，选择共享资源的载体，充分"互联网 +"的思维，运用新媒体手段，突破区域的界限。二是基于内容进行"二次开发"，广泛置于社会需求，研究学校课标定位，研究社会核心价值定位，突破单学科的界限。三是基于人才培养进行"横纵开发"，讲解员向研究员培养，科普员向科普辅导员培养，突破单业务条线的界限。

4. 结语

总之，通过与社会的融合，通过国与国的联通，通过科技化的手段，讲故事的方式，传播科学技术和人文思想，传播世界进步和中国文明，讲述车性化、人性化故事。"博物馆作为保护和传承人类文明的重要殿堂，是连接过去、现在、未来的桥梁，在促进世界多元文明交流互鉴方面发挥着特殊的作用。未来北京汽车博物馆将以车为媒，以中法汽车文化交流为示范，进一步推动国际间汽车文化交流，传承丝路精神，并赋予其新的时代意义，为"一带一路"文化建设贡献力量。

参考文献

[1] 郑奕. 博物馆教育活动研究 [M]. 上海：复旦大学出版社，2016：138.
[2] 葛宇春. 通过"二次开发"加强展品教育功能的有效性 [J]. 自然科学博物馆研究，2017（2）：42–43.

浅谈北京南海子麋鹿苑博物馆科普工作开展方式

白加德 ①

摘 要 北京南海子麋鹿苑博物馆以生态道德教育和环境教育为主线，通过创新科普创意载体、拓展科普传播途径、丰富科普活动等方式开展科普工作，不断地为"人文北京"的建设添砖加瓦，得到了来苑参观者的广泛赞誉。

关键词 科普创意　传播途径　活动方式

北京南海子麋鹿苑博物馆（以下简称麋鹿苑）是北京市首座户外类型的生态博物馆，承担着对生态知识、生态意识、生态文化等方面的科普教育及宣传任务。麋鹿苑以生态文明建设、生态道德教育为立足点，整合苑内科普资源，通过对科普载体的创新、扩宽科普传播途径、完善科普活动方式等途径，形成了一系列独具一格、发人深省的科普教育项目，在科普教育工作中取得了良好效果。

1. 不断研发交互式科普设施，创新科普载体

科普设施建设不但可以起到传播科学知识、播撒科学理念的作用，而且可以通过体验科技成果，促进参观者科学素养的提高。麋鹿苑积极建设国内外独具特色的生态博物馆，充分结合苑内自然生态优势，不断研发交互式科普设施，不断创新创意科普载体，形成了"麋鹿回归纪念园""动物家园""生态文明园""民俗文化园""低碳科普园"五个主题展示区。

麋鹿回归纪念园由科普栈道、科普围墙、麋鹿文化墙、唐诗麋鹿座椅、麋鹿科学发现纪念碑、乾隆狩猎图浮雕、贝福特公爵雕塑、麋鹿角石雕、麋鹿传奇展览、麋鹿大事记科普专栏等科普设施组成。麋鹿回归纪念园中的基础设施不仅为参观者提供了

---

① 白加德：北京南海子麋鹿苑博物馆馆长。邮箱：baijiade234@aliyun.com。

安全舒适的物质保障，更提供了丰富的科普知识。在栈道上常年布置有生物多样性宣传牌，科普围墙上则有有关麋鹿故事的小巧画作，麋鹿文化墙及唐诗麋鹿座椅则有古人记述麋鹿的相关词句。这样，参观者在苑内就可以通过不同载体享受到麋鹿文化的魅力。麋鹿科学发现纪念碑、乾隆狩猎图浮雕、贝福特公爵雕塑、麋鹿角石雕、麋鹿传奇展览等科普设施以时间为主线，全面记录了麋鹿被科学发现、中国本土灭绝、远渡欧洲、回归故里的画面，向世人讲述命运多舛的国宝历险记，更加佐证了"国家兴、麋鹿兴"。

动物家园科普设施区中的科普设施在设计制作时结合了动物的可爱造型凸显了"童趣"。这里的鲨鱼之家、燕子之家以鲨鱼鱼鳍、燕窝等雕塑向广大参观者讲述人类食用鱼翅、燕窝对自然带来的残酷破坏，也对人自身的健康埋下严重的隐患。鸟类迁徙地球仪、蜜蜂之家、壁虎爬墙、野狼钻洞、鸟笼等科普设施不仅让孩子们能够直观地体会到这些动物与人类生活息息相关，更为孩子们提供了玩耍的环境。"三不猴"则向孩子们展示了中国的传统文化："非礼勿视，非礼勿言，非礼勿听"。

麋鹿苑通过建设世界鹿类小展馆、湿地科普观鸟台、历届奥运会吉祥物动物雕塑广场、鹿剪影式中国传统护生诗画、灭绝动物公墓等形成了生态文明园。其中鹿类剪影以中国传统文化——"剪影"为表现手法，树立了麋鹿奔跑、跳跃、进食等多种形态，在剪影上还绘有"乌鸦反哺""羊羔跪乳"等护生诗画，这些诗画讲述了动物的感人故事，也教育人类要保护动物、爱护自然。灭绝动物公墓用多米诺骨牌为载体，记录了那些已经灭绝的动物和濒危野生动物的历史与现状，兼有中西方文化特色，给人留下深刻印象。

麋鹿苑通过建设南囿秋风石、万国欢迎石、观鹿台石刻对联、生肖青铜群雕、文化桥等形成了民俗文化园。生肖青铜群雕是典型的交互式科普设施，十二生肖以青铜铸成，生肖动物的兽首模仿了圆明园大水法的兽首制成。这些动物雕塑形态各异，手中还拿着不同的装饰"法器"。在生肖雕塑的水泥底座上刻有这些动物的生活规律，让人们在参观的同时能够了解中国的民俗文化。低碳科普园是麋鹿苑别具一格的主题展示区，在这里汇集有环保格言石椅教育路径、碳足迹和生态足印小径、低碳生活迷宫、碳计算日晷等科普设施。低碳生活迷宫以柏树为围挡，在其中设立有不同生活方式，如使用一次性筷子，这种生活方式破坏了森林资源，则此路不通，需要绕行其他路径。这种寓教于乐的方式，吸引着孩子不断摸索，不断深化低碳生活的观念。碳计算日晷是精心设计的碳排放量计算器，通过转动石盘，就可以清楚地看到不同的出行方式、生活方式带来的碳排放量，同时也能够算出消耗这些排放出的碳所需的森林资源。

近年来，麋鹿苑科普设施得到广大参观者的欢迎，其中，鸟类迁徙路径地球仪和麋鹿科学发现纪念碑获得国家专利局颁发的外观专利。这些科普设施设计灵巧、主题鲜明、创意十足，为参观者留下了深刻的印象，为麋鹿苑的科普传播工作奠定了扎实的物质基础。

2. 注重科普作品创作，拓展科普传播途径

科普作品创作是科研工作的继续，也是科学知识、科研成果等进行传播、推广的媒介和桥梁。注重科普作品的创作，拓宽传播途径，加大传播力度，是麋鹿苑博物馆服务人文北京的特色之一。麋鹿苑科普工作人员加强科普著作的撰写，加强科普展览的创作水平，借助网络平台进行科普传播，为参观者提供免费科普讲解等方式，不断拓展科普传播方式，打造麋鹿苑科普品牌。麋鹿苑充分调动科普工作人员的积极性，出版发行了《麋鹿与麋鹿苑》《麋鹿苑》《地球伦理》等系列科普读物 20 多部。科普人员撰写的百余篇科普文章通过报刊等媒体进行发表，深受广大读者的喜爱。麋鹿苑领导更是以身作则，从多角度多类型地筹划了多主题的科普系列讲座内容，并赴全国多个省市开展高水准的科普讲座。

麋鹿苑充分发挥网络传播平台的重要作用，对现有科普资源进行整合，在博物馆网站开通了麋鹿苑故事会、麋鹿苑印象、麋鹿苑影像和立体导览图等科普专栏。通过专栏，将描述麋鹿苑优美自然风景的抒情散文、展现麋鹿苑生物多样性的优美图片、讲述麋鹿身世的视频资料等科普资源向广大参观者开放，加强与市民进行科普文化交流，使大家能够足不出户领略麋鹿苑风光。麋鹿苑在每年的科技月、科技周及和平周都有针对性地布置灭绝动物主题展、生物多样性展、麋鹿传奇展览等大中型展览。这些展览展示了人类与自然的关系、生物多样性的重要意义、麋鹿的身世及生物学特征等科普知识。通过展示，促进了广大市民的科普知识增长，强化了保护生物多样性、保护自然的理念。这些展览也获得中国科学技术协会"科普日优秀活动奖"。免费讲解是博物馆工作人员为广大市民奉献的一道科普大餐，这项活动使参观者对苑内科普设施的含义有了更加清晰、更加深入的认识。通过工作人员动情的讲解，结合参观者所见、所闻、所感，使大家深深地感受到人与自然和谐统一的重要意义，体会到环境保护对造福子孙后代的深远影响。

3. 开展丰富多彩的科普活动，完善科普活动方式

麋鹿苑博物馆始终坚持立足服务基层、服务广大群众，不断丰富科普活动、拓

宽科普教育思路，梳理科普活动，经过系统整理，形成独特科普资源，增强对公众和社会的吸引力，推动公众科学素养的提升。麋鹿苑积极与周边社区和中小学校携手合作，开展了"科普进学校、进社区"活动，组织科普工作者赴北京市大兴瀛海一小开展"灭绝动物"主题展览及讲座，为四合庄打工子弟小学的学生组建了环保兴趣小组，与天宫院社区服务中心结对子开展"国际生物多样性日"知识竞赛等系列社区科普活。麋鹿苑积极与媒体合作，发挥媒体在科普宣传中的作用，精心组织活动，不断创新科普活动方式，与中国天气网合办了"保护麋鹿生存湿地"主题科普宣传活动，与新浪网亲子论坛联合开展系列亲子活动，与中央电视台科技频道《科技人生》栏目合作拍摄了《寻鹿记》，与北京电视台《这里是北京》栏目合作拍摄了《南海子麋鹿苑》。以上节目的录制，以群众乐于接收的方式传播科学知识，促使大家积极地接受各种科学信息，而且通过电视媒体提高了麋鹿苑科普品牌的影响力，提高了知名度。

麋鹿苑还以世界地球日、世界环境日为契机，大力开展未成年人生态道德教育活动，与"自然之友"合办"为灭绝动物扫墓、在麋鹿苑观鸟"活动，麋鹿苑创作的科普剧《动物联合国大会》和《麋鹿苑的夏天》参加了2012年"全国科普日北京主场活动"，深受观众好评，麋鹿苑还利用社会大课堂等平台开展大型公益活动，并推出"生物多样性数码摄影展""麋鹿沧桑""灵长类科普展"等若干小型专题展。

通过与国际环保组织共同开展科普活动，不断扩大自身影响力。近年来，麋鹿苑利用暑期与美国"地球教育研究所"合作，以苑区为教学场所，组织青少年开展了"地球守护者"科普活动。

总之，麋鹿苑以建设国内外独具特色的生态博物馆为抓手，以"创新科普创意载体"为物质基础，以"拓展科普传播途径"为教育载体，以"丰富科普活动"为提升助力，每年平均接待参观者40多万人，获得北京市"社会大课堂示范基地先进集体奖"，获得"国际科学与和平周贡献奖"，被评为"北京市优秀科普教育基地"。

浅析天津自然博物馆展览与科普活动开发现状

黄克力 [①]

摘要 天津自然博物馆立足馆藏标本资源，着力于展览与科普活动开发。天津自然博物馆在常设展览、临展及北疆博物院复原陈列展览中，锐意创新、充分挖掘馆藏资源、积极响应国家战略构想。在依托馆藏资源与馆内展览的基础上，打造主题科普活动，构筑精品科普活动体系。通过馆校共建的开展，使自然博物馆的科普教育活动形成内外呼应的长效机制。

关键词 自然博物馆 展览 科普 教育

天津自然博物馆的前身为始建于 1914 年的北疆博物院，由法国天主教神甫、博物学家桑志华创建。自 1957 年，始用天津自然博物馆之名。1997—1998 年，天津自然博物馆在马场道原址完成改扩建工程，2013 年再度迁建，2014 年初新馆正式面向社会开放。

天津自然博物馆新馆位于天津市文化中心，占地面积 5.7 万平方米，总建筑面积 3.5 万平方米。新馆以"家园"为总主题，由"家园·探索""家园·生命"与"家园·生态"三个专题、十二个版块构成常设陈列，包含科普剧场、自然科普教室、活体蝴蝶园、恐龙挖掘谷、4D 影院等活动功能区，形成集收藏与研究、展示与体验、文化交流与科普教育、文化旅游与智性休闲于一体的综合性自然博物馆。

天津自然博物馆于 2008 年被国家文物局认定为首批国家一级博物馆，2013 年获批全国首批中小学科普教育社会实践基地，年均观众量近 200 万人次。天津自然博物馆始终秉承以观众为本、文化惠民的公共文化服务理念，打造精品陈列与精品科普教育活动，为观众提供高品质的博物馆文化体验。

---

① 黄克力：天津自然博物馆馆长，研究馆员。邮箱：764715304@qq.com。

1. 天津自然博物馆展览展示特色与设计开发现状

1.1 常设陈列: "家园"主题陈列

天津自然博物馆是世界上唯一一家以"家园"为主题的自然史博物馆。

"家园·生命"展区以生命的起源与演化为主线,以生命演化中的大事件为线索,以时间演化为顺序,以古今结合为表现特色,从不同视角展示地球"家园"亿万年生命历史长河中的重大事件。同时,引入叙事情节手法讲述"家园"正在发生的故事,运用探索和发现的视角,借鉴国际上自然史博物馆的最新理念展示地球生命"家园"的波澜起伏与宏大壮阔。

"家园·生态"展区则以美国著名慈善家肯尼斯·贝林先生历年捐赠的200多件珍稀世界野生动物标本为基础,以大景观、大手笔为表现形式,利用人工造景及背景画,结合各种现代化的展示手段,生动展示动物生活的真实场景,再现世界上最具代表性的野生动物及其生态环境。

1.2 临展:《丝绸之路——自然大观》

2016年末,为积极响应国家"一带一路"倡议规划,天津自然博物馆推出主题展览《丝绸之路——自然大观》。《丝绸之路——自然大观》展览面积约600平方米,采用主题单元的展示形式与地理空间位置的叙述顺序,以古丝绸之路上重要城市为节点,从自然历史角度展示丝绸之路的风光与自然遗产。以骆驼元素串联古代丝绸之路上的各个节点,以现代丝绸之路上高铁、飞机、网络等现代交通通信工具为结尾,展示新时期"一带一路"建设的宏伟蓝图。

展览以展示丝绸之路沿线国家和城市的自然生态与自然遗产为主题,配合多场以"丝路上的自然"为主题的特色科普活动,旨在增进社会公众对当前国家提出的"一带一路"政策的关注与理解,为公众提供具象化体验,切实感受"一带一路"倡议构想的现实意义。

1.3 北疆博物院:《回眸百年,致敬科学——北疆博物院复原陈列》展览

位于天津外国语大学院内的北疆博物院是天津自然博物馆的前身,这座20世纪20—30年代曾闻名中外的博物院自20世纪40年代起就进入了长久的蛰伏。2014年,天津自然博物馆正式启动北疆博物院修缮复原工程,并于2016年初完成北楼及陈列室的修缮及陈列复原,并正式面向社会开放。

以北疆博物院修缮复原工程为基础整合开发的展览——《回眸百年，致敬科学——北疆博物院复原陈列》，在设计上以修旧如故、传承发展、创新提升为核心理念，依据复原陈列、人文历史、开放库房三大部分内容，复原展出各类自然标本及人文藏品近 2 万件，荣获 2016 年度全国十大精品陈列称号。

北疆博物院复原陈列以"复原"为核心，而又不拘泥于此。于"复原"层面上，设计布展精益求精，大量对比桑志华时代的影像文字资料，在建筑格局、风格、细节及展品展柜布置上忠实再现百年前北疆博物院之风貌。此次复原开放，将馆藏标本中的精品璀璨悉数展出，旨在使公众从丰富而富有层次感的展览中体味到科学发现的乐趣与科学探索的精神。

同时，复原陈列又特辟北疆历史人文展区，让北疆博物院馆藏 2 万多件人文藏品重见天日，将人文藏品与文献资源有机地整合在一起，进而引发人们思考理性科学精神西学东渐为中国近代科学发现和博物馆发展所带来的机遇与启示，充分发挥人文藏品与历史资料在解读历史与科学研究中的价值与作用。

此外，北疆博物院复原陈列创新性地将标本库房与研究工作区域全方位对外开放，着力发掘标本与文物的内在价值，充分将"让文物活起来"的指导思想贯穿于科普工作实践之中。使公众近距离接触展品的同时，也将博物馆人的工作以及藏品背后的故事呈现出来，以期达到"在参观中体验、在体验中思考、在思考中升华"的目的。

2. 天津自然博物馆科普教育活动开发现状

习近平总书记在党的十九大开幕式上指出，要坚定文化自信，推动社会主义文化繁荣兴盛，推动文化事业发展，完善公共文化服务体系，为文化事业工作者提出了新任务和新要求。作为国家一级博物馆、全国科普教育基地，天津自然博物馆始终坚持拓展教育服务功能的深度和广度，深入贯彻落实《博物馆条例》中博物馆教育服务的有关要求，将科普教育作为重中之重，力求实现教育活动的常态化、特色化与多样化。自 2014 年初，新馆开放以来，天津自然博物馆年均接待观众近 200 万人次，其中青少年观众超过 60 万人次，年均为公众提供讲解服务 1000 多场，开展形式多样、各具特色的科普教育活动 300 多场。

2.1　以展览为纽带，营造主题特色体验

馆藏资源是博物馆一切工作的基础，展览则是联系馆藏资源与教育活动的重要纽带。天津自然博物馆先后推出"金猴报春——猴年特展""富饶的南海——南海特

展"以及"丝绸之路——自然大观"等特色原创展览，并以此为依托开展了"金猴迎春——春节特别活动""海底生花——南海展特别活动""丝路上的自然"等一系列主题活动。

在做好原创展览的同时，我馆积极组织外展，先后与河北博物院、天水市博物馆联合开展"会飞的花——世界珍稀蝴蝶展"，创新性地将活体蝴蝶放飞与蝴蝶科普剧引入展览，拓宽馆际合作方式，丰富科普教育活动的形式与内涵，获得了良好的社会反响。

2.2 深入探索开发，构筑精品活动体系

我馆现有雨天课堂、活体面对面、特色科普冬夏令营、科普情景剧、博物馆奇妙夜等 10 多个精品科普活动，年均科普活动开展数量超过 300 场。在注重知识传播的同时，融入趣味性、新颖性与艺术性，尤其重视互动性的开发，充分运用博物馆资源，为公众尤其是青少年营造出一个寓教于乐的科学营地。

我馆自然科普讲堂每周开讲，结合自然科学、热点话题、学校课程等，邀请知名专家来馆，开展丰富多彩的主题科普讲座。科普情景剧则以更为生动直观的形式，传播科学知识、环保教育等理念，全年演出超过 200 场，尤其受到青少年观众的喜爱。雨天课堂、活体面对面等活动则为观众提供了更直观的参与体验，形式纷繁的科普教育活动互为补充，共同构建起了一个富于特色的精品活动体系。

尤值一提的是，2016 年我馆推出的"博物馆奇妙夜——与恐龙同眠"系列科普活动，开创了夜宿博物馆的先河，将自然科普、野外探险、亲子活动与体验探索相结合。打破传统推广模式，利用"互联网+"传播功能，创新性地增加了在线直播环节、实现线上线下互动。虽然活动参与者仅 40 多人，但线上线下互动人员过万，极大程度地拓宽了受众面，开辟了博物馆科普教育活动的新渠道。

2.3 开展馆校共建，建立科普教育常规化的长效机制

天津自然博物馆结合馆藏和专家团队优势，选择部分中小学作为试点，充分发挥博物馆作为学校教育"第二课堂"的社会功能，将自然博物馆打造为中小学生校外科普教育实践基地。专门设立了自然科普教室，并于其中开辟体验区、互动区、图书区、探索区及标本展示区，形成独特的多元体验模式。同时，科普教育团队结合新馆展览的特色，采用学校教师和博物馆科普人员结合的模式，针对中小学课业特点研讨、开发、设计校本课程。自 2016 年以来，我馆陆续与天津南开科技实验小学、恩德里小学等 12 所学校，以签约形式开展馆校共建，使学生能定期到馆内体验探索，在实践中学

习，感受自然魅力。

　　于博物馆之外，天津自然博物馆科普服务小分队先后走进40多所学校，按照学生年龄的特点，制作了鸟类、恐龙、蝴蝶、远古人类的学习手册、科普课、科普展览，以标本模型制作、科普剧表演、手工制作、互动游戏等形式融入到馆校活动中。通过"馆校合作"的方式达到了对青少年学生的积极引导作用。设计开发的系列课程，在保证内容独创的前提下，补充、完善了青少年的知识体系，充分发挥了合作双方的优势，实现共赢。

　　标本是自然博物馆工作的基础，教育则是自然博物馆的灵魂。天津自然博物馆始终将科普教育置于首位，深入贯彻"创新、协调、绿色、开放、共享"的发展理念，以提高全民科学素质为己任，以"一带一路"倡议带来的发展契机为助力，坚持馆际交流合作，不断完善科普教育与社会文化服务。

天文馆的科学传播与旅游体验
——墨西哥科苏梅尔天文馆

米拉格罗斯·瓦尔奎兹 ①

摘　要 科苏梅尔天文馆是墨西哥金塔纳罗奥州四座天文馆之一，位于加勒比海中的科苏梅尔岛上。科苏梅尔岛作为旅游胜地，每年都会接待无数慕名而来的游客。当地丰富而独具特色的旅游资源为天文馆的发展带来了机遇与挑战。在发达的旅游业氛围中，科苏梅尔天文馆的目标不仅限于服务当地公众，也期望吸引更多外国游客。科苏梅尔天文馆采取的一系列措施，使其不仅成了岛内的教育中心，同时也成了游客向往的观光目标。

关键词 天文馆　科学传播　旅游体验

目前，墨西哥拥有天文馆近 40 家，数量多于智利、比利时、丹麦等。金塔纳罗奥州有 4 家天文馆，是墨西哥天文馆最多的州。在金塔纳罗奥州科学和技术委员会的规划中，天文馆是本地区的科学、技术与文化普及场所，主要内容包括天文学、地区生物多样性以及玛雅文化遗产。

科苏梅尔岛位于加勒比海中，临近墨西哥尤卡坦半岛东海岸，与普拉亚德尔卡曼市隔海相望。旅游业是科苏梅尔的经济支柱，科苏梅尔是拉美地区重要的旅游目的地。除了白色的沙滩和碧蓝的海水，科苏梅尔岛最著名的景点之一就是天文馆。科苏梅尔恰安嘉安天文馆可以让国内外游客在浓厚的知识氛围下聚会、交流思想，也是开展多种活动的理想场所。

1. 科苏梅尔天文馆概况

1.1 天文馆的建设背景与历史

科苏梅尔是墨西哥第三大岛，人口密度居全国第二。岛屿南北长 48 千米，东西宽

---

① 米拉格罗斯·瓦尔奎兹：博士，墨西哥科苏梅尔天文馆馆长。Email：mvarguez@frutosdigitales.com。

14.8 千米，距离金塔纳罗奥州普拉亚德尔卡曼市海岸线 20 千米。科苏梅尔岛不仅以碧蓝清澈的海水以及绵延的白色沙滩闻名，而且拥有长度仅次于澳大利亚的世界第二长的沿海珊瑚礁。因此科苏梅尔是全世界公认的潜水和浮潜较好的地点之一，每年接待乘坐游轮的观光游客约 300 万人次，这使该岛成了墨西哥的旅游胜地。但科苏梅尔人所关注的话题绝不仅限于旅游，科学传播同样为科苏梅尔所热衷。

如今，科学普及已经成为社会关注的焦点，我们完全有理由对科技进步和创新进行长期的传播，以轻松的方式吸引青少年接触科学，并在此过程中与科学建立永久的联系，培养他们各方面的能力和科学素养。2015 年 8 月 20 日，科苏梅尔恰安嘉安天文馆首次向公众开放，目的在于架起科苏梅岛人民与科学之间的桥梁（彩页图 33）。

建设科苏梅尔天文馆的意义不仅限于本地，同时也是金塔纳罗奥州天文馆体系的重要组成部分。科苏梅尔恰安嘉安天文馆是本州天文馆体系中的第三家，之前切图马尔于 2011 年建成了尤克尔卡布天文馆，坎昆于 2013 年建成了卡尤克天文馆。州内第四家天文馆位于普拉亚德尔卡曼市，于 2015 年 12 月开放。通常，天文馆在建设和运行方面有很多相似之处，然而，每座天文馆也都拥有各自的特色，从而赋予每座天文馆独特的个性。科苏梅尔恰安嘉安天文馆的名称来自公众，"恰安嘉安"（Cha-an Ka-an）在玛雅语中意为"观察天空"。

1.2 天文馆的建设与设施情况

墨西哥国家科学技术委员会（CONACYT）斥资近 300 万美元修建了科苏梅尔天文馆。场馆总占地面积为 10000 平方米，建设用地由金塔纳罗奥州政府无偿提供，距离可停泊最大吨位游轮的普埃尔塔·玛雅港不远。

天文馆已建设成一座服务于科苏梅尔居民的活动中心，同时也成为该岛引人注目的新景观。天文馆的宗旨是：促进观众理解并支持科学与科技创新，保护金塔纳罗奥州的玛雅文化遗产和生物多样性，方便观众开展有益健康的活动，享受思想交流与探索实践带来的乐趣。

恰安嘉安天文馆不同于州内乃至整个拉丁美洲地区其他天文馆的特色包括：以考古成就为基础，将玛雅考古天文学、生物多样性和古代航海作为主题。馆内配置了一座三维沉浸式球幕影院。球幕影院设有 95 个座位，6 台投影仪组成的系统可以进行二维和三维的展示。影院有多部影片可供观众选择，其内容涉及天文学、环境保护、文化、高新技术以及探索发现等领域。除了这座拥有巨大吸引力的影院，馆内还设有一座天文台，两间可举办研讨会的教室，一个供研究人员与公众分享其科学工作成果的礼堂，以及一间虚拟现实体验室。馆内设施还包括一个水资源教室和一个自然环境展厅，这两处设施聚焦于保护自然资源的展示与活动。

2. 天文馆开展的工作及社会影响

在天文馆对外开放的两年中，已经接待了近 25000 名观众，他们至少在天文馆内参与了一项活动，例如，观看球幕影片、开展天文观测、参加各种会议和讲座，以及参加特别的天文活动等。虽然科苏梅尔天文馆已经获得了金塔纳罗奥州人民及国内观众的好评，但是为了吸引更多游客，仍须面对各种短期和长期的挑战，其中之一便是公众的需求正在逐渐提高。因此，需要为社会创造性地组织更具吸引力的活动，同时扬长避短，探索适合的发展途径。我们希望恰安嘉安天文馆不仅能为孩子，也能为广大青少年、成人、家庭主妇、企业家、教授、研究人员等社会各界人士以及来岛上观光的游客提供更具吸引力的活动。另一项重要目标是针对不同的观众群体分别组织适合他们的活动。换句话说，根据学生在课堂上学习的科目，提供对学校有用的教学内容，从而促使学生将在课堂上学到的理论用于生活实践。最后一项目标，是在讲解地理、历史、物理、数学等方面的问题时不只采用讲座的形式，还可以利用球幕作为演示工具。为了实现这些目标，天文馆期待与教育机构建立一种长期的机制，定期组织相应的观众群体参观天文馆。

在天文馆与科苏梅尔公众之间建立良好的沟通渠道十分重要。正在实施一些这方面的计划，如"隔壁公园中的天文馆（Planetario en tu parque）计划"。这一活动颇受欢迎，要确保制定更为行之有效的方案，不只是在最受欢迎的公园，而是尽量在科苏梅尔的所有公园中开展活动，以接触并吸引更多的观众。为了让人们有兴趣参观天文馆，还需要走出馆门举办一些"外展"活动，也就是将经常在天文馆内开展的部分活动搬到馆外进行展示。此举也可以帮助那些因为经济状况无力承担参观天文馆费用的人们接触到我们组织的活动。

由于游轮旅游业在科苏梅尔占据举足轻重的地位，我们重要的合作伙伴之一就是旅行社，以确保天文馆提供的各项活动能够列入科苏梅尔岛观光产品的清单。这使天文馆不仅作为一个文化交流中心，而且可以成为一个旅游景点，为众多游客提供在沙滩、阳光和岛上各种常规活动之外的其他选择。

天文馆团队肩负的另外一项重要使命是与其他机构建立联系。经过努力，我们已经与国家自然保护区管理局、科苏梅尔公园和博物馆基金会、科苏梅尔观鸟俱乐部等机构建立起合作关系，并联合开展了一些活动。与其他机构开展合作能够拓展科苏梅尔天文馆的观众群体，并使观众能够接触到除天文学之外的各种活动内容，这对岛上居民也同样具有重要意义。另外，我们的合作机构并不限于墨西哥国内，还包括许多其他国家具有业内领先水平的机构。与中国同行的合作就是如此。

2017 年 11 月底，北京天文馆和科苏梅尔天文馆签署了合作备忘录。这一协议正是"一带一路"倡议带来的重要成果。协议内容包括：在促进公众对天文学的认识和理解方面交流经验，组织工作人员、科学传播者、场馆管理者以及研究人员开展交流活动，联合举办活动、开展研究项目、座谈会或专题展览等。

3. 为外国游客提供科学服务

虽然科苏梅尔天文馆希望成为当地公众熟悉喜爱并乐于访问的场所，但仍有很多人没有去过天文馆，甚至不知道它的存在，这就要求我们不断制定和实施新的方案。但是，要同时满足不同观众群体（公众、学校团体和游客）的要求绝非易事。在以往的实践中，我们针对公众和学校这两个群体已经有了较为成熟的办法并开展了一些活动，但是针对外国游客的措施仍然有待探索。

游览科苏梅尔的游客主要是来享受沙滩、阳光和绿松石般湛蓝清澈的海水，天文科普活动对他们并不是很有吸引力，因此，合理的价格策略就成为激励游客前来参观天文馆的必要措施。我们最重要的策略是传播玛雅先民为我们留下的科学知识遗产。因此，天文馆之旅的主要内容便是让游客有机会了解更多关于玛雅天文学、考古学和导航技术等方面的知识。我们还开设了"玛雅数字""玛雅历法"和"玛雅数学"等讲座。针对游客开设的天文观测项目主要包括：玛雅人眼中的天空、玛雅星座和玛雅天体的故事。需要注意的是，观测活动会受到天气的影响，云雨天气有时会阻碍观察活动的正常进行。

仅凭讲座和博物馆之旅来吸引游客是不够的，我们必须开发一个完整的旅游产品系列，使外国游客获得独一无二的体验。为了实现这一目标，我们便开始着手制作球幕影片。在汇集了大量古代玛雅人的科学和文化遗产后，墨西哥首部动画球幕影片终于诞生，这就是《揭秘玛雅先民：宇宙的观察家》（彩页图 34）。

该影片由一家名为 Frutos Digitales，致力于科学传播的企业制作，得到了欧洲南方天文台（ESO）的支持，并制作了 4 种语言（西班牙语、汉语、英语和葡萄牙语）版本。问世以来，该片已经在世界上 30 多个国家的 150 多个天文馆中免费传播。《揭秘玛雅先民：宇宙的观察家》寻访了 6 个不同的玛雅文化遗址，以表现玛雅建筑学与某些天体运动间的重要关系。通过这部球幕影片，我们能为有兴趣了解玛雅文化的游客创作更有吸引力的科学文化产品，为旅行社、酒店等客户提供更多可供选择的旅游方案。

尽管以上措施反响良好，但我们仍然需要更多有针对性的产品，并确保这些活动项目能够被游轮上的游客预定，而这正是我们面临的最大挑战。

4. 结语

科苏梅尔恰安嘉安天文馆是一座可以进行科学交流的中心，也可以帮助公众在科学、技术、创新等重要的方面进行学习、辩论、发表意见、做出选择等活动的规划，与此同时强化科苏梅尔的科学文化氛围。

虽然科苏梅尔天文馆在当地已经享有很高的知名度，但仍有必要制定策略将其影响力扩展至更多人群。我们还需要持续开发更新颖的活动，以吸引更多公众参与。对于到访的游客，必须针对他们多变的需求开发相应的产品，并与旅游行业经营者，如旅馆、旅行社、出租车司机等建立联系。此外，对于科苏梅尔天文馆尤为重要的是需要吸引乘坐游轮前来观光的游客。要做到这一点，就有必要与游轮旅游的经营者建立联盟，以便通过他们实现科普旅游产品的预售。

对于发展中的科苏梅尔天文馆来说，挑战是巨大的，尤其是当科学似乎没有得到普及或失去影响力的时候。然而，必须利用所有可能对我们有利的条件，不管它们看起来有多么遥不可及。我们必须在公众与科学之间构建一种情感联系，正是通过这座桥梁，科学才能呈现出更具吸引力的一面。只有当人们意识到自己与科学之间的联系有多么密切时，才能对科学产生更为浓厚的兴趣并愿意接受科学知识和理念。

参考文献

[1] PEMBROKE. PINES FL. Méride, Mexico Targets $7 Million Boost by Hosting the FCCA Cruise Conference & Trade Show [EB/OL]. (2017-10-17) [2017-12-1]. http://www.f-cca.com/press/17-FCCA-Cruise-Conference-Boost-Business. html.

科技博物馆的"科普外交"
——澳大利亚国立大学和国家科技馆的"科学马戏团"项目

格拉汉姆·杜兰特 [①]　　威尔·格兰特

摘　要　澳大利亚国立大学和澳大利亚国家科技馆在以科学促进澳大利亚及全球各地民众之间的联系方面，因联手创建的推广项目——科学传播硕士/"壳牌科学马戏团"而闻名。30年来，这个项目在澳大利亚提供跨区域和偏远地区的科技实践活动，开发出一套切实可行的方法——利用便携式科学普及展览和简易器材，推出适于不同地点和观众的科普展示与体验。在澳大利亚举办科学马戏团的经验可供许多国家借鉴，并已经应用于"科学马戏团外交"中。

关键词　科技博物馆　科学传播　科学秀　科学马戏团　跨区域交流

1. 澳大利亚国立大学

澳大利亚国立大学是一所位于澳大利亚首都堪培拉的世界一流大学，在众多澳大利亚大学中有着独特的历史。这所学府于第二次世界大战后成立。澳大利亚现任副总理、诺贝尔奖得主、物理学家布赖恩·施密特曾指出：其成立标志着"现代澳大利亚的开端"。这所大学的办学宗旨为："培养澳大利亚的知识领袖，并给予这个国家开展研究的机会……以确保澳大利亚在世界文明国家中占据一席之地。"该校至今秉承一项独特的国家使命——为政府和整个国家提供充分开展研究和教育的环境。

---

① 格拉汉姆·杜兰特：澳大利亚国家科学技术中心教授。E-mail：Graham.Durant@questacon.edu.au。

2. 澳大利亚国家公众科学认知中心

澳大利亚国家公众科学认知中心（CPAS）是世界上最多样化、也是澳大利亚最古老的学术与科学交流中心。它于1996年被确立为澳大利亚国立大学的一个学术中心，其任务在于提高社区的科学意识、促进公众讨论与科学相关的话题以及提升科学家的沟通技巧，在全国和国际范围内促进现代科学惠及民众。通过开展以研究为导向的教育，使我们的学生在促进民众参与科学、技术或医学信息这些与他们密切相关的活动方面，成为娴熟的沟通者。该中心就科学在公共领域的传播方式开展调研，开发民众认识科学的新途径，并鼓励民众探索解决21世纪科学焦点问题的方法。

3. 国家科技馆——澳大利亚国家科学技术中心

作为国家科技馆的澳大利亚国家科学技术中心，旨在通过启迪式学习体验激发和激励学生。国家科技馆的资产包括位于堪培拉的馆属科学中心、伊安·波特基金会技术学习中心、一系列巡回展览、世界一流的非正式学习参与计划，尤其是富有创造力、忠诚、热情和多元化的全体员工（他们与澳大利亚一些领导机构展开合作）。国家科技馆的核心产品是可亲自动手体验的展品、科学秀、科技展，以及可提供高质量参与的人性化互动项目。国家科技馆开展的各项活动以堪培拉为基地，并在澳大利亚各地推广，每年通过多种合作形式使数以百万计的澳大利亚民众参与进来。国家科技馆也在科学中心领域的国际交流中发挥重要作用。国家科技馆的运营主要通过政府资助、门票收入、店铺销售、展览会租赁和赞助来维持。待全新的国家科技馆基金会建立后，便可依靠慈善捐赠将国家科技馆的服务覆盖范围扩展到更多的孤立社区。我们的愿景是：以科学技术为手段，为澳大利亚和全世界所有民众创造更美好的未来。

4. 澳大利亚国立大学、国家科技馆与壳牌公司间的合作

澳大利亚国家公众科学认知中心与国家科技馆是源自同一个目的的两个项目。国家科技馆最初是一个未启用的学校礼堂中的公共设施，通过所收藏的一系列动手演示展品来传授物理知识。直到1988年，国家科技馆搬入了澳大利亚和日本合资兴建的专用建筑中，成了澳大利亚国家科学技术中心。科学马戏团的历史可以追溯到1985年，由时任国家科技馆馆长的迈克尔·戈尔博士创设，他最先将馆内展品带到澳大利亚其

他地区进行展览。科学马戏团从那时开始运作，30多年来其足迹已经遍及澳大利亚的每个角落。科学马戏团是目前世界上运作时间最长、行程距离最远的科学中心外展服务项目。著名的科学传播学者布莱恩·特伦奇把这个项目描述为世界上最早的科学传播项目之一。2014年，"壳牌国家科技馆科学马戏团"被全球电信基金会教育风险投资评为促进科技事业的前二十大项目之一。

该项目成立伊始便开始从澳大利亚各地招收来自各科学专业的大学毕业生，通过让他们在澳大利亚国立大学接受研究生教育（目前是科学传播硕士教育），掌握关键的科学传播技巧，然后随"壳牌科学马戏团"赴澳大利亚各地参加巡回展览，利用超便携展品为广大学生提供科学秀并举办公开的动手参与型展览。

从一开始，这个全国范围的项目便极其重要：从全国各地招收新生，并确保该项目尽可能多地涉及全国各地。对于参与项目的学生而言，前往全国各地乡村、人迹罕至的地区是项目一大亮点。当新一届科学传播硕士／科学马戏团研究生于每年年初到达堪培拉时，他们便开始了一个理论与实践相结合的充实学年。

5. 科学马戏团的发展

麦金农和布赖恩特记录了科学马戏团的成果。虽然该项目的许多关键成果得以量化，如科学传播研究生的数量、开展科学秀的次数（超过15000次）、巡回展览的行程距离、造访的城镇数量（超过500个，包括90多个原住民社区）、接受启蒙的学生人数等，但科学马戏团所产生的影响想必比这个层面的记录重要得多。

科学马戏团对于科学传播这一学科在澳大利亚的发展而言是不可或缺的，而在全球也颇具影响力。麦金农和布赖恩特记述了早期科学马戏团研究生的故事。

他们委婉但有针对性地批评这个项目在准备不足的情况下便派他们出去。"你们对自己将要去做的事情心里有数吗？"女发言人问道。他们的回答是：他们只能在科学秀和科技展中接受培训，初次巡回展览想要聚集人气都要依靠他们的热情和人格魅力。布赖恩特和戈尔立对此做出回应，主要包括：增加公开演讲课程、简单的英语写作训练、展览设计，以及更多演示和学术背景方面的培训。这是澳大利亚在科学传播方面开展培训的首次尝试——并无"最佳方案"供参考。

这样的进展直接导致1996年澳大利亚国家科学公众认知中心的设立，进而言之，这也促进了过去20年科学传播产业在澳大利亚的发展。

科学马戏团在澳大利亚发展的30多年中，需要不断开发一些简单便携的动手实践展品、简单而有效的科学演示，以及无须复杂设备便可在边远地区举办的工作坊。目前，国家科技馆的科普宣传大篷车可在两个小时之内设立一个拥有50件展品的实践型科学中心。若要前往更为偏远的近海岛屿，科学马戏团的学者必须把所有的道具打包装上轻型飞机，运抵科学秀与工作坊的所在地。

澳大利亚的经验适用于世界所有的偏远地区：训练有素而充满热情的展示者、超便携式展品、激动人心的科学秀以及简便易行的工作坊——各要素可任意组合，成为了科学马戏团成功将巡回展览带到不同国家和地区的关键所在，同时也是"科学马戏团外交"的基础。有时只有展品进行巡回展览，有时会有少数展示者随行。总有一种科学马戏团的模式可以适用于某一情况和预算。

6. 科学马戏团外交

6.1 "科学动起来"（1996年和1997年）

为办好由联合国教科文组织、澳大利亚国际发展署和澳大利亚政府外交和贸易部资助的前往一些南太平洋岛屿，名为"科学动起来"的特别巡回展览，科学马戏团特地复制了一些相关展品。曾在科学马戏团工作的澳大利亚国家科技馆学者向所罗门群岛、瓦努阿图、斐济、基里巴斯、图瓦卢、马绍尔群岛、库克群岛超过42000名参与者提供了多元化的科学秀以及工作坊。1997年，他们又前往汤加和西萨摩亚，约18000人参与了本次活动。

6.2 东帝汶（2002年）

继成功地在9个南太平洋国家开展科学展览后，联合国教科文组织要求澳大利亚国家科技馆为东帝汶制订并实施一项科学教育试点计划。科学马戏团将展品安放在帝力的街道边，由科学马戏团的学者进行一系列科学演示。

6.3 印度（2005年）

国家科技馆科学马戏团协调员丽斯·霍格应邀出席由印度国家科学博物馆委员会组织举办的国际物理示范研讨会，并进行简单的科学演示，以庆祝2005国际物理年。

6.4 泰国（2006年）

国家科技馆受澳大利亚国际教育署的委派，为2006年8月在泰国举办的国家科学技术博览会贡献一些"科学马戏团"的展品，并有幸拜访了玛莎·查克里·诗琳通公主殿下。

6.5 韩国（2006 年）

在韩国科学节上，澳大利亚国家科技馆科学马戏团的展品受到了约 50000 韩国人的热捧。国家科技馆受到澳大利亚国际教育署韩国分部、韩国科学基金会、科学技术部、信息通信部、工商能源部以及韩国教育与人力资源部的资助，参加了韩国科学节。国家科技馆对韩国工作人员进行了科学展示方面的培训，并与韩国相关机构进行了会谈。国家科技馆通过"科学动起来"展览为澳大利亚国际教育署的学习促进项目提供支持。

6.6 中国（2010 年）

两位澳大利亚国家科技馆科学马戏团成员在北京、上海和广州巡回展览，并在上海世博会澳大利亚馆的科学周上进行展示，与中国科学技术馆、上海科学技术馆以及广东科学中心的同行分享科学传播的途径。科学马戏团成员为北京、上海和广州的共 6 所学校 1055 名儿童举办了 24 场科学秀活动。

6.7 阿布扎比科学节（2012 年）

澳大利亚国家科技馆科学马戏团的学者开展一系列的科学秀以及实践工作坊，支持阿布扎比科学节。

6.8 越南（2013 年）

作为澳越建交 40 周年庆祝活动的一部分，澳大利亚国家科技馆受澳大利亚外交贸易部之邀赴越南参加展览。2013 年 4 月，赴越科学马戏团先后拜访了河内、岘港和胡志明市，主要展览内容涉及四个要素："科学动起来"展览、作为补充的科学秀节目、面向岘港和河内教师的科学教育专业能力培训计划及其他交流活动，主要包括与越南科技传播中心科普部联手举办研讨会，与越南部长官员进行圆桌会谈，以及由国家科技馆馆长进行多媒体演示。国家科技馆成员与为展览会提供解说的越南学生一起工作，展品用越南语和英语标注。在为期 25 天的巡回展览中，有 7591 位参观者参与了展览和科学秀，74 位教师参加了专业研讨会。此后，澳大利亚大使馆组织人员将面向教师的科学马戏团建议演示手册翻译成老挝语和柬埔寨语，并分发给这些国家的教师。巡回展览及相关活动得到了广泛报道，在三周的时间内，有超过 100 篇媒体文章与读者见面。展览本身吸引了超过 7500 名观众，相关报道吸引了数以百万计的读者，带来广泛的影响。

6.9 日本（2014 年和 2018 年）

澳大利亚国家科技馆成立于 1988 年，这是日本政府和日本经济团体联合会向澳大利亚人民献上的慷慨赠礼。2014 年，赴日科学马戏团特别感谢了日本对澳大利亚国家科技馆的投资。此次巡回展览使日本政府与商界领袖欢聚一堂庆祝澳大利亚国家科技馆成立 25 周年，并决定在未来对澳大利亚长期投资。本次巡回展览的目的得以实现，促成日方形成了文件——《新世纪的澳大利亚》。这次巡回展览深入偏远社区，加强了澳大利亚国家科技馆与日本科学中心的合作。2014 年赴日科学马戏团在东京、南三陆町、盛冈、久慈和三泽等地进行了巡回展览，开展的互动科学秀吸引了 14000 人参与。在 2014 年 4—5 月，科学马戏团进行了为期 5 周的巡回展览和科学展示，造访了日本东北部经历海啸后正在恢复的社区。科学马戏团此次巡展得到了科技发现中心（珀斯）的资助，与日本科学未来馆展开合作，将把国家科技馆科学马戏团的成功科学传播外展模式介绍给日本。此次巡展还与政府、企业和教育机构进行了高层交流，包括举办国家科技馆、科学未来馆和日本东北地区三个区域科学中心之间的科普能力建设研讨会。

2018 年，日本将成为澳大利亚政府落实公共外交的重点国家。澳大利亚国家科技馆通过与大阪科技馆、科学未来馆和日本其他区域性科学中心联手合作，将在日本关西地区多地开展一次科学马戏团巡展活动。此次巡展旨在表达澳大利亚对作为国家科技馆创始投资者的日方的感激，同时也能传达这样一个信息：澳大利亚愿为今后 30 年的澳日合作铺平道路。

6.10 非洲（2015 年）

赴非洲科学马戏团项目基于十几年间在南非开展的一系列活动发展起来。赴非洲科学马戏团项目由公众科学认知中心领导，在澳大利亚国家科技馆的支持下正在不断发展，旨在帮助非洲国家的科学中心开发教育活动。赴非洲科学马戏团是澳大利亚政府资助的项目，旨在为南部非洲 5 个国家的学校、教师和社区提供有趣的科学知识，同时也着手于对非洲工作人员的培训与能力培养。主要的课程内容包括：科学秀、教师职业及专业进修（通过举办研讨会及分发书籍等资料）、动手实践型展览和员工培训。这是澳大利亚政府三大使团：国家科技馆——澳大利亚国家科学技术中心、澳大利亚国立大学、杜塞尔基金会与众多非洲合作伙伴之间的合作项目。该项目于 2015 年 5 月 3 日—7 月 15 日启动，共有来自 5 个国家的 41367 人参加，其中有 37392 名学生参加了科学秀，2843 个团体和家庭参加了科学秀或科技展览，451 个团体和家庭参加了研讨会，519名教师参加了职业发展研讨会，162名非洲科学传播工作者接受了为期数周的培训，发放了3200份教师资源手册，可以指导教师利用普通物品展现非凡的科学实验。由澳大利亚的AMF磁力公司捐赠的成千上万个磁铁被分发到数百所学校。

7. 科普巡展

在澳大利亚的各种海外援助项目中，国际发展项目的国际志愿者为科学马戏团专家在部分国家（包括纳米比亚、印度尼西亚和菲律宾）组织开展科普能力及科学演示方面的培训工作，创造了许多机会。曾经的科学马戏团成员、澳大利亚国家科技馆和澳大利亚国立大学的工作人员已经在包括缅甸、泰国、韩国、蒙古和文莱在内的国家实施了许多项目。

8. 科学马戏团的影响

壳牌科学马戏团的成功不断吸引世界各地的人前往澳大利亚观摩和体验正在运行中的科学马戏团活动。这是产学研（大学、科学中心和商业企业）三方合作的一个很好范例。此项目在其他国家中带动了类似的科学中心外展和科学传播培训模式，并已促使澳大利亚境内其他外展项目得以发展。科学马戏团的影响也是许多博士研究生和其他研究人员的研究主题。

30多年来，在澳大利亚跨地区以及在偏远地区开展的动手实践科学体验活动，促使一套经过验证、切实可行的模式发展成熟，即采用简单易用的材料，开展易于实施的科学展览，为不同场地的观众提供因地制宜的科学展示与体验。在澳大利亚开展科学马戏团项目的经验值得许多国家借鉴，并已经运用于"科学马戏团外交"中。澳大利亚国立大学和国家科技馆的愿景是：将科学马戏团举办展览、科学秀以及外交方面的奇妙经历发扬光大，令世界各地儿童、学生和成年人的生活焕然一新。

参考文献

［1］BRIAN. All staff welcome［EB/OL］.（2016-2-6）［2017-12-1］. http://www.anu.edu.au/news/all-news/all-staff-welcome.

［2］TRENCH B. Vital and vulnerable：Science communication as a university subject［M］. Netherlands：Springer, 2012：241-258.

［3］BRYANT C，Gore M M. The evolution of a masters course in scientific communication：Some reflections on experience at the Australian National University［M］．Quebec City，Quebec：Multimondes，1999：141-158.

［4］MCKINNON M，BRYANT C. Thirty Years of a Science Communication Course in Australia：Genesis and Evolution of a Degree［J］．Science Communication，2017，39：169-194.

集自然历史、世界文化于一身的
皇家安大略博物馆

摘 要 皇家安大略博物馆融合自然历史和世界文化于一身，藏品超过 1250 万件，大量开展国际范围的研究。2017 年参观人数超过 130 万人次。该馆设有 27 个永久的世界文化和自然历史展厅、2 个儿童动手实践探索展厅和 4 个用于临时展览的展区。该馆收集了两只鲸鱼的骨架，将其中一只鲸鱼的心脏做塑化处理后在主展厅展出，将在加拿大各地和其他国家和地区进行巡回展出。

关键词 皇家安大略博物馆 自然历史 世界文化 国际视角

皇家安大略博物馆是世界上较大的博物馆之一，肩负着通过研究、收藏和展示向公众传播科学和文化的双重使命。它位于加拿大人口稠密的多伦多市，是安大略省政府旅游文化体育部的一个下属机构。从历史上看，皇家安大略博物馆平均每年吸引近 100 万名观众，2017 年创下有史以来最高纪录，达到了 130 万人次，比上一年增长了 23%。大部分（58%）观众来自多伦多地区，而美国观众在外国观众中占比最高，达到 13%。

1. 历史沿革

根据安大略省议会通过的《皇家安大略博物馆法案》，皇家安大略博物馆于 1912 年正式成立，但在博物馆新建筑完工之后才面向公众开放。新建筑坐落于多伦多大学哲学家小径的东北角（表 1），其正门位于布鲁尔西街，这也是该市的主要街道。安大略省和多伦多大学平摊博物馆的运营费用，由多伦多大学负责管理，最初有 20 名员工。皇家安大略博物馆最初下设 5 个不同专业的博物馆，其内容集中涵盖考古学、地质学、矿物学、古生物学和动物学，每个博物馆都有单独的馆长。

---

① 伯顿·K·利姆：加拿大皇家安大略博物馆主任，博士。E-mail：burtonl@rom.on.ca。

表 1　加拿大多伦多皇家安大略博物馆重大历史事件编年表

年份（年）	里程碑事件	说　明
1912	设立《皇家安大略博物馆法案》	由省议会设立
1914	博物馆开馆	建筑位于多伦多大学哲学家小径旁
1933	首次扩建	在女王公园的侧翼位置开建一个 H 形的建筑
1951	修建容纳加拿大历史文化藏品的新建筑	由西格蒙德·塞缪尔博士捐资修建
1968	建造天文馆	由塞缪尔·麦克劳克林捐赠 200 万美元修建
1978	第二次扩建	新策展中心和露台展厅的翻新工程动工
1987	加德纳陶瓷博物馆	马路对面的陶瓷艺术博物馆归安大略皇家博物馆管理
1994	文物展厅	博物馆主馆中的加拿大历史文化新展厅开馆
2001	皇家安大略博物馆的复兴	2005 年水晶宫向公众开放时，复兴项目逐渐达到高潮
2017	女王公园入口	历史悠久的圆形大厅入口重新开放，作为布卢尔街入口的备选方案

　　该博物馆的设立主要得益于埃德蒙·沃克和查尔斯·柯雷利的宏伟愿景，人们公认他们是促成该博物馆得以设立的背后最大动力。沃克作为一位著名的银行高管和慈善家，热衷在多伦多建立一个科学文化机构。柯雷利作为一位艺术和考古学收藏家，收集了许多古代文物（主要来自于埃及），这些文物最终成为皇家安大略博物馆的藏品。柯雷利被任命为皇家安大略省考古博物馆的首任馆长，一直任职到 1946 年退休。其他馆藏品（如动物学方面的藏品）则与 19 世纪 50 年代中期建立的多伦多大学生物博物馆密切相关，这些藏品原本主要用于教学。此外，安大略省教育厅师范学校的藏品最终被划转至皇家安大略博物馆。1968 年，博物馆受多伦多大学管理的历史宣告结束，皇家安大略博物馆成了安大略省政府下辖的一个机构。

　　自 1933 年起，该博物馆经历了数次扩建，在东边的女王公园附近增设了一座新建筑物和入口，并与原有的哲学家小径建筑物相连，形成了一个 H 形构造。1951 年，西格蒙德·塞缪尔大厦以钢铁实业家的姓名命名，在省立法机构的对面修建了皇家安大略博物馆的南部建筑，用以容纳加拿大历史文化藏品，在获得藏品方面曾经发挥重要作用。1968 年，身为汽车业巨头的塞缪尔·麦克劳克林捐资修建了加拿大的首个天文馆，它位于安大略皇家博物馆的旁边，并接受其管理。1982 年，博物馆的建筑发生了重大扩张，在博物馆的南部修建了一座新建筑，用于容纳馆藏和进行研究（现称之为路易斯·霍利石策展中心），加上在原有的 H 形建筑北部新建的露台展厅，形成了彼此互连的矩形建筑体。扩张阶段的高潮发生于 1987 年，当时位于马路对面、已有 3 年经营历史的加德纳陶瓷博物馆由皇家安大略博物馆接管。但是，该馆在 1995 年由于天文馆关闭，规模开始缩小，建筑也被出售给多伦多大学。1997 年，加德纳陶瓷博物馆再次独立运营，采用捐赠者的姓名。西格蒙德·塞缪尔加拿大历史文化大厦于 2000 年被出售给多伦多大学，收藏品也被搬到了主博物馆。这种缩减的趋势在 2007 年发生了逆转，当时的一个改革项目将 1982 年的露台展厅变为迈克尔·李秦水晶宫（内设 6 个新展厅），并将入口移回布鲁尔西街（彩页图 35）。

2. 当代的皇家安大略博物馆

2.1　领导机构

一个由 21 名理事组成的理事会负责管理皇家安大略博物馆，理事会主要负责为安大略省的人民制定博物馆的政策、流程并监管资产。馆长和总裁负责领导高级管理部门维持皇家安大略博物馆的日常运作，副馆长和副总裁负责提供支持。乔舒亚·巴塞基是现任馆长，他正致力于将皇家安大略博物馆重塑为北美跨学科的艺术、文化与自然博物馆。除了兼职工作人员和 1300 多名志愿者，博物馆还有 300 多名全职工作人员。

2.2　愿景与挑战

皇家安大略博物馆的愿景是打造一个了解持续变化中的自然和文化世界重要目的地，并得到全球范围的认可。数字和社交媒体是博物馆正在涉足的新领域，旨在增加网络曝光率，从而帮助博物馆提高参观人数。博物馆网站（https://www.rom.on.ca）2017 年的访问量超过了 370 万人次，比上年增长 20%。社交媒体参与度增长 70%，在所有平台（脸书、照片墙和推特的）的收视次数（放置广告图像的网页每次显示，就是一次收视次数）达到近 15 万次。

如何更新老化的基础设施是一个持续的挑战。容纳馆藏和研究设施的策展大楼于 1982 年建成，目前它不但已经过时，而且扩建空间的大部分面积也已被使用，这给未来的发展带来了一个持续性的挑战。信息技术和基础设施相应需要更新，财务投资力度需要进一步加大，其中也涵盖了博物馆标本的数字资产领域和在线可用性。

2.3　馆藏与研究

目前，皇家安大略博物馆有超过 30 位策展人在从事标本的采集工作，进行艺术、文化和自然方面的研究。在过去的一年里，他们奔赴全球约 30 个国家进行了实地考察，所著的 100 多份作品被收入各种学术期刊和书籍中。

世界文化方面的一些收藏亮点包括加拿大历史文化藏品，它们汇集了来自加拿大的图像、装饰艺术和物质文化。来自中国的藏品堪称世界一流，为重要的综合性藏品。皇家安大略博物馆的中国藏品跻身世界前十名（指中国之外的博物馆，不含中国的博物馆）。在自然史中，伯吉斯页岩藏品堪称世界上产自加拿大的最大、最完整、最早的化石样本，这对于说明五亿年前那场著名的寒武纪生命大爆发中动物的早期演变极为重要。冷冻组织是增长速度最快的藏品，它与几个生命科学学科交叉，跻身世界前十名。包括 11 万多件用于 DNA 和遗传分析的样本，以研究珍稀濒危物种，评估生物多样性，并且帮助建立生命树。

2.4 社区

该博物馆与多伦多地区的周边社区有着广泛的联系。皇家安大略博物馆的社区接入网络旨在帮助突破社会、金融或文化障碍，确保尽可能多的人能融入进来。博物馆与 62 家非营利组织建立了合作关系，免费发放 10 万张入场参观门票。加拿大大型的公民科学项目之一是由皇家安大略博物馆与其他地方机构合作发起的安大略省年度生物多样性普查倡议，旨在记录多伦多地区的多种物种。约 250 名专业科学家得到约 400 名业余自然学家的协助，2017 年在祈德河地区记录了 1400 种物种。创意公共节目的规划是吸引不同观众参观博物馆的另一种创新方式，受欢迎的活动包括"星期五之夜"，该活动针对年轻人群，旨在将他们吸引到博物馆中来。

2.5 教育

2017 年，皇家安大略博物馆通过运用有讲解的参观、动手实践活动以及与安大略省教育课程相关的在线资源，接待了来自安大略省不同地区的超过 11 万名学生和教师。我们推出了一个能够走出实体博物馆建筑的外展服务项目，它包括一个充气的便携式穹顶的大众型移动天文馆，可以模拟星座的夜景，并可出借给学校、图书馆、社区中心以及其他博物馆。

2.6 会员与慈善事业

博物馆有近 15 万会员（个人及家庭会员），其中包括两类享受最高级专有权的赞助会员（可被允许"幕后参观"）。皇家安大略博物馆理事会还是皇家安大略博物馆基金会的筹款部门，它拥有自己的独立理事会和管理团队，旨在为博物馆的收藏、研究和展览提供支持。在 30 个以上的策展人中，有 8 个是讲席教授。

2.7　展厅

皇家安大略博物馆的常设展厅陈列着约 30000 件展品。有 30 多个展示艺术、考古和自然科学的展厅，其中有 17 个世界文化展厅、10 个自然历史展厅、2 个动手实践发现展厅和 4 个临时展览空间。去年最受欢迎的两个展厅分别是以儿童为中心的动手实践生物多样性展厅和探索展厅。这些展厅编排了一些节目，其特色是集成基于戏剧的各项活动以及主要由可触摸标本（也有一些活的动物）构成的多感官体验。

2.8　展览会

2017 年，皇家安大略博物馆举办了 7 个涵盖艺术、文化和自然的大型展览。能够例证这种多样性的三个展览为：给人带来视觉震撼的奇胡利玻璃艺术作品——其构成要素来自于沙子、火、美丽；拥有 5000 年历史的文身文化仪式——仪式、身份、痴迷、艺术；"来自大洋深处：蓝鲸故事"所表现出的大自然宏伟，它已成为皇家安大略博物馆新颖的展品。除了蓝鲸展品，"阿尼新纳贝格——艺术力和家庭相机"是内部展览，是为 2017 年加拿大联邦成立 150 周年庆典而开发。

2.9　巡回展览

原创巡回展览是皇家安大略博物馆始终积极追求的一个领域。如 2018 年的"第三种性别"展览——浮世绘（一种日本版画）中漂亮的年轻人成为广受纽约的日本社区好评的艺术展示，"国王与法老"展览——皇家安大略博物馆与中国的南京和金沙遗址博物馆合作，共同展出了古代埃及与中国的珍宝，吸引了 100 万人前来参观。

"来自大洋深处：蓝鲸的故事"是皇家安大略博物馆正在推进的下一个巡回展览。它主要针对模块化的展览进行开发，为了适应全尺寸空间且更具影响力，使其适合接展机构的实际和特定展览构件需求，能适应中等到较小尺寸的场地。整个展览包括 2014 年从纽芬兰鳟鱼河中打捞出的 26 米长蓝鲸骨架。此外，蓝鲸的塑化心脏标本令人印象深刻，高度达到 1.5 米，是所有已知动物中得以保存的最大心脏，肯定会激发观众的想象力和好奇心（彩页图 36）。

蓝鲸展览共有 9 个组成部分，其中包括一个沉浸式的多媒体三屏视频介绍，详细介绍了在加拿大圣劳伦斯湾流冰群中发现已死亡的 9 只蓝鲸的悲惨事件，以及皇家安大略博物馆以了解科学和教育为目的，将两只蓝鲸的骨骼恢复原状的过程。下一部分是对这只曾经存活的最重动物尺寸和骨架的大揭秘和互动讨论。在"生物和生物学"部分设置了一个声室，对蓝鲸在数百千米的海洋中用于沟通的次声波频率叫声进行了详述。博物馆陈列了一副全尺寸的头骨和带有真正鲸须的下巴，它是来自落基港的标本，用于解释这种体型最为巨大的动物在吞食一些最小生物（类似虾类的磷虾）时的滤食性行为。"进化"部分包括令人印象深刻的 5 副鲸鱼化石，观众可以完美地追踪陆地哺乳动物回归海洋的过渡。基因组使用交互式图形屏幕描述鲸鱼的遗传学特征，皇家安大略博物馆牵头开展有关读取蓝鲸身上 30 亿个 DNA 代码的研究，以及阐明其对我们理解鲸鱼成功适应海洋生态以及这些濒危物种生存的意义。"保护鲸鱼"部分简要介绍了有关捕鲸的历史，以及保护人员在救助这些庞然大物于濒临灭绝边缘过程中所付诸的种种努力。最后一部分是海洋保护专家的视频，他们热情洋溢地向社会宣传保持健康鲸鱼种群的益处和重要性。

3. 结论

皇家安大略博物馆是加拿大规模最大的自然历史和世界文化机构，每年参观人数超过 100 万。馆藏品和研究方面的优势最初集中在安大略省和加拿大其他地区，但现在其范围具备国际视角，涵盖许多不同的艺术、考古和自然领域。超过 30 个展厅展示了博物馆的标本、永久展示的手工艺品，以及临时展览。博物馆最近推出的一个举措是开发出巡回世界的展览，其内容包括"来自大洋深处：蓝鲸的故事"——主要基于博物馆从纽芬兰打捞并处理的鲸鱼，展示海洋哺乳动物令人敬畏的进化以及保护组织为挽救它们于濒临灭绝边缘所付诸的种种努力。

参考文献

［1］ROYAL ONTARIO MUSEUM. Annual Report 2016-2017［EB/OL］. http://www.rom.on.ca/ann-ualreport/index.php.

［2］DICKSON L. The museum makers：the story of the Royal Ontario Museum［M］. Toronto：University of Toronto Press，1993.

［3］CURRELLY C T. I brought the ages home［J］. 1956.

［4］DYMOND J R. History of the Royal Ontario Museum of Zoology. 1940.

［5］BROWNE K. Bold visions：the architecture of the Royal Ontario Museum［J］. 2007.

［6］Royal Ontario Museum. ROM business plan：fiscal 2017-2018［EB/OL］.（2017-7-13）［2017-12-1］. http://www.rom.on.ca/sites/default/files/imce/business_plan_2017_final.pdf.

在国际合作背景下促进科普教育
——科罗廖夫太空博物馆

维多利亚·切维塔克 ①

摘 要 乌克兰拥有约 5000 家博物馆。作为现代社会发展背景下的一种需求，博物馆能否对大众产生真正的吸引力？科罗廖夫太空博物馆作为乌克兰国内处于领先地位的太空博物馆，其作为教育和文化中心的发展战略恰能满足这一需求。科学教育能使博物馆，特别是科技博物馆与普通公众的联系更密切，并可以带给公众积极的影响。开展国际合作将有助于借鉴世界各地的经验，并能促进博物馆发展成为普及科学的中心。

关键词 科技博物馆 科学教育 国际合作 世界最佳实践

科罗廖夫太空博物馆是成立于 1970 年的一座乌克兰科技博物馆，位于乌克兰日托米尔市，苏联太空火箭系统首席设计师及国家太空计划发起人谢尔盖·科罗廖夫就诞生于此。在科罗廖夫所处的时代，苏联取得了一系列伟大的航天成就：发射了世界上首颗人造地球卫星、首个月球探测器、首个金星探测器、首个火星探测器、首次载人航天飞行以及人类首次在太空进行舱外活动。

如今，博物馆内设有两项常设展览：谢尔盖·科罗廖夫院士纪念馆和专注于世界空间技术成就的展厅。独特的馆藏包括 25000 多件物品，其中有科罗廖夫的个人物品和家庭纪念品、最初的航天器和相关藏品，其中包括联盟 27 号飞船返回舱、月球岩石及土壤样本，以及真人大小的精确模型。博物馆的首要任务之一是开展科普教育，促进空间探索以及与空间有关的活动，并鼓励游客发展行星际思维。实现这一目标的最好办法就是在国际合作与文化交流的过程中学习借鉴世界上最佳的工作方式。

---

① 维多利亚·切维塔克：乌克兰科罗廖夫太空博物馆外事部主任。E-mail：victoria.chetvertak@gmail.com。

1. 开展科普活动的初衷

据乌克兰独立新闻社（UNIAN）报道，乌克兰国内拥有约 5000 家博物馆，对于这个国家来讲，这是一个鼓舞人心的数字，人们可以从中真切地感觉到乌克兰博物馆的复兴。但博物馆对大众能否真正产生吸引力？换句话说，"要帮助公众最大限度地汲取这些藏品物件的内涵，让人们切实感受到博物馆正在成为他们生活中不可缺少的一部分"。博物馆已经成为现代社会发展的一种需求。科罗廖夫太空博物馆作为乌克兰国内处于领先地位的太空博物馆，其作为教育和文化中心的发展战略恰能满足这一需求。

博物馆的办馆理念通过其活动得以凸显，面向公众的教育和互动展示，常规以及定制的导览，节日、文化活动和展览，这些活动使博物馆能够吸引更多观众。

在制订工作计划时，我们始终牢记这样的理念："博物馆是服务社会及其发展的非营利性机构。它应该面向社会开放，为了教育、学习和分享之目的，收集、保存、研究、传播和展示各种有形和无形的人类遗产及环境。"[①] 这一对博物馆的定义建立于最佳博物馆模式之一的基础上，它形成于 20 世纪 80 年代后期的阿姆斯特丹。这种模式融汇了收藏（包括文物藏品 / 物品的搜集、保护和管理）、研究和交流（结合教育和展览）。

最近乌克兰的一些博物馆，其中也包括我们的博物馆都能证明，世界博物馆业的最新趋势为将博物馆作为与公众沟通的一种渠道。博物馆逐渐意识到自身的影响力，以及博物馆与其展品所产生影响的重要性，这种影响无法被任何故事或文字取代。举例来说，这样的影响对于儿童更为显著，我们可以这样理解："博物馆"为人们留下的第一印象，可以伴随他们的一生。

在这种背景下，科普教育似乎是使博物馆，尤其是科技博物馆与普通公众联系更密切的一种便捷方式。教育活动能够带来积极的影响，如令人难忘的学习体验，提高观众对科学技术的认知水平，促进不同年龄段的人和社会上的学习风气，增进公众与从事尖端技术开发的科学精英之间的信任和理解。

探索和知识建构始于"发现"，也就是一种需要观察、描述和询问的现象。在这个过程中，在现象得到证明之前，并无任何解释，信息从来不会如此轻易地获得。每个新发现都基于不同探究者个人的技能和知识基础，这就意味着没有哪两个过程完全相同。事实上，对现象的发现和解释由学习者主导，会被他们的年龄、期望和兴趣影

---

① 详细内容参见：http://icom.museum/the-vision/museum-definition。

响。通过这种方式，博物馆可以帮助观众了解科学研究的过程，即科学探究包含假说的形成、观察、测试、反复试验、控制、重复。此外，科学的认识方法还能帮助观众应对日常生活中遇到的各种问题。所以说，博物馆绝不只是展示一堆展品的地方，而是一个能够呈现科学及科学过程的复杂性和多元性，诠释诞生这些展品与知识的文化、社会和哲学理念的场所。

2. 工作方式及对国际合作的期望

近年来，博物馆积极开展各种科普宣传活动。尽管我们拥有近48年的博物馆运营经验，但在实现由单纯的馆藏向在实践中体现社会演变以及拓展观众兴趣转型方面，我们还是个初学者。在现代社会中运用更广泛的科学技术在我们的博物馆中更受青睐。

我们的第一项措施是通过怀旧之旅"复兴"我们的纪念展览馆。通过播放录音和纪录片、穿戴具有历史感的导览服装带给观众特别的体验，从而使历史变得触手可及。如今，观众可以沉浸于20世纪的氛围中，体验苏联航天总设计师最高机密的生活故事。

生日会为学习新颖、重要的内容提供另一个绝佳机会。一场令人兴奋的游戏可能会令你的探索更进一步。

另一项挑战是借鉴国外的经验。诸如"欧洲研究者之夜"以及"科学野餐会"等成功项目受到了观众的青睐。

对我们来说，把博物馆教育活动置于博物馆之外更广阔的背景下也非常重要。各博物馆在挖掘各自主题或展品方面的确表现出了非凡的创造力，在吸引更多观众的同时提供知识和信息。馆际交流活动之所以非常重要，是因为它们可使我们的博物馆与更多的博物馆及机构携手工作，共享信息，相互学习。从这个层面上来说，国际合作和经验可为我馆的进一步发展提供更多的机会。我们可以从国际合作中获得什么？借鉴世界各地博物馆的最佳惯例和做法，可以帮助我们更好地满足社会不断变化的需求。国际间各博物馆的合作与交流可以促进不同国家的博物馆分享经验，开展科普活动，开展专家交流，促进各博物馆的进一步发展。

3. 结语

科罗廖夫太空博物馆目前正以促进科学普及作为发展战略，开展一些引进的创新项目，同时开展广泛的国际合作，旨在借鉴世界最佳的做法和经验，促进博物馆发展成为科学普及机构。这是时代对我们的迫切要求，也是我们积极支持召开"一带一路"背景下科技博物馆国际研讨会的原因。我们相信：本次汇聚各国代表的研讨会所提供的国际交流机会，将促进更好国际合作方式的诞生，有助于本着互利原则制定发展战略，并为构建卓有成效的合作网络创建平台。

参考文献

［1］Ольга Фішук. Фішук О. україна-музейна Країна［EB/OL］.（2010-5-18）［2017-12-1］. https://www.unian.ua/culture/358865-ukrajina-muzeyna-krajina.html.

［2］Hudson，Kenneth. Museums：Treasures or Tools?［J］. Education al Facilities，1992：61.

［3］РОЖКО В М. Національна музейна політика：засади наукової діяльності музейних інституцій［M］. Prostir. Museum，2016.

［4］Jo-Anne Sunderland Bowe. The Creative Museum Analysis of selected best practices from Europe［EB/OL］. http://creative-museum.net/wp-content/uploads/2016/06/analysis-of-best-practices.pdf.

［5］Котвіцька，К. Показати з кращого боку：як українські музеї стають сучасними［EB/OL］.（2016-8-25）［2017-12-1］. https://zeitgeist.platfor.ma/museum-in-ua.

［6］MARIA XANTHOUDAKI，BRUNELLA TIRELLI，PATRIZIA CERUTTI，et al. Museums for science education：can we make the difference? The case of the EST，Journal of Science Communication［EB/OL］.（2007-6-6）［2017-12-1］. http://www.museoscienza.org/scuole/download/ESTarticolo_Jcom0602.pdf.

国家达尔文博物馆与俄罗斯自然历史博物馆协会

安娜·克林丽娜 ①

摘　要　创建于 1907 年的国家达尔文博物馆是俄罗斯历史悠久的自然历史博物馆之一，由亚历山大·科茨等人共同创办，该馆每年为不同的观众举办 50 ～ 60 个不同主题的展览。在国家达尔文博物馆中设立的俄罗斯自然历史博物馆协会发挥着联系俄罗斯各自然历史博物馆的作用。

关键词　博物馆　历史　现状　俄罗斯自然历史博物馆协会

位于莫斯科的国家达尔文博物馆是俄罗斯历史悠久的自然历史博物馆之一，它创建于 1907 年，2017 年迎来了其成立 110 周年纪念日。目前，它在俄罗斯自然科学博物馆中居领军地位，也是俄罗斯政府文化部下属所有俄罗斯自然历史博物馆的中心。许多俄罗斯博物馆归属于不同的政府部门，它们全部通过在国家达尔文博物馆中设立的俄罗斯自然历史博物馆协会和国际博物馆协会俄罗斯委员会维系在一起。

国家达尔文博物馆是俄罗斯唯一专门的进化史博物馆，总面积约为 20000 平方米，馆藏约 40 万件。它还收藏有关俄罗斯自然科学博物馆和展览方面的信息，并出版了介绍 462 家博物馆的参考书。博物馆每年都会举办各种科学会议、研讨会和圆桌会议，在过去 10 年里，举办了超过 330 次会议。

国家达尔文博物馆的展览在主楼中的占地面积达到 5000 平方米，包括 6 个展厅，每年举办 50 ～ 60 个临时展览。博物馆可向其他博物馆提供承办展览的业务。10 年来共计组织展览达 723 次。每年的参观人数超过了 65 万人次。国家达尔文博物馆的发展势头迅猛，全年不断地为参观者提供学习、探索、享受新事物的机会。

---

①　安娜·克林丽娜：国家达尔文博物馆馆长。E-mail：anna@darwin.museum.ru。

1. 国家达尔文博物馆的创建历史

在俄罗斯，许多博物馆都从私人收藏发展而来。国家达尔文博物馆由莫斯科高等女子学院的年轻达尔文主义教师亚历山大·科茨于 1907 年创建（彩页图 37）。

亚历山大·科茨自小就收集动物学藏品，大学毕业后，他受邀到莫斯科高等女子学院教授达尔文主义课程。他曾经用自己的收藏品来阐明课程内容，这就是国家达尔文博物馆的开端。

1911 年，亚历山大·科茨和妻子参观了许多欧洲的自然历史博物馆，萌生了一个建立新型博物馆的想法。科茨希望为百姓建造一座博物馆，但他既无容纳博物馆的建筑物，也没有资金，但他有一位他一直尊称为博物馆联合创始人的非凡的志同道合之士——他的妻子 Nadezhda Ladygina-Kohts，她负责在博物馆内进行科学研究，后来成为了科学博士。

第三位创始人瓦西里·瓦塔金是一位艺术家，后来成为艺术学院的院士。科茨认为艺术作品应该向动物标本注入活力，为展览增添美感。瓦西里·瓦塔金致力于博物馆工作的历史有 40 多年，现在的博物馆中展出了许多他的绘画和雕塑作品。

第四位共同创始人是标本分类学家菲利普·费德罗夫，他的毕生时光都在博物馆中度过，数以千计的博物馆展品均出自他富有创造力的双手。

国家达尔文博物馆是由这个伟大的团队组建，但却没有可供容纳博物馆的建筑物。亚历山大·科茨的一生都致力于向全国所有的领导者发出请求资助的呼吁，但均以无果而告终。亚历山大·科茨去世后，维拉·伊格纳季耶夫娜继任为第二任负责人，她设法让莫斯科政府决定为国家达尔文博物馆建造一座大楼，但她也没有在生前看到它的建成。博物馆大楼的建造花费了 20 年的时间。如今的国家达尔文博物馆拥有三栋大楼，包括主楼、展览馆和技术楼（彩页图 38）。

2. 国家达尔文博物馆的现状

每年，国家达尔文博物馆都会举办 50 ~ 60 场不断变化的展览，试图通过侧重于不同主题的展览，使不同的人可以找到自己感兴趣的东西。

每年有超过 65 万人参观国家达尔文博物馆，其中有超过 40% 的观众成了博物馆的常客。博物馆的参观者中有 85% 是家庭观众，只有约 13% 的参观者是组团而来的学生。博物馆可在一年内提供约 4000 次游览讲解服务。

玻璃陈设柜上贴有 QR 卡，任何参观者均可扫描二维码，用智能手机收听使用俄语或英语的游览讲解。

博物馆内设置了各种互动的综合设施，其中，有一个名为"行走于进化之旅"的全互动展览（彩页图 39）。

最近博物馆推出了一个名为"认识你自己——发现世界"的互动中心，在狭小的区域内设置了 52 件互动综合设施。中央大厅每天开启一项名为"生命行星"的灯光视频音乐展览，周末每天举行三次。此外，每年举办国际鸟类日、国际动物日、母亲节等 11 项公共节日活动。

博物馆每年出版发行 20～45 种出版物。在博物馆有限的区域内还创办了两个展览，其中一个展览的主题是植物，沿着"环境之路"的所有植物都标有俄文、拉丁文和盲文。另外还有植物学游览项目，包括针对盲人参观者的服务项目。第二个展览的主题是古生物学，主要包括曾经生活在俄罗斯境内已经灭绝的动物的雕塑。

博物馆也开展科学研究。博物馆的工作人员会开展各种野外调查。每年有 50～60 篇文章发表在各种科学和通俗杂志上。博物馆还举行联席会议、研讨会和圆桌会议。

博物馆的全部展览均设计得适合残障人士参观，残障人士可以单独或组团参观博物馆。2016 年，共有 8000 多名残障人士参观了博物馆。

3. 俄罗斯自然历史博物馆协会

1971 年，国际博物馆协会苏联委员会建立了自然历史博物馆分会。它在自愿的基础上，将自然历史博物馆的代表与当地历史博物馆中的自然部门科学机构以及政府各相关部门的代表联合在了一起。

1992 年，国际博物馆协会苏联委员会变更为俄罗斯委员会，自然历史博物馆分会在国际博物馆协会俄罗斯委员会框架范围内继续开展工作。

1996 年 9 月，在国家达尔文博物馆举行的自然历史博物馆会议做出了"依托国际博物馆协会俄罗斯委员会设立全俄罗斯自然历史博物馆协会"的决定。会上还通过了协会章程，选举出理事会。国家达尔文博物馆成了该协会的科学数据和方法论中心。协会在自愿的基础上联合各自然历史博物馆和地方历史博物馆中自然部门的代表，以及与自然历史博物馆直接相关的各科学机构、政府部门和代表。

全俄罗斯自然历史博物馆协会的主要目标是联合各自然历史博物馆员工的创造性努力和能力，而不论其从属地位。全俄罗斯自然历史博物馆协会肩负如下的各项职责：

（1）创建关于俄罗斯自然历史博物馆各种活动的信息库。2008年，全俄罗斯自然历史博物馆协会出版了一本介绍462家博物馆的专辑。

（2）促进博物馆工作人员的专业成长。

（3）支持自然历史博物馆的经验交流，包括普及国外博物馆的各种最佳做法。

（4）支持自然历史博物馆的工作人员发表科学著作。

（5）与国际博物馆协会自然历史博物馆和收藏国际委员会进行合作。

全俄罗斯自然历史博物馆协会每两年举办会议，对有关自然历史博物馆的各种议题展开讨论。自1996年以来，有超过1000人参加了这些会议。全俄罗斯自然历史博物馆协会与国际博物馆协会自然历史博物馆和收藏国际委员会（ICOM NATHIST）保持着密切联系。

2008年6月9—12日，国际博物馆协会自然历史博物馆和收藏国际委员会在俄罗斯国家达尔文博物馆召开年会。来自欧洲、亚洲、非洲和南美洲16个国家的50名代表出席了此次年会。

全俄罗斯自然历史博物馆协会组织开展的活动旨在构建一个共同的信息空间，为志同道合的人们进行沟通创造条件。定期举办各种兼具科学性和实用性的研讨会，旨在提高自然历史博物馆工作人员的专业技能。自1999年以来，共举办了19次研讨会，有来自246家博物馆的583名专家参加。此外还定期出版系统化的文献。每年公布国家达尔文博物馆的科学论文，作者不仅包括国家达尔文博物馆的员工，还包括来自俄罗斯和国外自然历史博物馆的同行。参与者们对协会提供的这种分享知识和经验、建立新业务联系的机会极为赞赏。

4. 结论

作为俄罗斯众多自然历史博物馆中的一分子，国家达尔文博物馆凭借具有奉献精神的员工努力并持续开展各项实践活动，如创造有趣的展览项目，为各类参观者营造互动中心和舒适环境，帮助他们发现和探索美丽的科学世界和野性自然，努力将博物馆建设得更加繁荣。

乌兹别克斯坦国家地质博物馆 ^①

A. sh. 艾麦德沙依枝　　H·A. 赛义夫 ^②

摘　要　乌兹别克斯坦堪称中亚独特博物馆和历史遗迹的宝藏。其中最突出的是国家地质博物馆，享有"地质宝库"的美誉。

乌兹别克斯坦地质博物馆成立于 1926 年，目前由乌兹别克斯坦共和国地质与矿产资源委员会负责管理。是由乌兹别克斯坦的一家小型地质博物馆发展来的。其藏品由地质组织、企业、机构、共和国和中亚著名勘探地质学家共同收集而来，目前，拥有大量藏品，为乌兹别克斯坦人民提供服务达 90 多年。

1. 地质博物馆馆藏资料

国家地质博物馆开展的各项工作及服务

（1）国家资助对古生物和独特地质资料石材（岩石、矿石、观赏石材和稀有矿物形成样本）进行收集和维护，以及获得有关具有科学、文化和审美价值的特色信息。

（2）向公众普及地质知识，丰富国有矿产原料基地，提高地下资源潜力。

（3）通过"Yosh Geolog"教育对地质感兴趣的儿童和学生，组织开展地质考察、游览和奥林匹克竞赛。

（4）编写有关记录，储存石质地质资料和古生物遗迹的说明。

（5）与新闻、广播电视媒体合作，共同促进地质工业取得成果。

（6）通过与乌兹别克斯坦国家机构"Uzbekkino"和乌兹别克斯坦国家电视广播公司合作，利用最新的信息技术（教育、科普、纪录片等），创作教育示范材料。

（7）博物馆的各种展品可使包括幼儿园孩子到专业地质学家在内的每位参观者都

---

① 国家地质博物馆。地址：11-a T.Shevchenko str., Tashkent, 100060；博物馆网址：muzeygeologii.uz；E-mail：muzeygeologii@umail.uz, muzeygeologii@yandex.ru；电话：(+99871) 256-11-92；传真：(+99871) 256-11-72.

② A. sh 艾麦德沙依枝　　H·A. 赛义夫：乌兹别克斯坦国家地质博物馆馆长。邮箱：muzeygeologii@umail. uz。

能找到与他们知识水平相对应的信息。经验丰富的导游、地质学家将为您解答参观博物馆期间提出的任何问题。

2. 科技出版物

2013 年，在 Kuraminsky 岭的北部山脚下，Akhangaran 附近发现了一具长鼻猿的遗骸。国家地质博物馆负责此次挖掘和安全工作。专家研究后认定这具长鼻猿的遗骸属于象科原脊象属最古老的代表。

2014 年，发表的科技文章为《关于在乌兹别克斯坦首次发现的首批古代大象遗骸（原脊象属）》，其发表在了名为"地质及矿产资源"（第 2 号）的科技杂志上。2016 年，《地质博物馆在共和国青年教育中发挥的作用》发表于名为"地质及矿产资源"的周年纪念刊物上（彩页图 40 ～图 43）。

2016 年 5 月 16 日，A.Sh. 阿赫迈德谢夫参加了由联合国教科文组织驻撒马尔罕办事处专门为世界博物馆日主办的国际科学与实践会议"博物馆科普教育：21 世纪全球和国家新战略"，并作了题为"在与参观者合作过程中，创新方法所起的作用"的报告。

自 2008 年以来，在乌兹别克斯坦共和国"教育法"获批通过，部长内阁颁布有关《乌兹别克斯坦青年才俊物质奖励法》，以及矿产资源委员会与教育部颁布联合决议的背景下，博物馆每年都会在学生中组织举办"Yosh Geolog"奥林匹克竞赛。

奥林匹克竞赛包括校级、市级、国家级 3 级，其目的在于培养学生对地质学的热爱和兴趣。以往奥林匹克比赛的经验表明，这一活动能够为解决地质人员短缺问题注入新动力。采矿和地质知识普及依据乌兹别克斯坦共和国范围内开展的从业人员培训计划实施，它覆盖了 2 万多名学生（彩页图 44 ～图 46）。

国家地质博物馆在相对较短的时间内不仅发展成为一个科学中心，而且成了矿产资源委员会的培训中心。既针对学龄儿童组织有校外学习，也为中、高级特殊教育系统地质及相关专业的学生开办有基础课程，并定期举办有关矿物学、岩相学、古生物学等讲座。在博物馆内部还专门为学生设置了一个特殊教育大厅。

博物馆举办各种专题展览，介绍乌兹别克斯坦矿产原料潜力，与外国合伙人和投资人举行商务会谈，讨论内容广泛，涉及矿产资源研究及开发层面构建方面的互利合作关系。在现代工业社会向信息化社会过渡的时代，地质博物馆面临的问题是如何向公众提供地质展品中所蕴含的科学信息。科学信息的传递、获得国家资助的评估主要依赖于各种媒体，包括初级—自然（地质标本）和各种次级（纸张、照片、电影、磁性、电子和其他媒体），其中，获得新科学知识和专家评估是最为重要且基础的。值得注意的是，地质标本最为主要，是揭示有关地质资料结构和组成的唯一可靠信息来源。博物馆大厅全景、展品摆放体系，以及博物馆导览资料中采用乌兹别克斯坦语、俄语和英语提供的展示资料和信息层次设计，帮助观众感知融为一体，加深对所见展品的理解。这将给每位博物馆的参观者留下难以磨灭的印象，并使参观者为其故乡矿物资源的自然性和独特性，古代探矿者和开拓者的无私工作，以及地下储藏室的现代发现者而感到自豪。

基辅国立理工学院博物馆

娜塔莉亚·皮萨雷大斯卡 [①]

摘要 国立理工学院博物馆是乌克兰的技术博物馆之一，也是乌克兰最大的大学博物馆。博物馆是乌克兰国立技术大学国立理工学院——伊戈尔·西科斯基基辅综合理工学院的一部分，同时，它还拥有建造于 1902 年的大学机械车间，以及建造于 1914 年的飞机机库。大学成立于 1898 年，是欧洲历史悠久、规模较大的技术大学之一。目前基辅综合理工学院的在校学生人数超过 2 万人。许多知名人士均曾在基辅综合理工学院工作和学习过。其中包括电焊的发明者尤金·佩顿、西科斯基直升机公司的创始人伊戈尔·西科斯基及火箭科学家谢尔盖·科罗廖夫。博物馆博览会涵盖 21 世纪各类技术，如广播、电视、计算机、电子、矿山与环境工程、机械和仪器、铁路、军火与军事通信、航空与航天探索，以及大学自身的历史发展。展览总面积达 2300 平方米，展 / 藏品 17000 件，覆盖时段从 19 世纪下半叶至今。博物馆举办针对年轻人的项目，邀请 20 世纪代表性人物参与活动。博物馆的使命是保护国家文化遗产并使其焕发光彩，引导青年选择职业。

关键词 技术博物馆 大学博物馆 教育 研究

1. 简介

乌克兰基辅理工学院是欧洲和乌克兰古老且规模较大的技术大学之一。

1898 年，在俄罗斯和乌克兰科技界和企业家的倡议下，为响应国家西南部工业建设的需要，乌克兰基辅理工学院得以创立，它在 19 世纪末发展迅速。

诸如 V. M. Kyrpychov, D. I. Mendeleyev, I. I. Sikorsky, S. P. Tymoshenko 以及许多其他世界著名科学家的生平和活动都和我们的大学有着密切的渊源。

基辅理工学院秉承在基础训练方面的深厚积淀，以及与毕业生实际工程技术发展和谐结合的举措，致力于在现有条件下，为学士、硕士、哲学博士及理学博士提供高

---

① 娜塔莉亚·皮萨雷大斯卡：基辅国立理工学院博物馆馆长。邮箱：tala1311@ukr.net

水准的培训。

目前，乌克兰396所大学和研究所中共有490个博物馆及展/藏品室。其中包括各种博物馆、教室、实验室、24个植物园、7个温室、17个观测台、2个天文馆及2个解剖剧场等。

2. 乌克兰基辅国立理工学院博物馆

乌克兰基辅国立理工学院博物馆是为纪念理工学院成立100周年，根据1995年5月29日乌克兰第360号内阁决议成立的，是乌克兰唯一的理工学院博物馆。展览会展示了乌克兰的工程发展历史，包括作为20世纪象征的新博览会（伊戈尔伊万诺维奇）西考斯基航空和航天学系，该博物馆于2008年开放。

3. 国立理工学院博物馆的展/藏品

乌克兰国立理工学院博物馆的藏品主要由各大院系和学术机构的分散展/藏品组成。大学里还有几个院系博物馆。发生于20世纪的各种命运多舛的事件毁损了这些博物馆的藏品。保护技术遗产的重要性未能引起国家和大学的重视。事实上，几十年来相关部门并未对技术遗产问题表示出适当的尊重。例如，欧洲独特轨炮、MESM以及第一台电子计算机有着极其珍贵的价值，但都没有被转移到其他博物馆进行收藏。在校长Zgurovsky的努力，国立理工学院博物馆于1998年成立，此后一直得到大学管理部门的支持。在博物馆建立后的一段时间里，集中收纳一些私人收藏主的展/藏品（不借助外力，便无法使藏品得到维护），促成了博物馆航空航天探索分部的成立。在国立理工学院博物馆的相关介绍中阐述了对乌克兰保存技术遗产历史的不同看法。

博物馆的建筑群，包括乌克兰国立理工学院博物馆的微型电动机、飞机和汽车车库。国立理工学院博物馆的展/藏品包括露天放置及在博物馆周围区域放置的航空博物馆展品，以及一些著名科学家和工程师的纪念碑。与此同时，它还具有国立和大学博物馆的地位。

博物馆和库房的展区面积分别为1700平方米和240平方米。

3.1 电子计算机展/藏品

就展/藏品的使用范围和独特性而言，电子计算机展品可谓在世界上没有任何与之类似的东西（近100个展示品），毫不夸张地讲，它有着与各种不同技术相关的独特零部件。早期计算机分部展示了一些首批乌克兰计算机，包括最早的 MIR-1（1965 年制造），其因于 1967 年被国际商业机器中心 IBM 收购而闻名。在乌克兰数学家及控制系统学家维克多·格鲁什科夫指导下的控制论研究所在推动计算机科学进步的过程中迈出了最初的几步。

3.2 空间分部的展 / 藏品

空间分部的展 / 藏品包括用于太空及为人类太空飞行做准备的各种展 / 藏品，它们在乌克兰可谓无与伦比（超过 1400 个样品）。此外，博物馆还收藏有由国立理工学院博物馆学生和科学家创造的现代纳米卫星"波利坦"。

3.3 武器展 / 藏品

博物馆所收藏的 19 世纪末至 20 世纪末的武器展 / 藏品（263 件展品），在欧洲也是无与伦比的。这些展 / 藏品揭示了小型武器作为技术装置从最开始到现在的复杂历史，着重介绍了设计和技术方面出现的各种革命性变化。该展 / 藏品汇集、展示了世界上一些最著名的武器系统（英国、法国、德国、美国、捷克斯洛伐克、苏联、中国）。

3.4 档案文件及藏书

博物馆的档案资料及藏书包括 19 世纪后期乌克兰国立理工学院博物馆的一些珍贵的且有价值的书籍和技术文献、原始图纸。特别有价值的展 / 藏品包括乌克兰国立理工学院学生在 E.O.Paton 等人领导下制作的各种桥梁总图和详图、关于画法几何学和数学的手写讲稿传真版。

4. 研究工作

博物馆的研究工作是其开展业务活动的一个重要领域，主要侧重于涵盖科学技术史、探索乌克兰科学家的创造性成果、评估科学家在创造和改进技术系统方面所做的工作，以及由科学家创立的学科和方向。这项工作包括搜索各种展 / 藏品、具有博物馆价值的纪念物品，以及对这些展 / 藏品、物品的进一步探索。博物馆展 / 藏品包括 17000 件各种形式由国家盘存的科技展品。此外还包括一些独特的展 / 藏品等。

5. 参观者及各种活动

博物馆累计接待参观者 3 万人次，并组织了 650 次博物馆游览活动。来自俄罗斯、中国、韩国、西班牙、马来西亚、美国、加拿大、芬兰、挪威、法国、德国、波兰、瑞士、日本、巴西、希腊、加纳等国的 95 个代表团有机会了解乌克兰的机器历史，科学家和技术人员的诞生地等。应大学各院系，特别是航空航天系要求而编制的专题游览数量有所增加。

博物馆与各种国际青年组织合作，举办各种节日（V Kyiv SteamPunkFest，TEDxKPI），极客日，循环进行乌克兰杰出设计师等科学阅读（2002 年以来，以阅读材料为基础，出版了 7 卷《乌克兰优秀设计师》），支持有趣的学生创新活动等。

6.结论

博物馆的宗旨是促进乌克兰青年成长为乌克兰技术、设计和科学精英的追随者。

博物馆的主要任务是有计划地构建现代大学的正面形象，同时考虑将社会结构、生活方式、社会、信息技术方面的变化作为青少年接受教育和职业的起点。博物馆尝试使用其唯一特殊的方法——通过展 / 藏品（展 / 藏品、媒体内容）来完成这一任务。

特种影视系统在科博场馆中的创新应用

李景霞 ①

摘 要 特种影视系统以其独特的技术优势在科博会馆建设中得以广泛应用,并发挥了其他展项不可替代的作用。本文系统地介绍了特种影视系统的概念、现状及需求,并试论述其创新应用。

关键词 特种影视　现状　前景　可持续发展　创新应用

1. 特种影视系统

特种影视系统是一种区别于传统电影的影视系统,可以将深奥、枯燥的科学知识以立体、互动、多维的方式呈现在观众面前。特种影视有三个特征,即用特种放映平台,在特种场合放映特种影片。因其数字技术手段的优势,近二十年来在各大科博会馆中得以广泛应用并发挥出了其他展项不可替代的作用。

1.1　特种影视的三特特征

1.2　常见的特种影视系统

科博场馆中常见的特种影视根据表现手段主要有梦幻剧场、情景式剧场、4D 影院、交互式 XD 影院、球幕影院、巨幕影院、飞行影院、VR 影院等。其最大的特点是

① 李景霞,宁波新文三维股份有限公司。联系电话 / 微信:13819805715,E-mail:346849489@qq.com。

传播科学内容的同时所展示出来的创新技术手段，使得影片更具冲击力和吸引力，并最大限度地增强观众的体验感和临场感。

2. 科普场馆对特种影视系统的需求

2.1 新媒体技术不断更新的需求

科普场馆在传播知识的过程中，往往通过一些常规的技术手段进行传播，表达的内容比较单一，展品之间相对孤立没有跨学科的联系。对于公众来说，新媒体技术越来越多地成为公众关注的焦点，传统的技术手段已经无法满足人们开拓的视野。对科普场馆而言，作为激发人们科学兴趣、启迪创新智慧的场所，必须具有最前瞻的创新能力，才能最大限度地体现其社会存在的意义。

特种影视集多种新媒体技术于一身，且不断融入新技术，符合时代特征。相对于传统的科普展品，它可以将多种深奥、零碎的知识碎片通过各种高科技创新手段综合性地进行传达，升华展品的展示功能，与人文、舞美艺术的高度融合使得它更符合大众的审美及参观需求。所以从某种意义上说，特种影视无疑是未来科普场馆中不可或缺的展示手段。

2.2 可持续发展的需求

自《全民科学素质行动计划纲要》颁布以来，各地的科技馆如雨后春笋般发展起来，科技馆事业在快速发展的同时，也出现了一些新问题和新矛盾，如可持续发展。那么如何使科技馆"永驻青春"，就需要我们定期对展品推陈出新，因此需要大量经费，但各地的资金有限，一旦跟不上，展品维护与更新就会停滞。特种影视是一种大投入、多功能、综合性的展示手段，如梦幻剧场，一次投资但是场地多功能，既可以进行梦幻剧演出，也可以用于儿童舞台剧表演，还可以进行科普实验或作为报告厅等。同时，节目内容还可以定期采购和配套，常换常新，从而达到可持续发展的目的。

2.3 免费开放趋势下的运营需求

科技馆展品的更新是科技馆可持续发展的重要手段。通常科技馆展品的更新率要求为每年 10% 左右。特大型的科技馆，如中国科学技术馆、上海科学技术馆等，拥有自己的展品研发队伍和研制场地，能保障每年更新超过 10% 的展品。但大多数科技馆经费有限，无法达到平均年更新率。如果馆内设置特种影视，收取相应的费用，可在一定程度上保障馆内资源的更新维护。

2.4 观众参观分流需求

对于一个大型的科技馆,展品数量为300～600件,日平均参观人数约为4000人,瞬时参观人数约为2500人,这大大超出了科技馆的承载能力。特种影视这种一次性可以容纳60～100人的大型展项,如馆内设置1～3件,可大大消耗科技馆的参观人数,是科技馆参观分流的一个重要手段。

3. 特种影视发展的现状及市场前景

3.1 中国特种影视发展的现状

自1959年中国首部立体电影问世以来,特种影视已经经历了58年的发展历程。其中,五花八门的电影形式层出不穷,或巨幕类、或感官类、或动感类,同时,拥有常规电影无法超越的视觉感受。虽然中国市场起步相对晚,相比而言,飞速发展的势头及庞大的市场需求不容小觑,尤其是世博会之后,更激发了国内市场,再加上大众对于特种电影的需求极大促进了特种影视行业的发展。截至2016年年底,中国拥有约34800块3D银幕,拥有球幕影院超过70家,拥有特种影视的主题公园超过100家。据了解,中影公司2011年一次性订购100套IMAX巨幕影院,可见特种影视市场之庞大。

国内科技馆中特种影视的占比也在逐年攀升,省级以上科技馆特种影视呈现数量多、投资大的特点,部分科技馆投资占比已超过20%,尤其是一些大型项目的引入,如飞行影院、传统骑乘、球幕影院等,但由于其场地要求大、投资成本高,在中小科技馆难以普及,为了更好地适应中小科技馆,大型的特种影视势必借助一些创新技术、创新手段,如VR技术、AR技术等,使其更加符合市场需求。

下面对部分科技馆特种影视的使用情况进行梳理见表1。

表1 部分科技馆特种影视的使用情况统计表

科技馆	开馆时间	特种影视
上海科学技术馆	2005 年	IMAX 立体巨幕影院、IMAX 球幕影院、四维影院、太空影院、梦幻剧场
广东科学中心	2008 年	3D 巨幕影院、数字球幕影院、4D 影院、虚拟航行动感影院、梦幻剧场
中国科学技术馆新馆	2009 年	球幕影院、巨幕影院、4D 影院、动感影院
临沂科学技术馆	2010 年	4D 影院、动感环幕影院、球幕影院、互动立体影院、动感立体影院、3D 全景声影院、科普剧场(激光剧场)、梦幻剧场
山西科学技术馆	2013 年	球幕影院、4D 三合一影院、梦幻剧场
辽宁省科学技术馆	2015 年	梦幻剧场、球幕影院、4D 影院、动感影院、IMAX 影院

3.2 中国特种影视发展的市场前景

对于特种影视定制市场而言，主要应用领域为科技馆、博物馆、主题公园，根据中国社科院发布的《2009年中国旅游发展分析与预测》(《2009年旅游绿皮书》)，自2010年起，中国将进入一个大型主题公园发展的新时期。科博场馆、主题公园市场规模的持续扩大将带动特种影视市场规模的持续扩大，尤其对于各省市级科技馆而言，特种影视作为新兴产品，对参观者极具吸引力，对发展科普文化、旅游文化具有带动作用，因而，市场需求更加旺盛。

从供应上来说，前些年，国内主题公园和各大科技馆、博物馆需要的特种电影大部分从国外购进，中国本土的特种影视供应商较少，并未从根本上解决国内市场的需求。近年来，国内一些从事特种影视的企业慢慢崛起，很大程度上已经掌握从设备生产到影片制作的全套技术，并在一定的领域展示出美好的前景。

4. 科博场馆中特种影视的创新应用

特种影视的创新应用很大一部分是由市场决定的。它因其科技含量高、视听效果出色、体验感强等特点深受各大科技馆青睐，成了各大科技馆的标配，也是广大观众争先体验的科普项目之一。其诸多技术如AR技术、虚拟现实、全息幻影、实时交互、VR技术等的引入，为特种影视的发展提供了新的突破，形成了一个新的领域，产生了如交互式影院、梦幻剧场、骑乘等各种高端的特种影视形式。

如辽宁科技馆梦幻剧场——"白垩纪之旅"（彩页图47）就是AR技术应用的典型案例，它是传统的科普剧结合多媒体演绎的一种综合性特种影视。它通过全息幻影技术、实景＋虚拟场景、真人＋虚拟影像、多层次画面互动等高科技手段，生动地演绎白垩纪时期的生长环境、动植物种类以及恐龙灭绝等知识点。

观众无须戴立体眼镜便可以看到栩栩如生的立体影像，多层次影像的穿梭、真人演员的神奇出现与消失、卡通形象与观众的实时互动等虚虚实实、亦真亦幻的特殊效果，使观众捉摸不透其中的玄机。

剧场充分利用了特种影视的特效手段突破空间、时间以及其他客观限制，是传统科普剧的一种全新尝试，它将晦涩、零散的知识点进行了综合性的表现、系统性的传达、艺术性的演绎，使观众置身其中学习科学知识、感受科学文化。

"传统骑乘"项目是集合最新科技的又一种特种影视形式，它因其超强的体验感在国内主题公园备受追捧，部分科技馆也引入了此项目，如杭州低碳馆"全球变暖"、佛山科技馆"梦想号"等,在科技馆中非常受欢迎,但目前尚未在中小型科普场馆普及,究其原因主要有三个:大场地、大投资、运营成本高。首先,在科技馆中此类项目空间要求是 2000 平方米左右。其次,投资 2000 万~ 3000 万元,这对于普通科技馆而言是负担不起的。(科技馆特种影视通常占地 200 ~ 500 平方米,投资 200 万~ 600 万元)。最后,运营成本高,成本包括维护成本、设备更新成本、环境更新成本、节目更新成本及人员费用,其维护费用及二次设计的成本很大程度上加大了科技馆运营的费用。

为了能适应中小型科技馆,符合大众品味,可从以下方面着手考虑。首先,减少大场景制作费用。将我们的视觉大世界缩小到眼前,在技术手段上可依靠时下流行的 VR 头显代替多个大场景,将人眼所不能极的物理空间通过 VR 眼镜实现,同时搭载自由度的动感平台,结合虚拟影像技术、全景 VR 技术、现场特效等高科技手段,实现 360 度全景体验,我们称之为"VR 骑乘"(彩页图 48)。其优势在于大大缩小场地。其次,节目内容可随时更换,且不受场地及布景的限制。最后,观众操作可以自助体验,1 名技术人员方可运营,大大减少了人员费用。

由表 2 可见,传统骑乘与 VR 技术的结合使得两者相得益彰,解决了空间、投资及其后期运营的瓶颈,进而开拓科技馆的科普教育模式,促进科技馆的可持续发展,为科技馆带来了新的亮点。当然,具有科普性、趣味性的影片开发也需要紧跟步伐,使其更加符合科技馆传播科学知识的宗旨。

表 2　科技馆传统骑乘及 VR 骑乘对比分析表

项目名称	场地面积(平方米)	投资	二次投入	工作人员
传统骑乘	2000 左右	2000 万~ 3000 万元	设备维护更新 5%=100 万~ 150 万元	4 ~ 7 人
VR 骑乘	50	100 万~ 300 万元	设备维护更新 5%=5 万~ 15 万元	1 ~ 2 人

5. 结论

总之,数字技术在特种影视中的创新应用,是未来发展的主流趋势之一,它不仅带来了全新的视听体验,还攻破了传统的特种影视在科技馆的可持续发展的难题。因此,科技馆必须要占领最新的前沿科技,才能使得越来越多的公众走进科技馆。

参考文献

［1］陈子俊. 前程似锦的特种电影［J］. 现代电影技术，2000（1）：9-16.

［2］岳鹏. 国内动漫及特种电影产业研究［EB/OL］.（2016-5-1）［2017-12-1］. http://wenku.baidu. com/view/cd620180b307e87100f6960c.html.

［3］许艳. 浅谈现代科普教育新形式——特种电影［J］. 科技与生活，2010（20）：1- 99.

［4］李红红，王志航，陈佳丹. 等，科技影视厅在科技馆可持续发展中的创新应用［C］. // 中国科 协年会. 2014.

中国科学技术馆的历史和发展

殷 皓[①]

摘 要 中国科学技术馆是中国唯一的国家级综合性科技馆,是实施科教兴国战略、人才强国战略和创新驱动发展战略,提高全民科学素质的大型科普基础设施。从 1978 年邓小平同意建设中国科学技术馆,到 1988 年一期建成开馆,到 2000 年二期建成开放,再到 2009 年新馆建成开放,中国科技馆历经 3 期工程,每个阶段都是一次历史性的跨越。近年,中国科学技术馆围绕强化实体科技馆展教示范作用、服务中国特色现代科技馆体系建设、推动中国的科学技术馆事业发展等方面,做出了积极的探索,取得了较好成绩。2018 年即将迎来中国科学技术馆开馆 30 周年,中国科学技术馆将担负新使命,拓展新思路,为提高全民科学素质贡献更多更大的力量。

关键词 中国科学技术馆 历史 发展

1. 中国科学技术馆发展历程

中国科学技术馆的筹建始于 1958 年,后因种种原因停工。直到 20 年后,1978 年 3 月,邓小平在全国科学大会上重申"科学技术是生产力"的观点,中国迎来了科学的春天,茅以升、王大珩等 83 位著名科学家在会上联名提出恢复建设中国科学技术馆的建议,同年 11 月得到邓小平圈定同意。从此,中国科学技术馆踏上了建馆创业的新征程,其主要经历 3 次工程。

自 1979 年始,历经十年艰辛,中国第一座"科学中心"中国科学技术馆于 1988 年 9 月 22 日建成开放(一期工程),建筑面积 2 万平方米,展厅面积 5000 平方米。

科技馆独特的展览教育内容和方式,逐渐受到越来越多公众的欢迎。中国科学技术馆乘势而上,二期工程于 1996 年 11 月立项,2000 年 4 月 29 日建成开放,建筑面积 2.3 万平方米,展览面积 1.5 万平方米。

① 殷皓,中国科学技术馆馆长。E-mail:yinhao@cstm.org.cn,地址:北京市朝阳区北辰东路 5 号,邮编:100012。

随着《中华人民共和国科学技术普及法》的颁布，公众对科普的需求日益高涨，中国科学技术馆新馆于 2005 年 4 月立项，历经 4 年努力，2009 年 9 月 16 日正式开放。

新馆总投资 20.3 亿元，建筑面积 10.2 万平方米，展览面积 4 万平方米。设有"科学乐园""华夏之光""探索与发现""科技与生活""挑战与未来"5 个主题展厅，展项 1000 余件（套）；设有 1 个短期展厅；设有公共空间展示区和球幕影院、巨幕影院、动感影院、4D 影院 4 个特效影院；设有多间实验室、教室、科普报告厅及多功能厅。开馆 8 年来，截至 2017 年 10 月底，接待公众 2720.67 万人次，超过一期和二期工程 20 年接待量 2100 万人次。2010—2016 年，平均每年 330 万人次。2016 年全年接待公众 383 万人次。2017 年暑期接待公众 175 万人次，2017 年 8 月 12 日单日接待 55866 人次，创新馆单日接待人次新高。

中国科学技术馆的建设和发展，一直得到国家领导人的亲切关怀。1958 年，周恩来总理批准筹建中国科学技术馆。1978 年 11 月，邓小平圈定同意建设中国科学技术馆，并于 1984 年为一期工程奠基题词。江泽民主席于 2000 年 4 月为二期工程建成开放题词："弘扬科学精神，普及科学知识，传播科学思想和科学方法。"胡锦涛主席于 2004 年 5 月视察中国科学技术馆并与少年儿童欢度节日，又于 2010 年 5 月再次来到新馆与少年儿童欢度节日。习近平主席于 2009 年 9 月来馆参加全国科普日北京主场活动，又于 2010 年 5 月再次来馆参加儿童节活动。

2. 主要工作情况

近年，中国科学技术馆按照"普及科学知识、弘扬科学精神、传播科学思想、倡导科学方法"的目标，秉持"体验科学、启迪创新、服务大众、促进和谐"的理念，围绕中国特色现代科技馆体系建设和全民科学素质提升，重点推进了以下工作。

2.1 强化实体科技馆展教示范作用

（1）积极推进展厅改造。最早的是 2011 年启动的"气象之旅"展区改造，后于 2016 年启动完成"太空探索"和"信息之桥"2 个展厅的整体改造。2017 年 2—9 月完成"华夏之光"整厅封闭升级改造。目前，"儿童科学乐园"等展厅改造正在推进。

（2）注重科技教育品牌建设。2016 年全年开展教育活动 4.2 万场，受众约 226 万人次。围绕中小学生需求，推出"中考串讲"等"定制你的科技馆之旅"系列活动。围绕增强馆校结合，推出"开学第一课"等品牌活动。举办"中科馆大讲堂"180 余期、听众 7 万余人次。策划"春江花月夜"等重点科学表演项目。原创首部大型互动科幻童话剧《皮皮的火星梦》于 2017 年 5—10 月，先后在馆内、香港航天科普展、山西等地演出 60 场，观众 2.34 万人次。

（3）深挖短期展览潜力。新馆建成以来共举办短期展览 75 次。主要有 4 种模式：一是自主研发，如"互联网展""心理学展"等；二是采用大联合、大协作方式进行合作研发，如"航天展""盐的故事"等；三是通过众筹的方式举办展览，如"虚拟现实""无人的力量"等；四是引进国际高水平展览，如瑞士"爱因斯坦展"、德国马普学会"科学隧道 3.0"展等。目前展出的"古希腊科技与艺术展"于 2017 年 11 月 3 日与希腊互换展览。

（4）发挥影视科普作用，提高影视原创能力。开馆 8 年来，累计放映各类特效影片 96 部，接待观众 390 万人次。原创制作完成 4D 电影 3 部、3D 电影 5 部、球幕电影 1 部。原创制作完成《平衡神器》等数十部科普影片，部分已在中央电视台播出。

2.2 服务中国特色现代科技馆体系建设

根据 2012 年 11 月中国共产党第十八次全国代表大会提出的"促进基本公共服务均等化"等要求，中国科学技术协会在整合以往工作的基础上，适时提出要建设中国特色现代科技馆体系，即在有条件的地方兴建实体科技馆；在尚不具备条件的地方，县域主要开展流动科技馆巡展，乡镇及边远地区主要开展科普大篷车、配置农村中学科技馆；同时开发基于互联网的数字科技馆网站。中国科学技术馆作为国家馆，在服务体系建设方面的主要工作如下。

（1）中国流动科技馆。2010 年启动研发，2013 年正式运行，以县城公众特别是中小学生为主要服务对象。截至 2016 年年底，共配发流动科技馆展览 295 套，巡展 1747 站，累计受益观众约 6757 万人次，圆满完成"四年基本覆盖县（市）"预期目标。2017 年 9 月，正式启动第二轮全国巡展工作，配发 69 套。截至 2017 年 10 月底，累计巡展 2216 站，参观人数达 8382 万人次。初步实现科技馆教育理念、内容、形式在县级城市的推广。

（2）科普大篷车。于 2000 年启动，以乡镇中小学生和农村居民为主要服务对象。截至 2017 年 10 月底，累计配发 1445 辆，开展活动 19.2 万次，行驶里程约 3377.1 万千米，受益人数 2.1 亿人次。其机动灵活的特点，很好地满足了基层公众的科普需求，被亲切地称为"科普轻骑兵"。

（3）农村中学科技馆。于 2012 年启动，以农村地区青少年学生和周边公众为主要服务对象，着力解决科普"最后一公里"问题。截至 2016 年年底，累计建设 293 所，受益青少年逾 137 万人次。2017 年底保有量将达到 539 所。该项目是政府、企业、基金会三类组织资源地有机整合，致力于"科技助力精准扶贫"的有益探索。

（4）中国数字科技馆。于 2005 年启动，目前定位是国家科技资源共享服务平台、现代科技馆体系建设枢纽、面向公众的科普网站。截至 2017 年 10 月底，网站日均浏览量超过 293 万，网站流量中国排名最高达 76 名，近期稳定在 100 名左右。官方资源总量 9.95TB。先后推出"榕哥烙科""科学开开门"等多项原创栏目。数字馆的建设，推动了优质科普资源的共建共享，正在打造一个"永不闭馆"的科技馆。

2.3 服务科技馆事业发展

（1）加强行业标准化、规范化建设。历经 4 年筹建，中国科普行业第一个标准化技术委员会——中国科普服务标准化技术委员会，2017 年 6 月得到中国国家标准委批复成立。同时组织修订《科学技术馆建设标准》，组织研究编写《绿色科技馆建筑评价标准》，为行业的发展确定标准和规范。

（2）加强行业交流研讨。以科技馆发展基金会、中国自然科学博物馆协会、博协科技馆专委会、科普场馆特效影院专委会等为依托，组织科技馆发展奖评选表彰、全国科技辅导员大赛、全国科技馆馆长培训班、科技馆馆长论坛等，编印《中国自然博物馆研究》杂志，编译美国《维度》杂志。

3. 中国科学技术馆国际交流与合作

3.1 基本情况

中国科学技术馆历来重视国际交流与合作。早在 1982 年，场馆尚未建成开放就已策划推出"中国古代传统技术展览"走出国门，远赴加拿大安大略科学中心展出。至今，该展览先后在美国、瑞士、英国、德国、比利时、意大利、泰国、新加坡、荷兰、希腊等 13 个国家和地区的 23 个城市巡展，参观人数 657.2 万人次。同时，早在 1983 年就引进了加拿大"安大略科学中心展览"。

近年，中国科学技术馆正努力通过考察、参会、培训、巡展等方式深化国际交流与合作。新馆开馆以来，截至 2017 年 10 月底，共派员进行 40 项 118 人次的双边或港澳台交流，参加 31 个 173 人次的国际组织年会或国际会议。

目前中国科学技术馆已加入国际博物馆协会、世界科学中心峰会国际程序委员会、北美科学中心协会、大银幕影院协会、亚太地区科技中心协会 5 个国际组织。其中，中国科学技术馆是亚太科技中心协会的创始成员单位之一，1998 年承办了协会成立（1997 年）之后的首届年会，2016 年再次承办 ASPAC 第十六届年会，164 名境外代表和 356 名中国大陆地区代表参会，大会取得较好效果。

3.2 未来计划

为落实习近平主席提出的"一带一路"倡议，进一步加强国际交流与合作，为构建人类命运共同体贡献力量，中国科学技术馆提出了以下4方面考虑。

（1）加大"华夏之光"系列主题展览全球巡展力度。推介"中国古代传统技术展""中国古代机械展"等已有的展览，策划推出"榫卯的魅力""做一天马可·波罗"等新的展览。

（2）面向中亚、东南亚地区推广中国流动科技馆、科普大篷车运行模式，推动实现常态化巡展。此次会议期间已与缅甸教育部签署了流动科技馆赴仰光等地巡展的合作意向书。

（3）采取"展览＋活动＋讲座＋科普影视＋文创＋互联网＋虚拟现实＋培训"集成方式，重点面向"一带一路"沿线国家和地区推出一批科学传播精品项目。

（4）计划于2018年中国科学技术协会主办"全球公民科学素质促进大会"期间，举办"一会"（科普场馆促进公民科学素质提升国际论坛）、"一展"（中国高新科技成就主题展览）、"一赛"（全球科学表演邀请赛），打造国际科学中心交流盛会。

习近平主席曾描绘："'一带一路'建设要以文明交流超越文明隔阂、文明互鉴超越文明冲突、文明共存超越文明优越，推动各国相互理解、相互尊重、相互信任。"中国科学技术馆非常愿意将此次国际研讨会作为与"一带一路"沿线国家和地区科普场馆增进交流与合作的良好开端，大力弘扬"和平合作、开放包容、互学互鉴、互利共赢"为核心的丝路精神，与各馆一道，努力推动研讨会提出的各项合作取得实质性进展，共同促进"一带一路"沿线国家和地区科普场馆共建共享、繁荣共赢。

从对外合作看北京麋鹿生态实验中心发展

宋 苑 ①

摘 要 麋鹿是中国特有的平原湿地动物，百年前曾因国家的衰败成为海外孤儿，幸得英国贝福特公爵搭救才免遭灭绝命运。北京麋鹿生态实验中心是 1985 年为麋鹿重引进回中国而设立的单位，成立 32 年来，中心的发展基本围绕着麋鹿展开。近年来，更是积极与海外学者、机构合作，研究麋鹿和麋鹿文化的发展之路、宣传生态文明建设理念。走出去，引回来，再走出去，麋鹿中心的发展离不开各方的协助，在"一带一路"政策引领下，麋鹿中心未来的发展会更好。

关键词 北京麋鹿生态实验中心 麋鹿 历史 发展

北京麋鹿生态实验中心（以下简称麋鹿苑），又名北京南海子麋鹿苑博物馆、北京生物多样性保护研究中心，是以麋鹿这一珍稀物种的保护为核心的生物多样性研究机构，更是一个围绕麋鹿保护主题宣扬生态保护、环境保护的科普场所，是集科学研究和科普教育于一体的综合型户外生态博物馆。麋鹿苑成立于 1985 年，为重引进珍稀物种麋鹿而建，1995 年起向社会公众开放。麋鹿苑的工作涉及了动植物研究保护及科普教育两大主要内涵。经过 32 年的发展壮大，麋鹿苑在麋鹿的保护与扩散、湿地生态系统恢复、生物多样性研究等科研方面成绩斐然，在生物多样性教育以及动植物保护环境教育等科普方面表现突出。随着生态文明建设理念的提出，"一带一路"政策的落实，麋鹿苑的科普工作日新月异，对外合作逐步增加。合作带来发展，如今，麋鹿苑内寓意深刻的科普设施、展览、活动鳞次栉比，对保护动植物、和谐共生的生态文明理论进行了良好展示和宣传。

---

① 宋苑：北京麋鹿生态实验中心科普部部长。研究方向：自然科学普及内容及方式。地址：北京市大兴区南海子麋鹿苑，邮编：100076，E-mail：yl_s2000@aliyun.com。

1. 麋鹿的三国情缘

麋鹿，同大熊猫一样，是中国特有物种，曾广泛分布在中国东部的平原、湿地上。它曾是中国皇权的象征，在商周时期就已被圈养在当时的皇家猎苑中，并被当时的皇族定为皇家御用祭祀物和猎物。中国有两句成语"鹿死谁手""逐鹿中原"，这里面的鹿说的就是麋鹿。由于麋鹿的生活环境在历史中屡遭战乱灾荒，清朝以后，仅北京南苑皇家猎苑里还存活有最后一个种群的麋鹿，其他地区的麋鹿均已灭绝。

1865 年，法国神父阿芒·戴维来中国传教，作为一个博物学家，他在北京南苑发现了麋鹿。通过南苑侍卫，戴维神父购买了两套麋鹿的骨架、皮肤等标本，并直接运回法国巴黎自然历史博物馆进行鉴定。至此，麋鹿在动物分类学上有了自己的一席之地，也开始被欧洲人接受。

因战乱和水灾，麋鹿于 1900 年在中国灭绝。幸而之前有一些麋鹿被运到了欧洲。这些麋鹿生活在欧洲的动物园里，但它们的繁殖率非常低，面临着灭绝的风险。此时，英国第十一世贝福特公爵因喜爱出资购买了分布在欧洲各个动物园的所有麋鹿，共计18 头，放养在乌邦寺庄园里。他的这一壮举挽救了麋鹿，让这一物种得以继续生存。

第二次世界大战期间，贝福特公爵家族唯恐战乱导致麋鹿的灭绝，又将一些麋鹿送出了乌邦寺，但当时并没有送回中国。直到 1985 年，麋鹿在中国消失了 85 年后，第十四世贝福特公爵才最终和中国政府签订了协议，将麋鹿重引进到它当初消失的地方——中国北京南海子。

2. 麋鹿苑的发展

麋鹿苑因 1985 年麋鹿的回归而建。自建立起，麋鹿苑就与英国贝福特公爵家族保持着良好的合作关系，公爵还特派了专家驻麋鹿苑协助开展麋鹿重引进的各项事宜。建立之初，麋鹿苑的工作职责只有一个，即保护繁育重引进回国的麋鹿。1995年，麋鹿苑开始作为科普单位对外开放，麋鹿苑走上了科普大发展之路。自 1999 年被批准成为"北京南海子麋鹿苑博物馆"后，麋鹿苑被先后批准为"北京市科普教育基地""全国科普教育基地""社会大课堂基地""爱国主义教育基地""北京市环境教育基地""北京市生态文明宣传教育基地"等。在这些称号获得的同时，麋鹿苑的科普工作如火如荼地发展了起来。

2.1 科普设施

作为北京地区唯一的户外博物馆，麋鹿苑首先给公众展示的就是麋鹿的生活环境——湿地生态系统。配合湿地生态系统，麋鹿苑建设了很多宣传动物保护、生境保护的科普设施，如警示世人的"滥伐的结局"、悬崖峭壁的"燕窝"和遨游汪洋的"鱼翅"，用雕塑的象形形态和诙谐的讲解词让公众理解没有买卖就没有伤害；发人深省的"地球号"（彩页图 49）方舟延引圣经中"诺亚方舟"的故事，告诉公众为什么要保护地球生态；"灭绝动物多米诺骨牌"（彩页图 50）则是利用一个个倒下去的动物墓碑以及阻止的大手告诫世人为何要保护动物，保护生态；宣扬"生态和谐共生"的科普设施"中华护生诗画"和"鹿字墙"则向公众普及了中华民族自古就有的"护生惜物"思想以及博大精深的鹿文化。参观者常以"长知识、增见识""创意无限""没看到就想不到"等评语来评价麋鹿苑的科普设施，正因为有这些科普设施的存在，麋鹿苑才和其他公园以及动物园等区别开来。

2.2 科普展览

2001 年，麋鹿苑第一个室内常设展"麋鹿传奇"（彩页图 51）向公众开放，这一展览打破了麋鹿苑作为博物馆却无室内展的窘境。2010 年，通过与德国私人博物馆的合作，麋鹿苑引进了近 3600 件关于鹿的标本。以这些标本为基础，2014 年，麋鹿苑布置了常设展"世界鹿类"（彩页图 52），这一展厅全方位向公众展示了鹿类知识，并同时深入挖掘鹿与人类的关系、人对鹿的影响等文化内容，成为科普教育的新亮点。2015 年，为纪念麋鹿回归 30 周年，鹿角大观展、麋鹿苑生物多样性成果展等展览也依次推出。在充实博物馆展览的同时，也提高了麋鹿苑生态文明教育的基础。随着展览的逐步推出，麋鹿苑的参观者数量也逐年递增，这些展览累计接待游客超过上千万，好评如潮。

2.3 科普活动

科普教育也是麋鹿苑作为博物馆的重要工作之一。通过这些年与中外自然教育工作者的交流，麋鹿苑逐渐形成了一套属于自己的科普教育体系。目前，麋鹿苑利用自身优势设计了大量科普课程与科普游戏，并多次组织社会大课堂、社会实践等活动。麋鹿苑的科普课程内容广泛，其中包含专门针对中小学生设计的系列自然体验课程"麋鹿苑春夏物候相册""玩在大自然""麋鹿苑生态导览图"等。这些课程把生态文明理论带进课堂，让学生在体验自然的同时感受人与自然和谐共处的理念。每个参加过麋鹿苑自然课程的学生都表示形式新颖，所感颇多。麋鹿苑还进行了自然教育科普剧的创作，通过大众喜闻乐见的形式宣传环保理念。2016 年和 2017 年，麋鹿苑推出了全新的"科普剧汇演"活动，将自主编演的科普剧推上了专业舞台，让更多的人可以通过欣赏科普剧领略自然环保的理念。这一形式的推出，让麋鹿苑的科普理念更加贴近大众，无论是社区民众、学校团体，还是亲子团体，都在看完演出后给予了极高的评价，认为这一形式接地气，在放松的同时受教育，很值得推广。同时，麋鹿苑自主开发的"自然故事大讲堂"系列活动，利用活动将科普讲座、课程、游戏、手工等科普形式串联到一起，让参与者在欢乐中感受生态文明教育的魅力和意义。

麋鹿苑还以多种科普方式积极参与各类社会公益活动，宣讲生态环保理念，普及生态文明知识，把麋鹿苑特有的生态文明理念普及给社会大众，让公众了解生态环保的重要性。近年来，麋鹿苑的科普工作者们走遍全国，为全国民众送去多道科普大餐，让生态文明教育之花在祖国处处绽放。累计共有近 20 个省、市、自治区的超万名民众参加过麋鹿苑的科普活动，获得的评价也极为中肯，如"孩子特别喜欢""形式独特，利于推广""大自然还可以这么玩，太好了"等。

3. 合作带来发展

近年来，随着麋鹿苑的发展，麋鹿苑对外合作的机会越来越多，吸引到苑参观的外宾也逐年增加。2015 年，在麋鹿回归 30 周年之际，"麋鹿与生物多样性保护国际研讨会"在北京胜利召开。此次会议不仅加强了麋鹿和生物多样性保护的成果交流、检验了麋鹿中心的科研成果，也拉开了建立长效国际交流合作机制的帷幕。目前，与麋鹿苑保持着长期合作联系的国家有澳大利亚、英国、德国等。这些合作给麋鹿苑带来了机遇，通过合作，麋鹿苑的科普水平得到提升，也让更多外宾了解了麋鹿这一传奇物种。

未来，麋鹿苑会继续加强对外合作，以麋鹿这一国宝为代表，以麋鹿苑的科普成果为基础，学习国外的先进经验和技术，输出自己的生态文明科普理念，走出国门，寻求更多合作。

浅谈重庆科技馆展览展示及教育活动的开发与研究

张 婕[①]

摘 要 "一带一路"倡议是我国政府根据时代特征和全球形式提出的重大倡议。科学教育交流合作作为共建"一带一路"的重要内容,是提升沿线国家文化、教育合作水平的重要领域,也是促进沿线国家文化相融、民心相通的有效途径,科技馆的建设与发展也是科学教育的一部分。本文以重庆科学技术馆为例,主要介绍重庆科技馆的发展历史、展览展示及教育活动方面的开展情况及经验教训,探讨场馆未来发展面临的困难及挑战,提出几点对策建议,希望能增进"一带一路"沿线国家在科技馆建设和发展中彼此间的了解、认知和互信。

关键词 科学技术馆 历史 展品展项 教育活动

科技馆作为实施科教兴国战略的基础设施、城市文明的重要标志、公众接受科普教育的活动场所,越来越受到社会各界的高度重视,全国科技馆的建设形成了高潮。随着人们对科技馆的认识不断深化和科技馆建设理念地不断提升,事业的蓬勃发展,使科技馆这种源于西方文明的事物在与中国科普活动相结合的实践中形成了一种文化——科技馆文化。

1. 科技馆的理念及功能定位

科技馆是科普教育的场所,其展示教育的方式和功能是通过直观视觉刺激和互动体验,来唤醒人们进行探索和学习的兴趣,并逐渐改变观念,形成科学的世界观和价值观。因此,它还是学校第二课堂、旅游热点、公民培训教育基地、先进的科学技术

---

① 张婕:重庆科学技术馆馆员,重庆科学技术馆科普教育专员。地址:重庆市江北区江北城文星门街7号,邮编:400024,E-mail:182262914@qq.com。

推广中心、科技服务中心、国内外学术交流中心等。

2. 重庆科学技术馆建设概况

2.1 重庆科学技术馆建设指导思想

坚持以人为本，以"创新·和谐"为理念，以"生活·社会·创新"为展示主题，以激发科学兴趣、启迪科学创新为教育目的，为公众营造从实践中学习科学的情境，通过互动、参与、体验等教育方式，引导公众进入探索与发现科学的过程，为提高公众科学文化素质、建设创新型城市、构建和谐社会服务。积极打造成"体验科学魅力的平台，启迪创新思想的殿堂，展示科技成就的窗口，开展科普教育的阵地"。

2.2 重庆科学技术馆概况

重庆科学技术馆为重庆市委、市政府确定的全市十大社会文化事业基础设施重点工程之一，是重庆市科协直属事业单位，是面向公众的现代化、综合性、多功能的大型科普教育活动场馆，是实施"科教兴渝"战略和提高公民科学文化素养的基础科普设施。

其位于长江与嘉陵江交汇处的重庆江北嘴中央商务区（CBD）核心区域，于2006年1月7日奠基，同年10月动工建设，2009年9月9日建成开馆，至今，已开馆8年。该馆占地面积2.47万平方米，建筑面积4.83万平方米（其中，展览教育面积为3万平方米），总投资额5.67亿元（其中，建安工程4亿元，展示工程1.67亿元）。截至2017年8月，重庆科学技术馆已累计接待观众超过1240万人次。

3. 重庆科学技术馆展览展示及教育活动概述

展览展示和科普教育活动是科技馆可持续发展的动力和源泉，也是场馆发展的核心竞争力，下面就重庆科学技术馆展览展示及活动开发做简要概述。

3.1 重庆科学技术馆展览展示特色和设计开发现状

科技馆的展教理念是科技馆的灵魂，科技馆的展览不是单纯的展品陈列，而是通过展览来启迪人们的思想，唤醒观众意识。重庆科学技术馆内容建设分为常设展览、短期展览、科普培训实验、科普交流活动、科普影视等方面。其中，展品主要由多媒体演示型、互动参与型、原形实物或模型展示型等组成。

3.1.1 重庆科学技术馆展品展项设计的原则

（1）注重结合。展览内容上注重科技与经济、社会发展相结合，自主创新与借鉴引进相结合。

（2）突出互动。展品以互动、参与和体验为主，综合运用多种形式，体验科技的神奇与美妙，营造适宜的学习情境。

（3）体现特色。采取主题展开的设计方法，以故事线、知识链等方式构建内容之间的有机联系，体现采取科技馆的自身特色。

3.1.2 重庆科学技术馆展览内容的规划布局及升级改造

（1）常设展览展示的规划布局。围绕常设展览"生活·社会·创新"共设生活、防灾、交通、国防、宇航科技和基础科学6个主题展厅，以及儿童科学乐园和工业之光2个专题展厅。展品涵盖军事、航空航天、微电子技术、虚拟模拟技术、生命科学、基础科学等多项学科领域，展品数目达400余（套）。

各展厅展示内容如表1所示：

表1 展厅展示内容统计表

楼层	展厅名称	展厅内容概述	主题展区	展品数量
二楼	生活科技	展示衣食住行等日常生活中蕴含的科学原理和科技成就，传播"生活离不开科学，科学改善生活"的理念，引导观众穿出品味、吃出健康、科学居家，梳理节能与环保意识，享受信息技术的成果，养成科学的生活方式，掌握科学的生活常识	穿衣打扮、饮食健康、科学家具、人体健康、能源与环境、网络与生活	145
二楼	儿童乐园	展示适合3～10岁儿童身心特点的科技内容，模拟儿童生活环境，让儿童在可爱的家中感受身边的科学，在爱心社区体验城市发展中的科技成果，在快乐工厂了解现代化工厂的运转流程，在梦想学校参与职业模拟体验，感受创造的乐趣，在科学公园领会自然的神奇，激发儿童探究科技的好奇心和创新精神，培养儿童对科学的热爱和社会情感	可爱的家、爱心社区、快乐工厂、梦想学校、科学公园	39
三楼	防灾科技	模拟狂风、暴雨、雷击、火灾、地震、泥石流等多种灾害，体验灾害发生的情景，了解灾害发生的科学原理和政府防灾减灾的账册措施，学会科学面对灾害，促进人与自然和谐相处	重庆防灾、气象灾害、火灾、地震、安全知识、灾害体验	39
三楼	交通科技	以重庆特有的山水和人文环境为大背景，让观众了解交通发展的历史，在参与互动中体验各类交通设施和交通工具，掌握交通科技知识，畅想高速、便捷、人性的未来交通	交通体验、交通设施、交通工具、未来交通	19
三楼	国防科技	结合重庆独特的军工资源优势，运用实物模型、多媒体等多种表现形式，模拟战争的"新、奇、险、特"，展示军事与科学、战争与和平方面的知识，让观众了解国防科技的历史、现实和未来。系国内科技馆中首个单设的国防科技展区	千里纵横、明日之师	14

续表

楼层	展厅名称	展厅内容概述	主题展区	展品数量
四楼	基础科学	把"像科学教一样思考"融入整个展厅中,集中展示数学原理及其运用,通过神奇的力、电的世界、美妙的光、声音的世界等经典物理展品,让观众感受到科技的美妙与神奇,启发观众去发现并获得开启科学大门的钥匙	区位数学、经典物理	86
	宇航科技	从宇宙大爆炸开始,介绍宇宙的起源、中国古代天文学成就和天体运行规律等知识,探索"宇宙之谜"。模拟人类对宇宙的探索活动,体验最新的宇航科技成果,了解人类探索太空的历程,畅想"宇航之梦"	宇宙之谜、宇航之梦	40
副楼一楼	工业之光	形象地展示重庆工业发展历程中的重要科技成果,突出反映重庆老工业、现代工业和店里能源、汽车摩托、装备工业、重化工和电子等支柱产业的科技成就,让观众了解重庆工业发展的历史,感受科技在推动工业化进程中的作用,领略工业在推动重庆经济社会发展中的风采	工业文明、支柱产业	
公共空间	科学艺术	展示国内外和重庆科技文明的成果,展示科学之美、科技之光,感受科学与艺术的魅力	科技之美、科技之光	

（2）短期展览内容规划。短期展览是对常设展览内容的丰富和补充,具有主题鲜明突出、内容相对独立、展示时间较短的特点。短期展览重点展示国内外重要科技事件、重大科技成果、最新科技进展和重要科技活动、有影响力的科技人物、公众关注的科技热点,有助于增强科技馆的人流量,扩大科技馆的影响力。

（3）其他内容建设规划。科普培训实验是对展览教育的扩展和深化,主要包括培训、实验、科普讲座、报告会以及青少年科技竞赛、科技俱乐部、冬夏令营活动。科普交流活动是展览功能的拓展和丰富,主要包括科技论坛、科技沙龙、科技问题辩论会、科学家与公众见面等科学文化交流活动。科普影视是科技馆特色的教育形式之一。设置 4D 动感电影、巨幕电影等科普影视项目。

（4）筹备并启动场馆展品展项的升级改造工程。随着新馆效益的褪去,场馆参观人流量骤减,大幅度地更新改造展品迫在眉睫。经过长远规划,重庆科学技术馆于 2015 开始启动展品 5 年改造项目,逐步对场馆展品分区域进行更新改造,以达到展厅常展常新,满足忠实"粉丝"的参观需求。前期改造筹备工作具体表现在:①场馆运行期间,注意观察展品动态,对易损坏、参与性不高的展品与抗损坏、受欢迎的展品进行对比分析,为展品的更新改造提供理论和数据基础。②加强科技发展前沿信息的积累,关注时事热点,了解最新的科技研究成果,调查搜集观众感兴趣的热点产品,对时效性强的展品进行及时更新。③搜集各展品制作单位的相关信息,并整理归类。

了解各制作单位擅长的方向，例如是技术、信用度还是展品的稳定性等，为展品设计和改造提供坚实的后盾。

3.2 重庆科学技术馆教育活动设计原则及典型科普活动

科技馆教育活动是科技馆以激发公众的科学兴趣为起点，以提高公众的科学素质为目标开展的一系列具有科学性、趣味性、参与性的活动。

3.2.1 教育活动设计的原则

教育活动的设计遵循"以人为本"的原则，发挥公众的主体意识，调动他们参与活动的积极性，并在活动中能轻松体会、探究科学的奥妙和原理。

根据不同的对象，在活动的开发中要求：体现年龄的针对性；按照《全民科学素质行动计划刚要（2006—2010—2020 年）》的总体要求，突出互动性和趣味性；体现活动的教育价值，体现与社会生活的密切关联性。

3.2.2 受观众喜爱的典型教育活动

通过学习借鉴国内外优秀科技场馆的科普教育活动，以及不断的实践与总结，目前重庆科学技术馆已形成了不少的固定品牌科普活动，根据不同的活动对象，现选择具有本馆特色的活动列举如下。

（1）"科技·人文大讲坛"大型公益科普讲座，活动对象为全市市民。为向公众传递科学精神、学术思想，传播科技、人文知识，重庆科学技术馆于 2010 年起举办"科技·人文"大讲坛，面相社会公众免费开放的大型公益科普讲座，以"公益、高端、精品"为定位，以开展交流共享为宗旨，致力搭建科学家、专家学者、社会名流等行业精英与普通公众双向沟通的桥梁，曾先后邀请到纪连海、王渝生、张召忠、毕淑敏等身受广大市民朋友欢迎和喜爱的名家做客，成为了重庆市科普文化战线一道亮丽的风景线。截至 2017 年 9 月，已成功举办 63 期，受众约 60000 余人。

（2）馆校结合综合实践活动，活动对象为青少年。该活动是在为解决科技场馆普遍存在的周二至周五人流量不足和不均、科普资源利用率不高的背景下组织策划的。其以满足学校需求为导向，以启迪学生好奇心、培育学生想象力、激发其创造力为目标，以科技馆展品资源为依托，用科技馆特色的教育方式，适应学校学科方向、课程标准、课时和班级组织形式，开发多层次、多样化课程体系，以满足学校 1～9 年级学生的不同需求。内容包括专题参观、主题活动、趣味科学实验、快乐科普剧以及梦工场科学小制作五大门类。项目配套《重庆科技馆馆校结合综合实践活动指南》为学校建立了明确的"菜单式"科普服务方式和机制，有效联通了馆校互相了解和互通有无的桥梁，成为馆校结合项目特色模式的基础。此项目实施性是以科技辅导员的教师定位为支撑，服务对象为义务教育阶段的学生。

截至 2017 年春季学期，重庆科学技术馆馆校结合综合实践活动已经实施第四期了。总共四期服务学校 66 所，服务班级 867 个，服务学生 39187 名；实施课程 343 门，授课 1679 堂；满意率保持 100%。

（3）Light on 亲子科学时间，活动对象为亲子家庭。Light on 亲子科学时间是家庭互动式科学活动，通过动手实验、手工制作等方式让家长与小朋友一起度过一段充满乐趣的亲子时间，近距离接触科学，学习科学，体验科学带来的乐趣，从而促进亲情沟通，启蒙科学兴趣，鼓励孩子成为终生学习者。2014 年该品牌推出至今，已开展"To be 达芬奇""大大泡泡 bang"等主题活动 5 个，共开展 79 场，近 3000 组家庭参与其中，并得到他们的认同和喜爱。

4. 重庆科学技术馆场馆发展遇到的困难、挑战及对策建议

4.1 场馆发展瓶颈与挑战

经过 8 年的发展历程，重庆科学技术馆在开展科普展教活动中积累了一定的经验，但仍面临着一些新的困难和挑战。

（1）常设展览开发和展品研制工作的力度不够，展品更新速度慢，自主研发能力不足。

（2）在科普活动的开发上创新力度不够，场馆自身特色还不够凸显。

（3）科技辅导员展教业务知识储备还不够等。

4.2 场馆发展对策建议

（1）在展览开发与展项改造方面，要加快完成"模仿→模仿创新→自主创新""展品形式设计创新→展品展示技术创新→展品原理内容创新"和"展品创新→展区创新→科技馆展览整体创新"的转变进步过程。要注重展区内部展品之间的内在联系，避免简单的展品罗列与堆砌。此外，要高度重视短期专题展览，引进广受观众欢迎的短期展览，将其作为提升观众量、扩大科技馆社会影响的有效手段。

（2）在科普教育活动的开发上，虽然"馆校结合"项目在全市乃至全国科技馆都有很好的示范效应，但影响力也会随着时间的推移、新鲜感的褪去而逐渐消失。因此，科技馆人仍需不断加强自身业务学习和国家政策的研究，寻找新的活动创新点，努力创造出更多更丰富的具有科技馆自身特色的科普教育活动新模式。

（3）人才培养方面，注重加强现有馆内科技馆专业人员的培训，尤其是理工科类、

科学技术史、社会发展史等的学习，力争培养适应性广泛的通用型专业人才。通过"补课"的方式加强员工的业务学习和学术交流，形成浓厚的学习氛围，建设学习型的科技馆，为场馆的可持续发展奠定人才基础。

5. 结语

"一带一路"倡导共商、共建、共享的原则，旨在建设命运共同体、利益共同体、责任共同体。秉承着团结互信、平等互利、包容互鉴、合作共赢的思路精神，这是"一带一路"建设未来发展的方向指引和重要保障。科技馆的交流合作也是"一带一路"人文交流的重要组成部分，是促进沿线国家文化相融、民心相通的有效途径。各国具有不同的特点、经验和比较优势，通过深化科学教育交流合作，相互学习借鉴，有利于整体提升沿线国家科学教育水平和公民科学素质。相信，科技馆事业在"一带一路"政策的指引，在沿线国家的不断深入交流与合作下，共建科学传播的丝绸之路一定会实现。

参考文献

［1］李晓亮．"一带一路"科学教育合作与交流［J］．中国科技教育，2017.07：24–27.
［2］曾川宁．关于特色科技馆体系建设的思考［J］．科协论坛，2014.11：25–27.
［3］郑念．全国科技馆现状与发展对策研究［J］．科普研究 2010（6）：68–74.
［4］张婕，朱海根．关于科技馆与中小学校相结合的教育活动设计和评估的思考［C］// 馆校结合．科学教育论坛，2011.

重庆自然博物馆在"一带一路"建设中的科普理念与实践

赵 笛 [①]

摘 要 本文通过对重庆自然博物馆的发展概况、展陈设计、教育互动以及问题与挑战的介绍,阐述了重庆自然博物馆在"一带一路"战略贯彻实施中和新馆建设开放期间所坚持的科普理念与实践。

关键词 重庆自然博物馆 一带一路 理念 实践

在贯彻落实"一带一路"国家战略的过程中,为加强公众从科普角度认识了解我国在自然历史科学领域所取得的巨大成就,令"一带一路"倡议更深入人心,重庆自然博物馆以其近 90 年的深厚历史积淀,开展了大量的理论探索与实践活动,并取得了一定的成绩。在此过程中,也遇到一些新的问题与挑战。现将重庆自然博物馆的发展概况与遇到的问题挑战探讨如下。

1. 发展概况

重庆地处"一带一路"倡议路线与长江经济带的交汇点,是"一带一路"倡议的中心枢纽城市,有着重要的战略支撑作用。经贸合作,文化助力。重庆自然博物馆作为重庆市的三大国家一级博物馆之一,也是重庆唯一的自然科学综合类博物馆,用行动诠释"一带一路"的理念。让公众在参观过程中学习到多项科学知识,培养科学思想、科学方法和科学精神。当好"一带一路"的科普大使,努力提高全民科技素养,助推"一带一路"倡议的顺利实施。

---

① 赵笛:重庆自然博物馆地球科学部馆员,北京大学考古学院考古学及博物馆学在职研究生。研究方向:古生物学与博物馆学。地址:重庆市北碚区金华路 398 号,重庆自然博物馆地球科学部;邮编:400071;E-mail:18384378@qq.com。

重庆自然博物馆新馆（彩页图 53）坐落于峰峦灵秀、烟云缥缈的缙云山麓，占地面积约 14.4 万平方米，总建筑面积 3 万余平方米，展区面积约 2 万平方米。其设计定位为"西部领先，国内一流"。开放第一年观众总接待量为 307 万人次，营业收入 1400 万元。

其前身是民生公司总经理卢作孚创办的中国西部科学院，以"研究实用科学，辅助中国西部经济文化事业之发展"为宗旨，是中国第一家也是中国西南地区唯一的民办科学院。抗日战争时期，中国西部科学院联络内迁重庆北碚的十多家全国性学术机关，共同组建了中国西部博物馆。1953 年并入西南博物院，后改名为重庆市博物馆，1991 年独立建制为重庆自然博物馆。2015 年 11 月 9 日，重庆自然博物馆新馆正式向公众免费开放，2017 年被中国博物馆协会评为第三批国家一级博物馆。新馆开放后的基本信息数据如表 1 所示。

表 1　重庆自然博物馆新馆开放基础数据表

年观众接待量		展览数量		活动数量		场馆规模		研究成果	
325 万		16 个		28 个		3 万平方米		78 篇（部）	
基展	临展	基展	临展	校园活动	下乡活动	展区	办公区	论文	专著
307 万	18 万	6 个	10 个	21 个	7 个	2 万平方米	1 万平方米	75 篇	3 部

说明：研究成果中包含英文论文 27 篇，SCI 论文 8 篇，25 万字英文专著一部。

在重庆自然博物馆的发展历程中，取得的成绩主要有：1941 年，杨钟健在中国西部科学院撰写出版了中国人研究恐龙的首部科学专著《许氏禄丰龙》，并于同年在中国西部科学院的中央地质调查所进行了许氏禄丰龙骨架的首次公开展览，这也是中国出土的第一具恐龙化石骨架；1943 年夏，由中国现代地震学的奠基人，中国地球物理勘探工作的重要开创者李善邦研制的霓式水平向地震仪问世，这是中国自行研制的首台机械记录式水平向地震仪；2017 年，重庆自然博物馆的"地球·生物·人类——重庆自然博物馆基本陈列"展，在中国博物馆协会、中国文物报社主办的"第十四届全国博物馆十大陈列展览精品推介活动"中获得"全国十大陈列展览精品奖"。

2. 展览设计概况

重庆自然博物馆的常设展览主题为"地球·生物·人类——重庆自然博物馆基本陈列"。其展品除少数馆内旧藏，绝大多数来源于新入藏珍品。广泛征集于世界 7 大洲的 8000 余件展品，按照展陈大纲设计，精心挑选整理，对展示主题予以主动关联，通过展品与展品间的组合方式构成核心内容支撑。

重庆自然博物馆的新馆展区，围绕展览主题分为三层，其间分布有恐龙厅、贝林厅、重庆厅、地球厅、进化厅和环境厅。六大展厅的内容体系既各有侧重又相互衔接呼应，以物为基，以人为本，以系统性的有序组合，展示出地球演变、生物进化、生物多样性以及资源和环境与人类活动的关系，涵盖地质学、生物学、生态学、古生物学、古人类学等学科内容。地球厅以新地球观为指导，运用最新理论成果，多维度、多层次揭示地球科学进展。进化厅厘清进化脉络，阐释进化机制，探究进化原理。恐龙厅以翔实的标本、科学的布局复原恐龙世界奇观，激发大众探密恐龙时代的兴趣。贝林厅突出生物在全球视野下的地缘化分布特色，诠释物种间的关系。环境厅回溯人类文明进程中的人与地球的关系，拓展对未来发展的广阔视角，引发对构建人类命运共同体的思考。重庆厅展示本地自然历史的变迁、人与环境相互依存的生存智慧、强化"绿水青山就是金山银山"的科学理念、宣传生态文明城市发展的主题。整个展览将展品作为展陈内容的支撑点，通过抽丝剥茧，环环相扣的叙事主线，运用视觉系的差异化表达和异形空间的造型艺术处理，实现科学与艺术相融合的展览风格。

以恐龙厅为例，与国内其他陈列展侧重于对恐龙埋藏地点与生态性复原的简单展示不同，本馆在视觉上将展厅的整层空间再分割为三层，展厅入口处布置地下层，模拟野外发掘现场，地面一层为化石修复室与学者书屋场景，最后为恐龙进化树与恐龙生活场景复原。从恐龙的岩层埋藏状态到局部结构的解剖，以及地面二层的恐龙同时代伴生古动植物。通过动态线性轴的方式呈现展示概念，揭示恐龙的生物行为与恐龙时代的生物多样性，并对公众的科学疑问进行启发性参与式解答。既形塑了单元空间的独立性，又拓展了与其他展区的关联性。

3. 教育互动概况

为将重庆自然博物馆搭建成传播和共享知识与文化、教育与科学知识普及的平台，新馆开放前后，努力通过各种形式，与重庆各级教育机构共同创办开展活动。近年来所举办的活动主要有以下几种。

3.1 参与重庆市文物局为完善博物馆青少年教育功能试点所开展的工作项目

以设立幼儿园、小学低年级、小学高年级和初中四个学龄段为架构，设计青少年科普教育课程 22 个，科普教材图书 12 种。与重庆市第一中学和重庆状元碑小学等 15

所中小学签订合作协议，进校开展《大自然的书签》《一帆风顺——胭脂鱼物语》《小恐龙历险记沙画教育》以及《我的家乡》等20多个主题教育讲座活动。覆盖青少年群体2万余人。

3.2　世界地球日、国际博物馆日、重庆市科普周、全国科普日主题展览及科普讲座

举办了《珍惜地球资源转变发展方式——促进生态文明，共建美丽中国》《走进非洲》《恐龙那点事》等专题展览。将《三叠纪古海洋中的鱼龙》《大自然的精灵》《二氧化碳与低碳》和《大熊猫的起源与演化》举办到重庆市第一中学、西南大学附属小学、重庆市北碚区消防支队和重庆市北碚朝阳社区等学校、部队和社区。参观人数多达3万余人。

3.3　重庆市委宣传部"梦想课堂活动"

将科普宣传普及到乡镇，为江津市嘉平镇中心小学学生发放宣传资料千余份。在宣传资料中融入江津地区的地形地貌自然环境，介绍当地主要的动植物与出土的古生物化石。将晦涩的科学知识普及到孩子们所熟悉的生活环境中，激发孩子对家乡乃至于国家民族的自豪感。

3.4　重庆市北碚区特殊学校"自然科学园"项目

为重庆市北碚区特殊儿童学校捐赠援建科普教育园地一间。针对特殊儿童在认知方面的特点与难点，利用声、光、触等手段，细化宣教科普知识，让特殊儿童跨越生理上的不便，尽可能地享受到和普通大众对于自然科学的同等认知。

3.5　科普小展览到乡镇

在重庆周边区县乡镇，举办《会飞的花朵》《低碳家庭》《泥石流灾害防治》《身边的昆虫》《恐龙的身世》《奇异的石头》《邮票上的恐龙》（彩页图55）等展览157场，观众数量达129800余人。

4. 问题与挑战

重庆自然博物馆在长期的科普实践中，坚持保护自然环境就是保护人类文明延续的科普宣传理念，也取得了一定的成绩。但是我们也清楚地认识到自身在与"一带一路"的愿景与规划对照中还存在诸多的不足。如何去解决这些不足对我们的工作是一

项长期的挑战。例如在打造"一带一路"科普教育品牌过程中，专项资金的保障，科普外交人才的培养，品牌的知识产权保护与科普场馆交流合作之间的矛盾等都需要一一去面对。在展览陈列方面，重庆自然博物馆还需要在展陈设计的核心思想，展陈内容的连贯故事线，展示收藏的本地化原则以及观众的真实性体验上做大量的研究与实践。

因此，加强"一带一路"科普场馆交流合作，不仅是共享各方在展览展示、科学教育等方面的资源，共建科学传播丝绸之路，同时也为不同文化背景、不同主题的自然科学类博物馆提供了互相了解、深入交流的机会，弥补单一博物馆在认知上的局限性，形成强强联合体。

参考文献

［1］刘晓. "一带一路"对外传播研究［D］. 湘潭：湘潭大学，2016.
［2］侯德础，赵国忠. 爱国实业家卢作孚与中国西部科学院［J］. 四川师范大学学报（社会科学版），2000（1）.
［3］胡昌健. 六十年来重庆文博事业发展概述［J］. 中国博物馆，1996（2）.
［4］李善邦. 霓式地震仪原理及设计制造经过［J］. 地球物理专刊，1945（3）.

科普互惠，协同共享

——浅谈河北省科学技术馆的发展

徐 静 [①]

摘 要 河北省科学技术馆新馆自 2006 年开放以来，逐渐成为周边人们提高自身科学文化素养和休闲娱乐的重要场所，除了常设展览的常展常新外，丰富的科普活动也是吸引公众的重要方面。自 2013 年起，中国科学技术协会在全国全面启动中国流动科技馆巡展工作，河北省科学技术馆承接了多套流动科技馆，在各区县开展科普宣传，不仅扩大了河北科学技术馆的影响力，也带动了相应区县人们学科学的热情，掀起了学科学、爱科学的热潮。本文主要从河北省科学技术馆实际出发，谈河北省科学技术馆目前的发展以及今后的展望，并希望将"一带一路"作为科普事业发展的一个契机，更好地让科普互惠，将各个地区优秀的资源协同共享。

关键词 科学技术馆 科普活动 发展

习近平总书记在 2013 年 9 月和 10 月分别提出建设"新丝绸之路经济带"和"21世纪海上丝绸之路"的战略构想，强调相关国家要打造互利共赢的"利益共同体"和共同发展繁荣的"命运共同体"。"一带一路"贯穿欧亚大陆，东边连接亚太经济圈，西边接入欧洲经济圈。无论是发展经济、改善民生，还是应对危机、加快调整，许多沿线国家同我国有着共同利益。其中作为我国科普教育的主阵地——科技馆，在科学知识的普及、科学方法的倡导和科学精神的弘扬上发挥着独特的作用，同时科学技术馆也是全民科学素质教育的主要阵地。加强"一带一路"沿线国家和地区科普场馆的沟通合作，推动场馆的发展和进步，使之成为当地经济社会发展的重要力量，是我们每个科普人的责任。

---

① 徐静：河北省科学技术馆展览教育部副部长，文博馆员。研究方向：展览设计和科普活动开发。地址：河北省石家庄市长安区东大街 1 号，邮编：050011，Email：xujingkjg@163.com。

1. 河北省科学技术馆概况

河北省科学技术馆由旧馆和新馆 2 个馆组成,自 2006 年新馆建成开放后,旧馆则成为了办公区域。为适应社会发展的新形势,谋求河北科普工作的新发展,满足公众对学习科技知识的新要求,充分发挥科技馆的教育优势和重要作用,河北省委、省政府于 1998 年投资兴建了河北省科学技术馆新馆。2006 年 3 月正式对外开放,主要由常设展览区、科普电影放映区、会议休闲区等构成。常设展厅共有四层九个展区,内容涉及数学、机械、生命等多个领域。2010 年更新了三楼动手园区,改为安全展区;2013 年更新了二楼应用技术展区,建成一个小的儿童展区;2015 年更新三楼临展,改为机器人展区。2017 年 9 月更新完善二楼儿童展区,现已开始进入布展环节。

河北省科学技术馆作为一个省级大馆,在国内中小型科技馆建设方面具有一定的导向作用。河北省科学技术馆新馆自 2006 年开馆以来,开展了丰富的展览活动,吸引了大批观众走进科技馆参观,了解科学知识的同时感受科技的魅力。据河北省科学技术馆的不完全统计,自 2006 年新馆开放,常设展览开放 300 余天,每年接待观众量约为 20 万人次,穹幕电影和天象节目播放 600 余场,4D 影院播放 500 余场。自 2015 年 6 月免费开放以来,共接待观众 40 万人次,社会各界团体 98 个,每年开展科普讲堂、科学表演和科普剧演出 70 余次,举办各类科普宣讲活动近 20 场。其中 2017 年暑假期间馆内共接待观众 93266 人次,比上年同期增长 165%,越来越多的人选择在休闲时将河北省科学技术馆作为提高自身科学文化素养的重要场所。此外,河北省科学技术馆还通过科普实验、科学秀、科普剧等多种形式开展科普活动,吸引了更多的观众参与。开展科普临展多场,比如"大观世界"主题科普展览、"科学养生之如何应对慢性病"科普展览等,累计接待各界观众近 15 万人次。同时还创作编排科普剧和科普实验 10 余部,每年进校园活动 10 余次,组织特色天文科普活动近 10 场,有效地促进了科普工作与学校教育及传统文化的结合。

河北省科学技术协会于 2002 年 11 月购进科普大篷车一辆,并专门成立了河北省科学技术馆科普大篷车工作队。科普大篷车携带的展品可供观众动手操作,并配有以"崇尚科学、反对迷信""航天知识""科技、文明、小康"等为主要内容的几十块科普展牌,可按照地区和面向群体的不同进行多样化的科学传播。自 2003 年以来,科普大篷车多次参加我省有关部门和省科学组织的科技、文化、卫生"三下乡""科普活动日""四进社区"等大型系列巡展活动和专题活动,巡展范围北至夏都承德,南下燕赵古都邯郸,东到渤海之滨秦皇岛,西至坝上草原张家口。到目前为止,科普大篷车已经跨越河北省 11 个设区市,数十个县,行程 15000 多千米,接待参观观众 20 多万人次,得到基层群众的热烈欢迎和一致好评。

2. 场馆特色

2.1 信息化建设初见成效

2015 年,河北省科学技术馆加大力度,创作了一大批优秀的科学实验、科普剧作品,并通过网络放到了科技馆网站平台,供参观观众在网站浏览观看,并将作品制作成《感知科学》《感知科学 II》系列光盘。2016 年河北省科学技术馆"互联网 +"公众服务科普平台搭建完成。该平台基于信息化技术和互联网媒体传播,主要包括优化河北省科学技术馆官网建设、开通微信公众号和手机客户端,加强对新媒体的建设、运用及规范管理,是河北省科学技术馆实施"互联网 + 科普"行动,努力促进科普资源线上线下有效结合的工作新途径。河北省科学技术馆还荣获了"2015 年度全国科普教育基地科普信息化工作优秀基地"荣誉称号。

2.2 原创科普活动精彩纷呈

河北省科学技术馆每年举办社会公益活动约 30 次,包括科技馆活动进校园、免费天文培训课、天文户外观测、公益科普讲座等,都是紧追时下最热的科普话题。如《时空的波动——引力波》天文讲座、《精彩纷呈的中国载人航天》天文讲座、《创客与创客时代》公益科普报告会等。同时还开展了"大手牵小手,爱心暖童心"留守儿童公益活动、"寒假科技创想一日营"活动、"倡导低碳生活,共建绿色家园"环保主题活动、"童眼看世界,科学过六一"的活动等。

从 2013 年的首届全国原创微型科普剧剧本创作大赛到今年的第五届,河北科学技术馆报送的多部原创剧本和科普剧均获多项大奖,如《PM$_{2.5}$——谁才是凶手》《龙卷风》《霾老大之殇》等剧本还自编自导成科普剧,在科技馆内上演几十场,深受观众的喜爱。除此之外,原创科学实验《安静的小球》《玩儿转干冰》《魔法盛宴》《身边的摩擦力》等作品也在节假日给孩子们带去了科学知识与欢笑。

2.3 流动科技馆掀起学科学、爱科学热潮

自 2013 年起,在中华人民共和国财政部的大力支持下,中国科学技术协会在全国全面启动中国流动科技馆巡展工作。河北省科学技术馆也承接了巡展工作,2015 年,同时有 8 套流动馆在 18 个区县开展活动,展出 18 站,累计参观人数达 33.56 万,受

到基层群众,特别是青少年的广泛欢迎,取得了良好的社会效益。2016 年,在我省的流动科技馆已经增加到 10 套,在全省 10 个设区市巡展 42 站,接待社会观众 75 万余人次。自 2017 年初至今,承担了 17 套流动馆,展出 90 站,到 2017 年年底还有望增加。

3. 现状及发展规划

2015 年,中央财政投入 3.5 亿元补助资金,推动全国 92 家试点科技馆于 2015 年 5 月 16 日前实现对公众的免费开放。河北省科学技术馆是第一批免费开放的场馆之一,自免费开放后,来馆参观的公众数急剧增加,尤其是节假日给展厅造成了很大的压力,如何在观众量猛增的情况下提供更加高效的服务是河北省科学技术馆面临的主要问题之一。

3.1 宣传力度不够

河北省有 11 个地级市、22 个县级市,仅有河北省科学技术馆一个综合性科技馆,来科技馆参观的群体大多是周边群众和学校团体,偶尔会有旅行社参观,有不少本市市民都不知道河北省科学技术馆是做什么的,更不要说其他地方的公众,这说明科技馆的宣传力度还有待进一步加强。

3.2 参观体验有待提高

科技馆运行过程中最先关注的就是公众的参观体验,包括参观前、参观中和参观后。河北省科学技术馆比较重视公众参观的体验,如对于展品的讲解、表演科学实验、和观众的互动等,而对于参观前和参观后的体验忽略较多。参观前的体验除了通过各种途径获取的科技馆的相关信息外,还包括公众到达科技馆后但还未进入参观时应获得的相关信息,后一方面河北科学技术馆做的还不够。参观后的体验包括线下互动、参观收获、心得等。河北省科学技术馆的网站在 2016 年 5 月改版,对比之前单纯以图片和文字的形式进行展示,现在的网站又增加了"乐园漫游""特效影院""作品展示"等内容,其中作品展示都是河北科学技术馆自编自演的科普实验和科普讲座,可以让那些不能现场观看和观看后意犹未尽的公众再次感受科学的魅力。

3.3 人才队伍建设有待进一步提高

河北省科学技术馆的展览教育部主要负责展厅科普讲解、科普实验、进校园和部分流动馆的讲解、表演工作,工作人员的素质能够代表馆的软实力,但是目前这支队伍存在两个显著的问题：一是人员素质起点不高，展览教育部共 24 人，初始学历为本科的占 50%，硕士 1 人，剩下近 50% 为专科；二是专业不对口，工作人员所学专业里面文学类、管理学类和经济学占超过一半。这表明，河北省科学技术馆人才队伍不仅需要质的提升，还要培养复合型科普人才。

自 2006 年至今，河北省科学技术馆新馆已经走过了 12 年，在这个信息大爆炸的时代，科技日新月异，河北省科学技术馆无论是从规模还是设施都日渐落后，因此河北省科学技术馆与直属领导单位河北省科学技术协会也在积极地筹划建设新馆事宜，以期落成后能够辐射更广的地区，给更多的人带去更多更好的科普活动，将河北省科学技术馆打造成一个品牌。

4. 建议

我国提出的"一带一路"倡议通过政策沟通、道路联通、贸易畅通、货币流通、民心相通这"五通"，将中国的生产要素，尤其是优质的产能输送出去，让沿"带"沿"路"的发展中国家和地区共享成果。其中的民心相通指促进不同文明和宗教之间的交流对话，推进教育、文化交流，发展旅游等。教育交流已经走在了前列，"新丝绸之路大学联盟"已有五大洲 30 多个国家和地区的 124 所大学加入，在校际交流、人才培养、科研合作、智库建设等方面与联盟内高校开展形式多样的合作；丝绸之路经济带大数据云服务创新研究院、中国西部科技创新港均在筹建科技智库；国际汉唐学院和中国书法学院在土库曼斯坦、俄罗斯、波兰等国的研究中心建设紧锣密鼓……

科普场馆也可以建立一个科普联盟，在"一带一路"中发挥更大的作用。如可以加强科技合作，共建联合实验室（研究中心）、国际技术转移中心、加强科技人员交流、合作，开展重大科技攻关，共同提升科技创新能力等。不同国家和地区之间的合作都蕴含着无限的机遇。"一带一路"倡议提出四年间，世界见证了沿线各国在教育合作、科技交流、创新引领方面诸多的发展成就。科普场馆也应扛起"一带一路"创新大旗，为沿线国家和地区的发展做贡献，连接"中国梦"与"科技腾飞的梦想"。

参考文献

［1］晓霞. "一带一路"贡献"陕西智慧"——陕西着力构建科技教育中心［N］. 陕西日报，2017-09-08.

内蒙古科学技术馆的发展

郭 瑜 [①]

摘 要 内蒙古科学技术馆是内蒙古自治区政府的代建项目,是自治区"十二五"期间的重点民生工程之一,2010 年 8 月开工奠基,2013 年底土建工程竣工,2016 年 9 月 20 日正式开馆。科技馆是提高自治区全民科学素质的大型科普基础设施,同时也是自治区唯一的综合性科普场馆。

关键词 科学技术馆 历史 发展

1. 场馆概况

内蒙古自治区科学技术馆旧馆于 1983 年 12 月 5 日正式开馆,是全国最早建立的科学技术馆之一。内蒙古科学技术馆新馆是自治区"十二五"期间新建的综合性科普场馆,2009 年 10 月立项,2010 年 8 月 18 日开工奠基,总建筑面积 48300 平方米,展览教育面积 28830 平方米,建筑总投资 6.083 亿元。主要功能包括展览教育、公共服务、业务研究、管理保障等。2016 年 9 月 20 日建成对外免费开放。

新馆的外形以"旭日腾飞"为创意主题,造型寓意着马鞍哈达沙丘等地域特色和内涵。球幕影院,配合着绿色草地的背景,从远处看,像是一轮东升的旭日,赋予草原升起不落的太阳的视觉憧憬。

内蒙古科学技术馆将科技创新、科学普及作为实现创新发展的两翼,以弘扬科学精神、传播科学思想、倡导科学方法、普及科学知识为己任,以"世界眼光、时代特征、内蒙特色、创新发展"为建设目标,建成一座"国内领先、西部一流"的具有民族特色的现代化科技场馆。

内蒙古科学技术馆常设展览由 5 个展厅和公共空间室内标志性展项组成,以"探索·创新·未来"为理念,5 个展厅中设置了"探索与发现""创造与体验""地球与家园""魅力海洋""生命与健康""科技与未来""宇宙与航天""智能空间""儿童乐

---

① 郭瑜:内蒙古科学技术馆馆员。邮箱:276774029@qq.com。

园"9 个主题，共设展项展品 457 件，展品以互动、体验为主要的展示方式，体现了科学性、知识性和趣味性。此外，还设有数字立体巨幕影院、数字球幕影院、4D 动感影院，专题展览厅、科普报告厅等。

2. 特色活动

在内蒙古科学技术馆中有一个独具特色的展区——蒙医药主题展区。这一展区是内蒙古科技馆历时三年，在自治区蒙医药相关单位的大力支持下，结合幻影成像、体感互动等国内外最先进的展示手段，集中展示了蒙医的发展及理论知识、诊断方法、蒙医药知识、蒙医特色疗法等详细内容，介绍了蒙医的精髓。蒙医共有三种诊疗方法：望诊、问诊、触诊。而在蒙医的特色疗法中也包括艾灸疗法、酸马奶疗法、正骨术等。除此之外，内蒙古科学技术馆中还存放着一尊蒙医针灸铜人像，它是一尊佛像疗术铜人。它的原版现收藏于内蒙古医学院所属蒙医药博物馆内，该铜人高 61 厘米、质量为 21 千克，全身标有 611 个穴位，穴位标注精准，躯体上标明了针刺穴位、灸疗穴位、放血穴位等。在口口相传的医学传授年代，它不仅指导蒙医大夫运用医术治疗疾病，同时也是研究蒙医针灸最宝贵的文物之一。

今年我们正着手建立科技馆实验室，以美国的 STEAM 教育理念为基础，建立有关能源、机器人、数学等主题的科学实验室，以期更好地引导青少年选择研究课题，动手进行科学实验，培养他们创造性解决问题的能力。除此以外，我馆还在积极研究制定校本科学课程，课程的目标是合理、充分地利用科技馆资源，提高青少年的科学素养。课程内容的选择遵循以科技馆资源为基础，以发展学生的科学素养为目的，以学生已有的知识和经验为依据，符合学生的认知特点与兴趣，选择能够反映现代科技、体现 STS 的内容的基本原则。

在教育活动开展方面，我馆进行了许多尝试，获得了较好的社会反响。2017 年 1 月，由内蒙古科学技术馆、内蒙古电视台新闻频道主办，内蒙古睿学教育承办的 2017"雏鹰计划"青少年成长安亲体验营在内蒙古科学技术馆魅力海洋展厅安营扎寨。孩子们夜宿魅力海洋展厅，体验了不一样的科普之旅。2017 年 3 月，内蒙古科学技术馆又进行了为期 3 个月的馆校结合活动，共与呼和浩特市 18 所小学建立了联合伙伴关系，通过升级科技馆教育功能实现馆校共赢，让科技馆真正成为中小学生校外的第二课堂。馆校结合主要包括主题参观、展厅主题活动、观看科学秀及科普影片等，学生们的反响普遍较好。2017 年 7 月 10 日，"少年派的西北漂流记"内蒙古科学技术馆深度研学五日夏令营正式开营，预计将开展 6 期，本期小营员有 150 人，分别来自湖北、山东、浙江、江苏和湖南等地，孩子们通过参加此次夏令营了解了内蒙古的民族知识并且收获了前所未有的体验，在科技中获得收获，在科技中快乐成长。

3. 场馆发展所遇困境及展望

内蒙古科学技术馆正式开馆至今刚满一岁，在这一年中，我们发现了很多不足，也遇到了很多困难与困境。一方面，内蒙古科学技术馆地处我国中西部，享受到的资源与发达地区相比仍存在差距，而且内蒙古科学技术馆起步较晚，虽说在硬件设施方面我们不算落后，但在软件条件方面相对落后，对于科普课程开发及展品设计与开发的能力还很薄弱。加之现在内蒙古科学技术馆正在进行一系列的改造与建设，正是应该向其他科技馆汲取经验的时候，加强场馆间的合作与互访显得尤为重要。从另一方面来看，内蒙古科学技术馆中有许多具有地方特色的展项，这些展项所展现的科学知识也应该借其他各地区科技馆的平台传播出去，加强场馆间的合作与互访，互相借鉴互相补充。

4. 对"一带一路"科普场馆交流会的建议

第一，各地区的科技馆可以进行"一带一路"专题展览，对于本身处于丝绸之路经济带的各地科技馆可以结合自身特色与历史文化，丰富专题展览的内容。如内蒙古与俄罗斯、蒙古两国有着 4221 千米边境线，有通往俄罗斯、蒙古境内的古丝路、古茶路、古盐路等历史道路，在开设相关专题展览时也可以与俄罗斯、蒙古两国相关科技机构进行沟通交流，共同丰富和完善这一专题展览，同时促进科技馆共同发展。

第二，可以定期举办一些科技馆交流会，互相分享各科技馆间的教育活动开展情况及先进理念，更深入一步之后，可以与其他国家的科技馆进行联动，博采众长，尤其是向科技教育方面发展比较成熟的美国等国家多进行沟通和交流。

第三，对于欠发达地区科普人群进行界定。在开馆这一年中，我们发现很多观众认为科学技术馆只是专门为儿童提供的游乐场，来到科技馆主动学习科学知识的人比较少，家长带孩子来很多只是抱着玩的心态，其实这也是由于家长对科技馆功能及作用的不了解导致的，我们更应该让成年人具备科学思想、科学精神、科学思维，进而使他们去引导孩子，毕竟家长对孩子的教育在孩子的成长过程中占有举足轻重的地位。

相信通过这次的研讨会，能够推动各地区场馆的发展和进步，使之成为当地经济社会发展的重要力量。并能够通过提供"科普互惠共享资源"，共享各方在展览展示、科学教育等方面的资源，共建科学传播丝绸之路。

发挥宁夏重要区域特色
打造特色科技传播活动

柴继山 ①

摘 要 "一带一路"倡议是党中央、国务院根据全球形势深刻变化，统筹国际、国内局势做出的重大战略决策，对于开创我国全方位对外开放新格局、推进中华民族伟大复兴进程、促进世界和平发展具有重大意义。共建"一带一路"是我国顺应世界多极化、经济全球化、文化多样化、社会信息化的重要举措，更是"中国方案、全球治理"新模式的积极探索，为打造人类利益共同体和命运共同体增添新的力量。宁夏是西部大开发重点区域和丝绸之路经济带重要支点，加强"一带一路"沿线国家和地区科普场馆的沟通合作，推动场馆的发展和进步，使之成为当地经济社会发展的重要力量。

关键词 一带一路 科普场馆 沟通合作 发展

1. 引言

宁夏地处我国西北内陆，位于祖国西北部的黄河中上游地区，东邻陕西省，西部与北部接内蒙古自治区，南部与甘肃省相连，总面积 6.64 万平方千米。自古就是古丝绸之路的必经之地和商埠重镇。改革开放以来，宁夏通过国务院"西部大开发战略""关于进一步促进宁夏经济社会发展的若干意见""宁夏内陆开放型经济试验区规划""宁夏空间发展战略规划"等一系列战略举措，在保障宁夏经济社会文化长足发展的同时，为当前深入和全面对接国家"丝绸之路经济带"战略奠定了深厚的基础。实现"两个一百年"奋斗目标和中华民族伟大复兴的中国梦，需要以科技实力为支撑，以自主创新为动力。实施创新驱动发展战略，核心是推动科技创新与经济社会发展紧密结合，关键是提升自主创新能力，基础是提高全民科学素质水平。宁夏科学技术馆

---

① 柴继山：硕士，高级经济师，在宁夏科学技术馆从事科技馆科普教育研究。地址：宁夏银川市金凤区人民广场西路，邮编：750011，E-mail：3134923407@qq.com。

作为全宁夏回族自治区最大的综合性科普宣传教育基地，科普展教资源是科技传播的重要载体，对提升全民科学素质具有不可替代的重要作用。

2. 注重功能定位，打造一流科普场馆，为公众提供最优质的科普资源和科普服务

2.1 发挥区域优势，构建特色场馆

宁夏科学技术馆新馆是自治区成立 50 周年重点献礼项目，投资 2.5 亿元，占地 3.88 万平方米，总建筑面积近 3 万平方米，常设展厅 16101 平方米，于 2008 年 9 月建成开放。多年来，宁夏科学技术馆新馆以"体验科学，启迪创新；服务大众，促进和谐"为理念，积极发挥全区综合性科普宣传教育场馆作用，把场馆建设与回族文化、宁夏区域特色深度融合，为促进民族团结，提高少数民族地区公众科学素质做出了较大贡献。

2.2 场馆建设发展稳步推进，功能日益完善

按照中国科学技术馆建设标准，宁夏科学技术馆新馆共设置序厅和 13 个主题展区，拥有展品、展项近 500 件，矿物、动物、植物、古生物标本 1500 余件及 4D、穹幕两个特效影院，集科普展览教育、科技培训、科技影视制作与放映、科普报告、青少年科技活动等社会功能于一体的全区最大的综合性科普场馆，教育服务功能日益完善。

2.3 科学规划，展品常展常新

针对运行实际和公众诉求，从场馆建筑特点及展品布展实际出发，宁夏科学技术馆聘请全国科技馆业界专家，经过反复测算和论证，制定《宁夏科技馆展厅改造五年规划》。根据规划，依托中央财政科技馆免费开放专项资金，自 2013 年至今，累计投入 6200 万元进行展厅升级改造，实现了展品常展常新，使科技馆场馆服务能力明显增强，阵地作用更加突出。

3. 多措并举，"小馆大科普"格局形成

宁夏科学技术馆立足现有实际，在普及科学知识、倡导科学方法、传播科学思想、弘扬科学精神中，积极开展大联合、大协作，充分发挥科普资源优势，拓展科普覆盖

面，形成了"小馆大科普"格局。

3.1 创新展教理念、展教能力和水平不断提升

宁夏科学技术馆结合地域优势和自身场馆特色、在抓展厅升级改造等硬件设施提升的同时，面向广大青少年常年开展系列主题科普教育活动、科学小讲堂、科学趣味实验、科普剧表演等表演类活动，结合展品展项深度开发特色科普教育活动等软件来为公众提供科普服务，充分发挥科技馆作为中小学生"第二课堂"的作用，吸引青少年参观，调动他们的好奇心和探究精神，激发其对科学的兴趣，为青少年营造学科学、爱科学、用科学的良好学习氛围。

3.2 积极探索创新，青少年科技教育活动向纵深发展

宁夏科学技术馆肩负着全区青少年的科普教育重任，在创新青少年科技教育模式，促进青少年科技活动深入蓬勃开展方面做出了卓有成效的成绩。多年来，先后组织开展全区青少年科学节、青少年科技创新大赛、机器人竞赛、"大手拉小手，科普进校园"宁夏行、青少年科学调查体验活动、明天小小科学家、科学影像节等综合型教育活动；举办寒暑假"小博士"培训类活动；2016 年成功举办首届宁夏青少年科学节，历时 36 天，"区、市、县、校"四级联动，覆盖全区 5 市 22 县（区），开展了 300 多项内容丰富的青少年科技教育活动，累计参与青少年超过 20 万人次，有效引领全区青少年科技教育活动联动发展。

4. 现代科技馆体系建设取得显著进展

宁夏科学技术馆坚持把贯彻习近平总书记"科技创新、科学普及是实现创新发展的两翼，要把科学普及放在与科技创新同等重要的位置"的重要指示精神作为工作的出发点和落脚点，做到科普教育服务有内涵，展览教育活动有特色，青少年科技教育活动有亮点，科技馆体系建设有创新，着力青少年科技创新能力培养和公众科学素质提升，初步形成了实体科技馆、数字科技馆、流动科技馆、科普大篷车等为主要建设内容的现代科技馆体系。

4.1 常设展厅实现免费开放，观众接待与日俱增

常设展厅平均每年向公众开放 300 天，自 2014 年实现科技馆免费开放后，观众流量大幅增长，全年接待观众 55 万人次，较以往同期增长 10%，展厅日均接待观众近

2000 余人，接待总人数超过 290 万。

4.2 流动科技馆发展势头喜人，成效显著

流动科技馆在全国范围内率先实现宁夏全区覆盖，科普惠及民生。宁夏作为全国首批流动科技馆试点，自运行以来，足迹遍布全区 5 市 22 个县区，科普惠泽老、少、边、穷地区 100 多万人。

4.3 推进数字科技省区馆建设，搭建网络科普共享平台，"互联网 + 科普"取得明显成果

2011 年，我们以网络为依托，在宁夏科技馆建设网上数字化科技馆，被中国数字科技馆确定首期建设试点，建立了中国数字科技馆二级子站宁夏站点。2013 年，建设了拥有专利权的宁夏数字科技馆，2014 年以来，与企业联合建立展品线上线下数字化互动综合平台，最大限度地将科普资源惠泽于民，提升公众科学素质。积极推进"互联网 + 科普"，利用各种新媒体手段，促进科普碎片化整合，强化科普教育功能，扩大科普受益面。

5. 深化科协系统改革，推进科技馆持续健康发展

5.1 注重人才培养，为科技馆建设发展提供强有力的人才支撑

宁夏科学技术馆建立了健全的监督、考核、奖励机制，把工作实绩和工资待遇、岗位任用和评优评先等挂钩，有效地激发了全馆人员的工作积极性和创造性。采取"请进来、走出去"的做法，拓展馆际合作、加强交流学习，提升人员专业素质和管理水平。着力培养青年业务骨干，为他们创造实践锻炼机会、制定职业发展规划。近 5 年来，中层管理人员已完成近半数的新老交替，迅速成长起来的青年骨干为科技馆管理队伍注入了新鲜血液，引入了新思路、新动力，有效地促进了科技馆工作的创新性可持续发展。

5.2 打造科普精品，提高科普品质

随着国家、地方对科普事业的重视，公众越来越关注以传播科学知识、启迪科学思想为主要职能的科技馆建设。面对科技馆事业发展新的历史机遇，宁夏科学技术馆将在深化科协系统改革中，以建成中国特色现代科技馆体系为目标，以转变展示设计理念、转变展教工作重点和转变科普教育方式为主要任务，打造科普精品，提高科普

品质，发挥好宁夏科学技术馆在全区创新驱动发展中的重要作用，不断满足公众需求，为提高全区公民科学素质服务。

6. 科技馆发展面临的机遇与挑战

随着科技馆事业的蓬勃发展，科技馆的教育功能不断完善和拓展，面对公众日益增长的对科普教育需求层次的提升，满足为公众提供精准科普服务的迫切需要，科技馆展览创新面临着瓶颈和新的挑战。尤其是科技馆在实现免费开放以来，面临着参观观众激增，科技馆运营成本增加、运行压力增大，展品损坏日益严重，展品维修周期增长、更新缓慢等诸多不利因素，要想保证科技馆对公众持久的吸引力和关注度，要做到常展常新是一件非常不易的事情。因此，在展品设计中就要进行深入研究适合本馆实际和发展的展品展项，在既保证互动参与、体验操作的同时还要确保观众安全，展品便于维修保养等综合性因素，在展品常展常新上，宁夏科学技术馆进行了有益的探索和尝试。根据《全民科学素质行动计划纲要》的要求，宁夏科学技术馆以科普展览教育和青少年科普教育活动为主要内容，以规范化建设和职工素质培育为支撑，不断加强馆校合作和科普创新，突出科普教育社会效益，不断满足公众对科普的需求，实现创新、可持续发展。对科技馆的展览教育功能进行全方位地认知和定位，不断创新展览教育理念，不断提升展览教育水平，提升展览教育服务能力，进一步发挥科技馆在提升公民科学素质中的重要作用，任重而道远。

7. 科技馆建设发展的对策和建议

习近平总书记在党的十九大报告中指出："创新是引领发展的第一动力，是建设现代化经济体系的战略支撑。为加快建设创新型国家，实现为建设科技强国、质量强国、航天强国、网络强国、交通强国、数字中国、智慧社会提供有力支撑。"要以"一带一路"建设为重点，坚持引进来和走出去并重，遵循共商共建共享原则，加强创新能力开放合作，形成陆海内外联动、东西双向互济的开放格局。创新型国家建设离不开全民科学素质的提升，科技馆要紧紧把握住这一难得的历史发展机遇，不断研究创新建设发展理念，发挥好科技馆科普宣传教育阵地作用。科技馆要注重面向广大社会公众，尤其是青少年普及科学知识，传播科学思想和方法，弘扬科学精神，启迪智慧，培养创新能力，为培育创新型科技人才做出积极贡献。科技馆要注重体验探究式学习在场馆中的普及和运用，要注重展品资源更深层次的挖掘，结合当前国内外最新科普教育理念，开展多种形式的主题科普教育活动，永葆科技馆生机和活力。要加快科普信息化建设，完善数字科技馆建设，以数字信息技术为平台和载体，以实体科技馆为中心，利用"互联网+科普"平台和科技馆数字化科普设施，辐射带动数字科技馆、流动科技馆、科普大篷车及青少年科技教育工作逐步实现信息化建设发展，扩大科技馆科普覆盖面。推进科技馆免费开放工作管理科学化、制度化、规范化和标准化，加强绩效考核和监督，有效推动科技馆建设创新升级。要加强"一带一路"科普场馆之间的交流与合作，打造共建共享平台，促进科技馆建设发展理念深度融合。

浅析青海省科学技术馆发展现状

翟　咏 [①]

摘　要　青海省科学技术馆始建于 1987 年，新馆于 2011 年 10 月落成开放，是青海省规模最大、设施最齐全的科普类场馆。青海省科学技术馆充分利用现有资源，积极丰富展教内容和形式，持续扩大科普服务覆盖面，取得了良好的社会效益。与此同时，随着科技馆科普业务的增加和知名度的提升，运营中一些问题逐渐凸显出来，一定程度影响了科技馆职能的有效发挥。本文通过介绍青海省科学技术馆基本信息和教育活动开展情况，对其目前的运营现状展开分析，包括教育活动实施内容、取得成绩、存在问题等。

关键词　青海省　科学技术馆　教育　困难

1. 基本概况介绍

青海省科学技术馆始建于 1987 年，新馆于 2011 年 10 月 24 日落成开放，位于青海省西宁市城西区五四西路 74 号，建设总投资 4.3 亿元，占地面积约 3.67 万平方米，建筑面积 33179 平方米，展厅建筑面积 14000 平方米，设有 7 个主题展区、3 个特效影院、序厅、青少年科学工作室和教育培训中心。常设展品 280 余件，公众可动手参与的展品占 90% 以上，内容涉及基础学科、环境、生命、能源、交通、安全、信息、航空航天以及青海独特的自然环境、高原生态和社会经济发展中的重点科技内容。作为青海省规模最大、设施最齐全的科普类场馆，青海省科学技术馆承担着面向广大社会公众普及科学知识、传播科学思想、弘扬科学精神、倡导科学方法的重要职责，是青海省贯彻实施《全民科学素质行动计划纲要》的主要实施单位之一。截至 2017 年 9 月中旬，已累计接待公众 448.2 万余人次，其中青少年 197.6 万余人次，占总人数的 44%，有效促进科普活动的传播速度和广度，很好地发挥了省科技馆的

---

①　翟咏：青海省科学技术馆副馆长。邮箱：459970338@qq.com。

科普前沿阵地作用。

2. 科普教育活动开展实施情况

2.1 依托重大节假日，开展主题科普活动

近年来，青海省科学技术馆坚持以法定节日、公众假期、重要纪念日等为载体，年均开展主题科普活动 10 次以上（元旦、寒假、春节、劳动节、儿童节、暑假、国庆、圣诞、科技周、科普日等），取得了良好的社会效益。自 2015 年起，青海省科学技术馆以"科普你我同行，助力生态文明"为主题打造系列科普活动，不断扩大教育活动的深度和广度，品牌效应日渐凸显。目前，青海省科学技术馆举办的大型主题科普活动已在省内公众中形成重要影响，成为公众假期参观学习、了解科技知识的固定平台和窗口。

2.2 聚焦青少年人群，策划专题科普活动

青海省科学技术馆始终将提高青少年的科学素质作为科普教育的重中之重，每年开展丰富多彩、针对性强的青少年活动。一是通过"科技馆活动进校园"的形式，定期组织力量走进校园开展科普活动，丰富学生的课余生活，促进校内外教育融合。二是开展各类青少年科技营活动，如寒暑假的科普探秘营、户外生存能力训练营、高校科学营等，组织一系列参观、考察、露营等活动，活动地点包括馆内、互助酒厂、贵德科普基地、德令哈天文台、可鲁克湖、外星人遗址以及北京、上海、广东等地 211 重点高校等。三是加强馆校合作，开展新奇有趣的"科技课堂"教学活动。青海省科学技术馆新馆开馆以来，依托馆内资源，以稳步推进馆校合作为前提，加强课件研发力度，完善课程体系建设，努力打造科学教育一流品牌。截至目前，青海省科学技术馆已自主研发 11 个系列 1185 个课件，与西宁市 9 所小学建立合作关系，有来自 13 所学校的 82088 名学生参加了工作室的科学课程，极大地激发了他们学科学、爱科学的热情。

2.3 普惠基层群众，开展流动科普巡展活动

自 2011 年 6 月以来，青海省科学技术馆以中国流动科技馆巡展活动为契机，不断整合优质科普资源深入偏远地区进行巡展，有效改善了基层科普资源相对薄弱的现状。2015 年年底，流动科技馆巡展顺利完成全省县域全覆盖的目标，2016 年开始第二轮覆盖，并尝试从县域到乡镇延伸。截至目前，流动科技馆巡展已在全省各州市县共 64 个

站点开展，累计行程近 7 万千米，普惠基层群众 75.83 万人次。经过几年来的摸索实践，流动科技馆巡展工作已成为我省科普工作的一大品牌，不仅在省内广受赞誉，而且得到了业界的一致认可，达到全国一流水平。此外，青海省科学技术馆还积极开展贴近群众生活的"科普下基层"活动，将下基层活动与"科普之冬""科普与三区建设同行"等主题紧密结合，通过喜闻乐见的活动形式，向基层群众传播科学知识，发挥了很好的宣传和引导作用。

3. 存在的问题和困难

青海省科学技术馆自 2011 年开放以来，运行平稳、管理有序，受到社会各界广泛关注与一致好评。随着青海省科普事业的进一步发展，科技馆业务范围逐步扩大、工作内容持续增加、运维成本不断攀升，工作中遇到的问题和困难日渐凸显，主要表现如下。

3.1　运维成本高，经费不足

多年来，青海省科学技术馆经过不懈努力，在省内外知名度显著提升，得到了公众的认可和青睐。但是，持续增加的参观公众导致了展厅展项损坏率高、老化加速等问题，运维经费逐年上升。自 2016 年以来，科技馆陆续对部分展区进行改造升级，进一步增加了经费的开支。

3.2　人员流动性大，人才储备不足

长期以来，青海省科学技术馆一直存在人员流动较大、高素质人才匮乏的问题，究其原因主要有以下几点：一是临聘人员比例较高。临聘人员相较在职人员，在收入、保障、归属感等方面存在差异，工资偏低、工作时间长、工作量大、节假日不能与亲人团聚等因素进一步导致人员流动频繁，影响了人才队伍构建的稳定性。二是工作特质因素。青海地域辽阔，人口分布分散，公众科学素质水平偏低，青海省科学技术馆作为我省主要的科普工作实施单位，需要经常深入农牧区开展科普宣传活动，加之高海拔、高寒等特殊地理条件，进一步增加了科普宣传工作的难度。艰苦的工作条件和周期性下乡是造成人员流动性较大的因素之一。三是缺乏专业人才。青海省科学技术馆现阶段人才构成以本科和大专学历为主，专业构成以文科居多，无科技馆相关专业学历人才，缺乏理工科、高学历人才，工作以经验指导为主，加之上述人员编制和工作特性原因也造成部分人才流失，导致科技馆的专业人才队伍欠缺科学性和稳定性。

3.3 科技馆辅导员职称无自主系列

目前，科技辅导员队伍具备年轻化的趋势。一方面，他们热情、积极、敢于创新；另一方面，他们也对职称评审有较高需求。科技辅导员的工作具有教育性质，长久以来，科技辅导员"教育者"的定位并不明确，在职称评定层面较为凸显，从全国来看，除教育系列、馆员系列及工程师系列外，科技辅导员没有独立的职称评定系列和统一评定标准，导致多数科技辅导员无法参与职称评定，影响了他们的工作积极性，不利于他们的职业发展和晋升。此外，非在编人员无法参与职称评定，同时收入较低，工作积极性和主动性不强，间接导致员工工作效率下降、人才流失加剧。

4．对"一带一路"科普场馆交流合作的建议

青海自古以来就是丝绸之路的途经之地，在"一带一路"开放战略中更是具有不可替代的重要位置。"一带一路"国家科普场馆发展国际研讨会作为促进国际、国内科普场馆互惠互通、共建共享的科普盛会，对我国科普场馆尤其是"一带一路"沿线地区科普场馆的建设和发展具有重要意义。青海省科学技术馆希望以此为契机，深层次融入"一带一路"建设，与国内外科普场馆共同推动沿线国家及地区的科普事业向前发展。

一是希望主办单位能搭建一个平等、合作、交流、共享的平台，为各场馆提供互通有无、相互学习的渠道。

二是关注沿线地区科普场馆的发展现状和发展需求，了解经济欠发达地区科普场馆存在的发展困境，并给予相应的政策扶持。

三是梳理优秀科普场馆在管理、运行、人才培养、展览展示、科学教育等方面的典型做法，旨在促进处于不同发展阶段的场馆间的学习交流，以缩小差距，共同发展。

四是希望"一带一路"国家科普场馆发展国际研讨会能够切实成为国内外科普场馆共享资源、共谋发展的纽带，有力推动西部欠发达地区科普场馆建设，促进我国科普事业均衡发展。

普惠共享资源　打造科普阵地
——绍兴科技馆创新发展历程回眸与展望

顾尧根　丁耀栋 [①]

摘　要　在"一带一路"的时代框架背景下，绍兴科技馆以打造"聚人气、创品牌，高水平打造地市级一流科技馆"为中心目标，立足本职，创新管理、探索前行，开馆三年来，在推广科普知识、培养科创意识、提高全民素质方面成绩显著，得到社会各界的高度肯定。

关键词　绍兴科技馆　科普阵地　一带一路　发展历程

习近平总书记在全国"科技三会"上强调："科技创新、科学普及是实现创新发展的两翼，要把科学普及放在与科技创新同等重要的位置。"绍兴科技馆深入贯彻党的十九大精神，紧紧依靠各级党委政府的支持和社会各界的关心，以打造"聚人气、创品牌，高水平打造地市级一流科技馆"为中心目标，团结干群力量，深挖优势潜能，实现当年建设及开馆的工程建设记录，把最优质的服务、最优质的设施、最优质的资源展现给广大公众。开馆三年来，绍兴科技馆创新管理、探索前行，在推广科普知识、培养科创意识、提高全民素质方面成绩显著，以一流科普铁军力量助力绍兴"一带一路"重要枢纽城市建设，得到社会各界的高度肯定。2017 年 8 月 17 日，时任中国科学技术协会党组书记、常务副主席、书记处第一书记的尚勇同志视察绍兴科技馆时，对丰富新颖的展项展品和科学教室开设的课程表示赞赏。

1. 绍兴科技馆基本概况

绍兴科技馆位于绍兴市镜湖新区洋江西路 528 号，2009 年列入"绍兴市三年建设

---

① 顾尧根，绍兴科技馆办公室主任。研究方向：科技馆教育理念及实践。联系方式：13757581221，地址：浙江省绍兴市镜湖新区洋江西路 528 号，E-mail:35490827@qq.com。丁耀栋，研究方向：科技馆教育。地址：浙江省绍兴市镜湖新区洋江西路 528 号，E-mail: dingyaodong1205@foxmail.com。

计划"重点项目，2011 年 3 月动工，2014 年 5 月完成土建移交，2014 年 12 月 29 日开馆，总用地面积约 5.586 万平方米，总建筑面积 3.1 万平方米，其中地上面积 2.5 万平方米、地下面积 0.6 万平方米，建筑高度 24 米，土建工程总投资 2.5 亿元（彩页图 56）。新馆分为南北两区，北区为"彩虹儿童乐园"、科普影院区；南区为一层序厅、临展区，二层为"地球与生命""探索与发现"展区，三层为"科技与生活"展区。还有绍兴籍院士、青春期教育、禁毒教育、绍兴水生命 4 个公共展区，一个中小学生科技教育实践基地和室外小小科普植物园。

2. 绍兴科技馆近年来取得的成绩与特色亮点

2.1 做强阵地，彰显科普实力

2.1.1 常设展厅与时俱进

着眼于观众科普需求和展品（展项）创新，最大程度发挥展厅的社会价值，以新颖的方式展示给广大公众。如在"科技与生活"展厅，通过结构剖视图的方式把新能源汽车的内部结构呈现出来，贴近百姓生活；"地球与生命"展厅采用院地合作方式，通过与中科院古脊椎动物与古人类研究所合作，由中国科学研究院负责展厅内容的策划设计，由科技馆完成项目的实施，这样的合作在全国尚属首次。近期，在该展厅增添了 AR 设备展示古生物，以现代手法吸引更多青少年参观体验。

2.1.2 免费开放广受欢迎

经过积极争取，绍兴科技馆被列入全国科技馆首批免费开放试点单位，自 2015 年 5 月 16 日起实行常设展区、科普讲座及其他配套服务免费的惠民政策。免费开放后，极大丰富了广大市民的科技文化生活，提升了绍兴公民的科学素养，加强了社会效益的发挥力度。开馆 3 年多来，参观人数已逾 150 万人次，青少年观众约占 75.2%，年均接待观众 50 万余人次。

2.1.3 短期展览常展常新

短期展览能有效保持科技馆的新鲜感，绍兴科技馆根据广大观众需求，结合国内外最新科技进展和成果、社会关注的热点问题、有影响力的科技人物和事件等，不断丰富短期展览内容。充分利用一楼 2200 多平方米的临展厅，多渠道引进大中型短期展览，3 年来，已举办短期展览 15 次，吸引 50 万余人次驻足参观。2017 年 5 月举办"漾舟信归风——古代船模展"期间，承办"'一带一路'上的绍兴古桥"图片展，全面展示了绍兴古桥在"一带一路"建设中起到的作用。2017 年 7 月开展的"穿越达尔文星球"VR/AR 新媒体科普短期展览一度出现"一票难求"的盛况。

2.2 突显特色，激发科普活力

2.2.1 品牌活动好"戏"连台

以"爱科学、玩科学、秀科学"为主题，打造部门联动、社会参与的科普品牌活动 50 余场。如举办"六一科学嘉年华""科普夏令营""科学实验秀""科普手偶剧"等主题活动。主推爱上科技馆、大喇叭小讲堂、专家进展厅等系列活动，连续三年举办绍兴市校园科学达人大赛，通过积极探索，将"校园科学达人"大赛升级打造为省"科学玩家"青少年科学才能挑战赛，2017 年与浙江省青少年科技活动中心、省电视台举办第二届浙江"科学玩家"青少年科学才能挑战赛。通过经常性地开展活动，让科教基地持续保持活力。此外还主动出击，将"科普大篷车"开进社区、学校、农村，让偏远山区的群众也能在自己家门口感受科技的无穷魅力，2017 年"科普大篷车"的行程就超过 3000 千米，足迹遍布绍兴多个角落，让科普之光照亮更多人。

2.2.2 寓教于赛摘金夺银

分别针对青少年学生和信息学辅导教师进行信息学强化培训，根据中小学学生自身水平，量身定做安排培训课程，在为备战全国青少年信息学奥林匹克联赛的选手进行专题培训的同时，为全市中小学生普及信息学知识。同时，组织全市信息学优秀教师开展研讨，邀请绍兴市内外的资深专家作经验传授，研教结合，有效促进信息技术在教学中的应用。2011 年至今，绍兴已累积获得国际信息学奥林匹克竞赛金牌 5 枚、银牌 2 枚，位列全国第一，2017 年，在第 34 届全国信息学竞赛上，绍兴学子更是获得了 4 金 7 银 9 铜的历史最优成绩，通过信息学竞赛保送名牌大学的绍兴学子达近 60 人。2015 年至今，举办、承办、参加各类科技竞赛 50 余次，青少年科教氛围浓厚。2016 年，在第四届全国科学表演大赛中综合成绩位列全省首位，选送的 6 个节目均获奖，其中，全国二等奖 3 个。2017 年，原创微型科普剧本《杯酒"释"英雄》在第五届全国科学表演大赛中荣获一等奖，展现出了科技馆良好的精神风貌。

2.2.3 院地合作引贤进馆

在新馆开馆之日挂牌成立"中科院古脊椎动物与古人类研究所科普教育基地"。2015 年 10 月，与中国科学院院士、瑞典皇家科学院院士张弥曼团队签约成立院士专家工作站，这在全国科技馆行业属首开先河，院地合作成效显著。3 年来，院士专家频频来到绍兴科技馆，广泛开展特色鲜明的公益性科普活动，相继邀请郝跃、曹春晓、张弥曼、傅睿思等一批知名专家学者开展"科学大讲堂"，每年开展 10 余次。

2.3　做亮基地，打造科普名片

2.3.1　展教联动体现魅力

2015 年 1 月，绍兴市教育局和绍兴市科学技术协会联合发文在绍兴科技馆建立绍兴市中小学生科技教育实践基地。基地面向全市中小学四、五、六年级及初一、初二学生，开放时间为每周三到周五。活动"展教合一"安排 1 天，半天在"科学梦工场"开展科学实践活动，半天在展厅开展科普活动。其中"科学梦工场"共有 16 间教室，开设了 10 多个科学实践项目，能同时容纳 400 多名学生听课，包括科学教室、信息学教室、航模探究 box、建模探究 box、机器人体验室、3D 打印创作室、3D 创新实验室、思维训练室、创意百拼室、心理探索室、"科学梦工场"演播室等，这样的科技实践场所在全国科技馆中尚属首创。

2.3.2　馆校合作扩容提质

实践基地运行以来，学校和社会反响热烈。2015 年年初印发了建立基地的通知，3 月正式开始建设，9 月初建设完成，10 月 21 日基地新建的"科学梦工场"正式开班授课。运行两年多来，进一步扩大范围，共接待来自市内越城区、柯桥区、上虞区 39 所学校 9 万余名学生，通过建立校外科技教育实践基地这一创新做法，全市更多的学生参与到科学教育中来，实现了校外科技活动场所与学校课程的有效衔接，作为学校教育的补充，已成为全市中小学生科技教育"第二课堂"。

2.3.3　创新打造品牌栏目

与绍兴电视台合作推出深受中小学师生和家长欢迎的青少年科学栏目《科学梦工厂》，其通过新颖的展示手段推介科技馆品牌，立体展现科技馆环境设施、展品展项，以发展小会员和志愿者的形式，提高活动知名度，做亮中小学生科技实践基地。栏目共分为 4 个版块，"科学探索馆"中的所有科学实验全部由绍兴科技馆青少年活动部的科技辅导员设计开发，主持人和小嘉宾一起动手，共同探索科学的神奇和奥秘。"小飞马俱乐部"通过主持人进教室，让孩子们在一问一答中学习科学知识。"彩虹乐园"结合个人游戏和团队游戏，既有利于培养亲子关系，也增加了栏目的趣味性。"奇思妙想剧场"是展示科技小发明、科学文艺节目、科技创新成果的平台，旨在增强同学们的创新意识。

2.3.4　精准发力提档升级

进一步完善提升实践基地，为全市学生提供更加优质的科学实践活动。一是课程开发与时俱进，新开发科学实践课程 30 多门，每周定期试教，提升整体教学水平。二是教学环境不断改善，"科学梦工场"对部分教室进行改进。三是制订全年青少年科普计划，与教育部门联合发文组织全市青少年参加各类竞赛。四是联动学校全面推介，年初组织学校分管校长到科技馆现场召开实践基地工作推进会，研讨基地科技实践活动方案，商讨与学校交接注意事项，进一步完善机制促进基地有效运行。与此同时，努力把实践基地打造成绍兴市乃至全国创新型科普教育的名片。

3. 发展中遇到的困难及对"一带一路"科普场馆交流合作的建议

3.1 当前我馆存在的困难与挑战

在高质量办好全年特色展教活动的同时，我们创新开展各类青少年科技教育实践活动，在全市形成了良好的科普教育氛围，在地市级同类科技馆行业中已经走在了前列，但在实际运行中也存在着制约科技馆进一步发展的问题。一是新馆搬迁落成早于镜湖新区开发成熟期，目前场馆与居民集聚区相距较远，周边配套设施尚未完善，如缺少餐饮配套、公共交通不便、停车车位不足等。二是干事力量配备不足。按照《科学技术馆建设标准》规定，我馆人员编制数应达到155人，而编办实际核准总数仅为111人（事业编48人，编外63人），与省内规模相近的浙江省科学技术馆（30452平方米）、杭州低碳科技馆（33656平方米）分别有工作人员140人和145人。三是队伍稳定性低。我馆编外员工年收入总计4万元，待遇低于省内同类岗位水平，且多数员工必须在双休日、法定节假日继续上班，休息时间的冲突和偏低的待遇，导致编外员工自开馆以来离职率高达120%，严重影响我馆事业的持续发展。四是尚无单独的职称评审体系，馆内事业人员符合专业技术职务晋升条件时需到教育系统、文博系统等单位参加职称评审，因各种原因，评审通过率较低，由此成为阻碍事业人员晋升的绊脚石，不利于激发干事积极性。

3.2 着力破解发展难题，推动创新发展

科技馆发展事业是社会公益服务和公民科学素质建设的重要基础。一是积极争取市委市政府的领导和支持，将科技馆发展事业和公民科学素质建设的目标任务纳入地方发展规划，着力解决当前我馆存在的编制人员不足、周边设施不相配套等困局。二是要把科技馆青少年科技教育与学校教育有机结合起来，会同科协、教育等相关部门，密切配合，形成合力，共同抓好绍兴市中小学生科技教育实践基地建设。三是要建立健全激励考核制度，创新科学普及和青少年科技教育工作的组织管理机制，对在工作中涌现的先进个人给予精神和物质奖励。四是加强员工队伍建设，抓深抓细抓实"两学一做"学习教育常态化、制度化，深入开展"不忘初心、牢记使命"主题教育，以

党建引领全馆事业发展。五是加强志愿服务队伍建设，完善志愿者招募、培训、考核、管理制度，充分发挥学生、专家、市民等三支志愿者队伍的作用，建立健全与周边高校联动开展科普志愿服务工作机制，在有效降低运营费用的同时，推动公益服务文化在科技馆的落地生根、开花结果。

3.3　对"一带一路"科普场馆交流合作的建议

2017 年首届"一带一路"科普场馆发展国际研讨会《北京宣言》的提出，为"一带一路"建设注入强大推动力。绍兴作为"一带一路"重要枢纽城市，对我馆事业发展既是机遇，也是挑战，我们愿意与各兄弟场馆一道，加入到"一带一路"科普场馆交流合作的"朋友圈"中，共同促进科普事业发展，在此提出几点拙见。

3.3.1　补齐短板

在地区科普教育和青少年科技创新、科技教育事业发展上，还存在着不平衡不充分的短板，如对偏远山区、农村公众的科学普及力度还须进一步加大，青少年科技创新能力培养还须进一步加强，"馆校结合"实践还需进一步深入推广，馆际交流合作还需进一步深化。国内的科普场馆展陈内容在一定程度上呈现整齐划一、千篇一律，有的场馆布展略显粗糙，要进一步加强与台湾地区、国外知名科技博物馆的沟通交流。

3.3.2　建立平台

《北京宣言》提出要打造运转高效、信息共享的"一带一路"国际协同创新平台。笔者认为，促进"一带一路"科普场馆交流合作的有效途径是建立信息交流、数据共享平台，如由中国自然科学博物馆协会牵头，各专业委员会、工作委员会协作，在"一带一路"沿线划分片区成立若干个以交流互通、共赢共享为目标的地区协会，抱团取暖，合力推进沿线地区科普场馆事业发展。

4. 结语

习近平总书记指出，"中国科学技术协会各级组织要坚持为科技工作者服务、为创新驱动发展服务、为提高全民科学素质服务、为党和政府科学决策服务的职责定位。"绍兴科技馆作为市科协直属事业单位，始终牢记职责使命，着力强化科普阵地建设，充分发挥科学普及和青少年科技教育等重要作用。在新常态下，绍兴科技馆将继续深挖潜力，着眼未来，发挥科普辐射作用，将一如既往以活动带动科普工作，提升科技馆的人气和活力。同时，进一步完善数字科技馆建设，以"互联网＋"展现科普活动新魅力，带动科普工作新发展，为绍兴"一带一路"重要枢纽城市建设贡献科普动力。

武汉科学技术馆的发展

王锐利 [①]

摘 要 武汉科学技术馆运营 27 年来经过三次大的建设，展览理念与时俱进，场馆建设契合公众需求，科普活动丰富多彩，"参与、互动、体验"的参观形式，广受公众欢迎。未来，将根据场馆现状与公众需求，不断拓展与完善科普形式。

关键词 科学技术馆 历史 发展

武汉科学技术馆已经走过了 27 个春秋，它由新、老两座场馆组成，位于江岸区赵家条 104 号的老馆自 1990 年对公众开放，位于江岸区沿江大道 91 号的新馆于 2015 年 12 月 28 日开馆。作为青少年科普教育基地，两馆将并驾齐驱，共同为公众服务（彩页图 57）。

1. 武汉科学技术馆的建设

1.1 武汉科学技术馆初建成

1978 年，全国科学大会以后，国内出现科技馆建设的热潮，部分省、市科技馆相继投入建设。武汉科学技术馆从 1990 年 3 月 18 日正式开馆到 2006 年 12 月 30 日改扩建后重新对外开放，走过了不寻常的 15 年。

"七五"期间，武汉市委、市政府将武汉科学技术馆建设列为全市人民办的 20 件好事之一。1984 年，武汉科学技术馆选址在江岸区后湖乡，占地 38739 万平方米。1986 年 8 月 18 日，举行了武汉科学技术馆奠基典礼。

1989 年武汉科学技术馆建成，占地面积 26623 平方米，建筑面积 13374 平方米，规模在全国省会城市科技馆中位居前列。1990 年 3 月 18 日，武汉科学技术馆开馆暨武汉市首届科技博览会开幕。

---

① 王锐利，武汉科学技术馆助理馆员。地址：武汉市江岸区沿江大道 91 号武汉科学技术馆，邮编：430010，E-mail：381397238@qq.com。

武汉科学技术馆开馆后在经费不足的情况下，坚持科普教育的方向，先后举办"首届科技博览会""大型'活体'恐龙暨珍奇水生动物展""'嫦娥一号'航天科普知识展""'我们在现场'抗震救灾大型图片展""坚持科学发展，建设生态文明""2009国际天文年暨中国武汉日全食观测系列活动""辉煌科技60年""走近机器人""武汉市百万市民学科学——'科学防癌·健康生活'首届健康武汉系列活动""地外行星""'聆听大地的声音'生物多样性之旅展览""食品安全主题科普展""探索宇宙展览"等专题展览34场次、大型巡展11场次。与此同时，还积极拓展工作范围，积极开展各类青少年科技活动、承办武汉市青少年科技创新大赛等青少年科技类大型科普赛事，为提升未成年人科学素质做了很多工作。

进入21世纪，公众学习科学知识的热情日益高涨，党中央、国务院对科普场馆建设高度重视，使全国科技馆事业迎来了前所未有的发展机遇。武汉科学技术馆展厅展示面积不足、展览内容单薄、科普活动形式单一的现状难以满足公众学习科学知识的需要。在武汉社会各界地呼吁以及市领导的关心和重视下，武汉科学技术馆在运行15年后于2005年8月对主楼进行改扩建。

改扩建期间，武汉科学技术馆一边筹备展品，一边利用科技馆科普资源，采用"走出去"的方式，结合科普大篷车把科普活动送到学校、社区、乡镇、军营，在近两年的时间里，共接待观众22万余人次。

1.2 武汉科学技术馆改扩建后重新开馆

2006年12月30日，武汉科学技术馆改扩建后重新对外开放。武汉科学技术馆通过改扩建，扩大了展示面积、更新了科普展品、拓展了科普功能，之后又相继建成4D动感影院、全市最大的科普画廊、科普宣传电子屏和天象馆以及与天象馆配套的天文知识展区，使武汉科学技术馆的整体形象、教育功能有了很大提升。主要功能有科普展教、学术交流、科技培训、青少年科技实验、科普影视、科普画廊和科普电子屏等（彩页图58、图59）。展示面积由原来的1600平方米扩大到6400平方米，建筑面积扩大到15435平方米。科普展品260多件（套），90%以上可动手操作，内容涵盖数、理、生命科学等，展示和介绍应用科学技术和日常生活中的科学知识，如工业、农业、医疗卫生、气象、环保等；展示、介绍现代科学知识，特别是高新技术及其产业，如卫星与航天技术，计算机与信息技术、生物工程、基因工程、新材料和新能源、激光等。随着硬件设施的提升，武汉科学技术馆的办馆宗旨和教育理念更加明确，展教工作更加突出两种教育方式。一是展览的参与性，强调参观者在直接参与操作展品的过程中，亲身感受科学技术带来的乐趣，在体验中学习科学知识；二是科学的思想性，更加关注公众对科学思想和方法的掌握。每年接待观众 25 万余人次。

1.3 武汉科学技术馆新馆建成开放

随着社会的发展和科技的进步，在公众对学习科学知识的热情更加高涨的同时，科技馆科普工作重点也逐渐从对公众进行科技知识普及向提升公众科学素养方向转变。武汉科学技术馆老馆的硬件设施等客观条件已难以满足当下科普工作的需求。2010 年 7 月，杨淑子、李培根、李德仁、叶朝晖、赵梓森、邓秀新 6 位在武汉工作的中国科学院和中国工程院院士联名向武汉市委、市政府提出建设武汉科学技术馆新馆的建议。2010 年 8 月，武汉市委、市政府做出"将武汉港客运楼改造成科技馆"的决策，并将新馆建设列入我市"十二五"规划，连续两年写进《政府工作报告》。

武汉科学技术馆新馆馆址在美丽的汉口江滩，2011 年 11 月 20 日破土奠基，2015 年 12 月 28 日建成对外开放，总建筑面积约 3 万平方米，主楼改造及展示工程总投资 5 亿余元，是一座集多功能、综合性、智能化于一体的特大型科普教育活动场所，也是武汉市着力打造的"江汉朝宗"文化旅游景区群中的重要组成部分（彩页图 60）。截至 2017 年 9 月，新馆共接待观众约 290 万人次。

武汉科学技术馆新馆的顶层设计由国内科普大家主创，凝结了众多科学家的集体智慧。在展览理念上，坚持"见物见人见精神见智慧"的"四见"原则[①]，将科学的发现与人文精神、科学家的故事连贯起来。在展陈形式上，由国际、国内知名设计公司参与设计，创新展品占全馆展品总数的 40% 以上。此外，还有选择地从国外采购了一批经典展品。

本馆常设展览以我国古代著名诗人屈原的《天问》为开篇，隐喻人类自古以来的求知探索精神。展馆分设"自然板块"的宇宙、生命、水展厅，"创造板块"的光、信息、交通展厅，另设了数学、儿童展厅，展品数量达 600 余件，既将自然科学与工程科学有机地融合，又突出了鲜明的地方特点。"参与、互动、体验"的参观形式，广受公众欢迎。

1.3.1 地理位置独特

武汉科技馆新馆地处武汉中心城区，具有优越的区域环境的特点，可以表达为"三临、三老"，"三临"指临街、临江、临滩，"三老"指"老城区、老车站、老码头"。

---

① "四见"建馆理念即将展品背后关联的科学家、科学家所处的时代和科学发展脉络、科学的发现与人文精神、科学家的故事与科学的生活运用等连贯起来呈现给观众，让观众从中感受科学的乐趣，培养科学探究的兴趣。由中国自然科学博物馆协会名誉理事长、清华大学博导、武汉科学技术馆新馆建设专家组组长徐善衍提出。

新馆依江而建，是"江汉朝宗"旅游景区（正在申报国家 5A 级景区）的核心景点之一。特殊的位置意味着公众对这座场馆软、硬件设施有着更高的要求，这对武汉科学技术馆既是机遇，也是挑战。

1.3.2 建筑有特色

武汉科学技术馆新馆是目前全国最大的利用既有建筑改造成的科技馆。它既保持了原武汉客运港这座曾经江城地标性船形建筑的外形，又注入了科技元素，体现了环境保护和科学发展的新理念，寓意科普之舟扬帆远航，使这座历史建筑重抖精神、再放异彩。

1.3.3 展示有亮点

一是新馆实现了科技、历史、文化充分融合，新馆的建筑特征、所处地域特点以及馆内"琴台遇知音""古代人类智慧"等颇具历史和文化内涵的展示内容特色，彰显了新馆科技、历史、文化相融合的鲜明特征。二是展示内容具有鲜明的地方特色，交通、信息展厅充分展示了地方产业特色，水展厅充分展示地域资源特色。三是新馆展品具有注重创新、兼顾本地特色、互动性强的特点。其中，天问展项是本馆重点展项，创意来自楚国诗人屈原的《天问》，展项造型独特优美，观众可以用回答问题的方式与之进行互动，在全国具有独创性。

1.3.4 高端的智能化建设

新馆拥有国内领先的智能化系统，该系统包括公众服务、传播交流、运营管理三大功能，共 15 个子系统，涉及安防、票务、观众导览与公共服务、观众管理系统、科技馆网站、智能办公、展馆管理等方面（彩页图 46 ～图 62）。

2. 开展形式多样的科普活动

2.1 组织全国青少年科技创新大赛——武汉市选拔赛

武汉科学技术馆成功承办了第 23–32 届全国青少年科技创新大赛——武汉市选拔赛，每届大赛全市有近 5 万名学生参加。

2.2 举办科普夏（冬）令营和科技馆培训班

暑（寒）期举办"科学如此神奇""梦想助我成长"等夏（冬）令营，让青少年轻松学科技、快乐过暑（寒）假。每年组织各类夏（冬）令营活动 10 余期。武汉科学技术馆还针对不同人群定向举办各种培训班。常年为科技创新大赛的辅导老师举办武汉市中小学科技辅导员培训班，为市民和中小学生开办"声乐培训班"和"少儿思维训练班"等。

2.3 推出科普剧表演

2009 年，武汉科学技术馆先后推出自编、自导、自演的《家电夜话》和《魔法实验师》等科普剧，2017 年推出科普剧《不同物品的物理性质》《静香的生日宴会》《逃离火星》《海绵宝宝海底音乐会》《"擦亮"天空》，科学实验剧《"气"功》，科学表演秀《气球历险记》《神秘的宝藏》，这种寓教于乐、耐人寻味、充满新鲜感的科普形式深受公众的喜爱。

2.4 启动馆校结合工作

本着"密切合作，资源共享，共同发展"的原则，2009 年，武汉科学技术馆启动馆校结合工作，先后与江岸区、江汉区、汉阳区的 11 所中小学签订共建协议，开展丰富多彩的科普活动。

2.5 建立中国科学院·武汉科学家科普演讲团

2010 年 3 月，由中国科学院武汉分院和部分高等院校的院士、专家组成的武汉科学家科普演讲团成立，同时，以科普报告进学校为主要内容的"武汉市百万市民学科学——院士专家进校园活动"正式启动，通过这项活动建立学生与科学家面对面互动交流的平台，扩大了学生的知识面和科学视野，共举办 23 场报告会，累计听众达8200 人。

2.6 建立流动科技馆

2009 年，武汉科学技术馆对大篷车展品进行改造更新，结合馆内大型专题活动，开展大篷车"四进"（进广场、进学校、进社区、进机关）活动。流动的科技馆年均巡展50 余场，为更多的江城人民送去科技知识的同时也送去欢乐（彩页图 65、图 66）。

2.7 举办专题展览

武汉科学技术馆新、老馆都坚持举办专题科普展览，其中武汉科学技术馆新馆设有 1630 平方米的临展厅，开馆至今已举办"野生动物标本科普展""'海洋精灵'水母主题展览""未来织物""海洋权益与军事""'南海之美'——海洋生态与保护主题展览"等大型专题科普展览。临展内容贴近市民的生活，为提高全民科学素质起到积极作用，截至 2017 年 9 月，累计服务观众 1000 万余人次。

2.8 开办武汉科普讲坛

武汉科学技术馆新馆开馆，即着力打造"武汉科普讲坛"品牌活动，已经先后举办《科普的时代特性与责任》《"走读大武汉"之摄影常识》《日食的观测与拍摄》《领略科技魅力 携手走向未来》《机器人总动员》《4.22 世界地球日——垃圾分类公益讲座》等科普报告会，为广大市民送去了丰富的科普大餐。

3. 武汉科学技术馆展望

2016 年 5 月 16 日，武汉市市长到武汉科学技术馆调研，明确了"新馆和老馆在展示内容和科普活动上互相补充、形成差异化发展"的思想，提出了"（展览）按照既有相对固定、又有常展常新，扩大知识面、展示或介绍最新科技成果，展望未来科学发展方向"的要求，并且建议市科协会同市教育局，研究制定把科技馆老馆改造为科学体验中心及科学探究室工作方案。

两个场馆同时运营，差异化发展是必然选择。武汉科学技术馆新馆侧重展览展示教育，"自然板块"有宇宙、生命、水展厅，"创造板块"有光、信息、交通展厅，另设了数学、儿童展厅及专题临展厅，自然科学知识与工程科学知识有机地融合。老馆侧重动手实验、探究式学习，初步确定将建设十大科学探究室（STEAM、机器人、传感器、"小小科学家"、模型运动、天文、机械剧、数学、光学、创新教育十大创新性的主题科学探究室和一个社区科普大学教学基地）及科普影院等。两馆各有侧重，同时互相补充，从不同的角度推进科普工作，更好地为公众服务。

4. 武汉科学技术馆对"一带一路"科普场馆交流合作的建议

"一带一路"倡议实施以来，各国和地区间已达成越来越多的共识，"一带一路"沿线国家在自然环境、社会政策、经济和文化等方面存在巨大差异，大家都面临着复杂的发展挑战。为了共谋科技馆事业的发展，"一带一路"国家和地区共享发展理念、发展模式、发展成果，最终实现共赢。

4.1 线上模式

在科普领域，"一带一路"沿线各国和地区科普场馆可借助互联网、移动终端等新兴媒体，搭乘"互联网 +"的快车，互通有无，共享办馆理念及科普活动开展方式。

4.2 线下模式

通过学术论坛、科普展览、夏令营等多样化活动，促进"一带一路"沿线各国和地区科普场馆互相交流学习，达到良好的科普效果。

参考文献

［1］武汉市科学技术协会. 武汉科协 50 年［M］. 武汉：武汉出版社，2012.
［2］王刚. 武汉科技馆"满血复活"背后是院士博导团队［N］. 长江日报，2016-02-29.

突出重点，覆盖全省，不遗余力开展科普活动
——云南省科学技术馆科普活动开展案例

向纹谊 ①

摘 要 云南省科学技术馆在对公众进行科学普及时，以常设展览"科学的探索"和"体验科学"中国流动科技馆云南巡展为两大平台，加之馆校、馆社、馆企等合作形式，不遗余力地把科普资源洒遍云岭大地；以耳熟能详的身边科学为切入点，以"前沿科技"的演示、体验为吸引点，以青少年学生科学精神的启迪为着力点，以能吃苦不怕累的工作状态为保障，目的是为了让边疆地区学生，特别是少数民族学生、群众、干部等，能近距离感受"紧跟时代步伐"的科技展品，激发云南省公众对科学的热情，提高全省公民科学素质，更进一步地增强民众爱国热情和民族自豪感。

关键词 科普 边疆地区 前沿科技

1. 云南省科学技术馆简介

云南省科学技术馆作为中国科学技术协会和云南省政府授牌的科普教育基地，是实施科教兴滇、贯彻落实《云南省全民科学素质行动计划纲要（2016—2020 年）》的重要阵地，是全省示范性科技馆，是面向青少年和公众开展科普宣传和教育培训等科普活动的公益性教育机构。2011—2012 年，云南省科学技术馆对常设展览进行了科普功能提升改造，原"科普乐园"更名为"科学的探索"，自 2012 年 9 月再次面向公众免费开放，截至 2017 年 12 月，接待观众突破 120 万人次。展厅面积为 3550 平方米，互动展品 121 件，模型 175 件，以科技发展史为主线，分为"科学的启蒙时期""从实验室到工业化""从微粒到宇宙""低碳与发展"4 个主题展区，和"儿童天地""影视及科学表演区""梦工场""健康诊所""军事科技高地"5 个特色展区，为公众免费提供生动有趣的科学体验。

---

① 向纹谊：云南省科学技术馆体验教育副部长（主持工作），中级工艺美术师，参加工作 11 年，一线科普活动经验 9 年。地址：昆明市翠湖西路 1 号，邮编：650031，E-mail：15860043@qq.com。

2. 开展科普活动的背景

2.1 省内情况

云南是一个人口大省，也是全国少数民族种类最多的省份，世居少数民族就有 15 个。同时，云南地处中国西南边陲，属山地高原地形，山区面积占全省总面积的 94%，海拔落差大，地形复杂，这样的自然条件导致山区交通不便、信息闭塞。对于少数民族和边远山区来说，科普资源严重不足。为切实改善这一情况，努力提升我省广大群众，特别是山区、少数民族群众的科学文化素养，云南省科学技术馆坚持举好"科普"的旗帜，致力于免费对全省公众进行各种科学普及活动，不遗余力地将科普资源惠及全省公众。

2.2 地域情况

云南与越南、老挝、缅甸三国相邻，边境线长达 4060 千米，是中国与邻国接壤最多的省份，地域上、历史上曾用唇齿相依、一衣带水，山连着山、水连着水来体现云南与周边国家的友好关系。发源于中国的澜沧江——湄公河一江连接六国，从地理上把中、老、越、缅、泰、柬六国紧密联系在一起，正如习总书记指出："无论从地理方位、自然环境还是相互关系，周边对我国都具有极为重要的意义。"云南与邻国形成了地域共同体、命运共同体、利益共同体及战略共同体，这些因素注定了云南在国家修复邻国关系，"一带一路""海上丝绸之路"国家战略建设中有着无可替代的重要作用。

"我国同周边外交的基本方针，就是睦邻、安邻、富邻。突出体现亲、诚、惠、容的理念。发展同周边国家睦邻友好关系，是我国周边外交的一贯方针。要坚持睦邻友好，守望相助；讲平等、重感情、常见面、多走动；多做得人心、暖人心的事，使周边国家对我们更友善、更亲近、更认同、更支持，增强亲和力、感召力、影响力。要诚心诚意对待周边国家开展合作，编织更加紧密的共同利益网络，把双方利益融合提升到更高水平，让周边国家得益于我国发展，使我国也从周边国家共同发展中获得裨益和助力。"（习近平：《习近平谈治国理政》"坚持亲、诚、惠、容的周边外交理念"）

3. 科普活动方案设计的原则

（1）以我馆现有的各类科普剧、科学秀、科学实验等为基础。

（2）融入紧跟时代步伐，代表时代科技发展的成果。

（3）围绕社会热点话题组织科普活动。

（4）配套活动尽量不与常设展览及"流动科技馆"常规展品类似。

（5）活动方案必须具备科学性、易操作性、趣味性。

4. 活动开展方式

（1）利用好我馆两大平台（"科学的探索"科普展览及"体验科学"中国流动科技馆云南巡展），由我馆科普辅导员开展丰富多彩的科普活动。

（2）联合社会资源，挖掘热心科普事业的团体或个人，共同研究开发，开展如科普讲座、趣味活动、前沿科技体验等的科普活动（表1、表2）。

表1　云南省科学技术馆馆内外活动一览表

地点	内　容	形　式	目　的
馆内活动	音阶管、竹蜻蜓、万花筒、民族蜡染、拓印、再生纸、小机床、空气动力车、浮与沉、叶脉书签	利用给定材料，动手完成相关活动内容	通过我们开展各类科普活动，起到"弘扬科学精神，传播科学思想，倡导科学方法，普及科学知识，提升科学素质"的目的。学生们在动手制作的过程中，经历面对困难和问题时，提出解决困难和问题的设想、巧妙利用工具、动手操作解决所遇困难和问题等，使学生养成综合思考问题的习惯，激发学生"探索"兴趣
	空气炮、小小机器人表演、科普剧表演、科普影视联播	通过眼睛看、动脑筋，发现问题并提出解决问题的设想	
馆外活动	天文观测 宇宙剧场 科学教育机构	联合社会科普机构 1. 用天文望远镜观测 2. 球形幕布播放科普影视 3. 科普教具资源共享	
	云南开放大学 云南师范大学附属小学 昆明市第二幼儿园	利用主题活动联合开展活动。如：全民终生学习活动周、学校科技节/科技周	
	昆明城区 石林县科学技术协会 玉溪市科学技术协会	全国科技周、科普日期间及全民健身活动中联合开展科普活动。科普剧表演、3D打印演示、机器人表演等	
	中国流动科技馆	云南各县级站点巡展启动时，开展配套科普活动，如"驰骋之梦"体验、航拍体验、VR技术体验等	
	少数民族地区双语科普	联合行动。前往边疆少数民族地区开展前沿科技展示、科普大篷车活动	

表 2　云南省科学技术馆各类活动年度效益指标及完成情况统计表

开展内容	年度效益指标	完成情况
"科学的探索"科普展览特色活动开展场次	50 次以上	全年开展 64 场
"科学的探索"科普展览特色活动观众参与量	15000 人次以上	参与人次 17620 人次
"科学的探索"科普展览免费开放天数	280 天以上	全年开放 312 天
"科学的探索"科普展览观众总参观量	200000 人次以上	全年达到 321669 人次
"科学的探索"科普展览展品完好率	95% 以上	保持在 95% 以上
"科学的探索"科普展览观众回头客率	20% 以上	2017 年展览调查问卷观众回头客率 50% 以上
重大科普活动组织开展重点活动的数量	达到 8 个及以上	开展 12 个
重大科普活动参与活动的公众人数	参观人数达 3 万人次及以上	48600 人次
重大科普活动展览展示的科普资源数量	35 ~ 49 项	45 项
重大科普活动制作发放的科普宣传品种类及数量	2 种	5 种
重大科普活动活动参与单位的数量	参与活动的单位、学会、协会及社区达到 15 家及以上	参与 21 家
重大科普活动媒体对活动的宣传报道数量	数量达 20 篇及以上	宣传报道 35 篇
中国流动科技馆 13 套展品巡展 52 个站点完成率	巡展站点数量不低于 52 个	完成 53 个县级站点启动
中国流动科技馆观众覆盖当地中小学学生的比例	中小学生覆盖率达到 70%	各站点均由当地教育局联合发文，要求各学校有序组织中小学生参观体验，中小学生覆盖率超 70%
中国流动科技馆各项实验、表演活动观众参与量	观众参与活动人数超过 41600 人次	参与人次为 53000 人次
中国流动科技馆展品完好率	完好率达 90% 以上	保持在 95% 以上
中国流动科技馆巡展站点老少边穷县所占比例	70% 为老少边穷县	完成
中国流动科技馆带动州市县科普展览配套经费投入	90% 州市县有相应经费投入	各地均有相应经费投入
中国流动科技馆州市县领导班子及全民科素办领导带头参观体验	100% 带头参观体验巡展	各地均做到州市县领导班子及全民科素办领导带头参观体验
中国流动科技馆配套科普活动开展	90% 州市县巡展时有开展配套活动	完成
中国流动科技馆观众参观好评率	90% 好评率	好评率 99%
中国流动科技馆各县（县级市）愿意在"十三五"期间进行第二轮巡展	100% 愿意	100% 愿意
农村中学科技馆成功建成并按时开展的数量	80% 成功建成并按时开展	完成

开展内容	年度效益指标	完成情况
农村中学科技馆展品完好率	≥85% 完好率	展品完好率为 90% 以上
农村中学科技馆建设学校位于国家级贫困县	90% 位于国家级贫困县	完成
农村中学科技馆科技辅导员参加培训率	90% 建成学校参加	均参加
农村中学科技馆观众总参观量	≥150000 人次	172839 人次
农村中学科技馆观众参观好评率	80%～90% 好评率	好评率 95%
少数民族地区双语科普	联合行动 2 次以上。前往边疆少数民族地区开展前沿科技展示、科普大篷车	联合开展 6 次

5. 科普活动：从探索到坚持，最终形成品牌

5.1 馆内动手体验、制作活动

通过前期调查、准备，结合本馆实际情况，制定符合我馆开展的动手制作活动，本处以"小小智慧树"中的"万花筒"制作为例。

万花筒制作活动方案细则

活动主题：体验感受、手工制作

活动对象：九年义务教育的学生

活动人数：20～30 人

活动时间：2 小时

引　言

万花筒是一种光学玩具，只要往筒眼里一看，就会出现一朵美丽的"花"。将它稍微转一下，又会出现另一种花的图案。不断地转，图案也在不断变化，所以叫做"万花筒"。万花筒诞生于 19 世纪的苏格兰，由一名研究光学的物理学家发明。2～3 年后，几乎同一时期传道了中国和日本。19 世纪初，中国的很多玩具进入日本，其中就有万花筒。当时，作为利用光学的游戏，新鲜而有趣，万花筒成为糖果店吸引孩子的招牌玩具。

万花筒的图案是如何来的呢？原来是靠玻璃镜反射而成的。三面玻璃镜组成一个三棱镜，再在一头放上一些各色玻璃碎片，这些碎片经过三面玻璃镜的反射，就会出现对称的图案，看上去就像一朵朵盛开的花。

活　　动

1. 活动准备

（1）材料准备：镜子，纸板，不干胶，剪刀。

（2）分组准备：3～5人一组，每组一份材料。

2. 活动过程

（1）按照裁剪线（黑色实线）进行裁剪。

（2）撕掉镜子的保护膜。

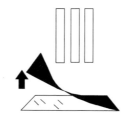

（3）利用双面胶把三面镜子粘贴到纸板所示位置，镜子反射面向上。

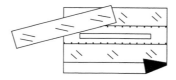

（4）按照虚线进行折叠，并用双面胶粘贴，制成等边三角体，镜子在万花筒内侧。

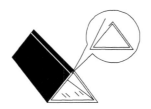

（5）使万花筒一端对着显示器或其他物体，眼睛从另一端观察，慢慢转动万花筒。

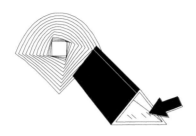

活动记录：

_____

_____

_____

_____

_____

……

经过长期开展活动，我们不断探索学习，相继制定了多种动手制作方案：音阶管、竹蜻蜓、民族蜡染、拓印、再生纸、小机床、空气动力车、浮与沉、叶脉书签。

5.2 馆外科普活动

5.2.1 少数民族地区双语科普

2017 年，我馆与少数民族科普工作队联合行动 6 次，前往边疆少数民族地区开展前沿科技展示、科普大篷车活动。全省拥有大篷车 75 辆，每辆车每年进乡开展活动不少于 10 次，每次受众人次不低于 500 人次。在联合开展的 6 次活动中，前沿科技展示、体验 18 次，共发放科普资料 32 万份，其中少数民族双语科普资料 10 万份，前往地区有楚雄州、临沧市、版纳州、宁蒗县等（彩页图 67 ~ 图 75）。

5.2.2 天文观测

云南省科学技术馆与云南省天文爱好者协会联合开展科普活动。初始阶段，利用"中秋节""科技周""科普日"时机，在馆内开展白天观日、晚上观月活动，把各类天文望远镜带到市民身边，免费观测、普及天文知识。近几年，随着中国流动科技馆云南巡展活动的开展，把天文望远镜带到边远山区，为当地公众讲解天文知识，让他们亲自观测星空。

5.2.3 "宇宙剧场"活动

"宇宙剧场"是一种球幕立体宇宙剧场，直径 18 米，倾角 15 度的标准半球内配备球幕布作为银幕，剧场内播放超高分辨率的细腻画面，使画面艳丽感人，3D 效果卓越超群。主要影片有"恐龙灭绝之谜""宇宙太空舱"等，播放时间为 10 ～ 15 分钟。

因在试点时的效果突出，"宇宙剧场"配套活动得到中国科学技术协会和中国科学技术馆的认同，配备于之后的展品中，更名为"球幕影院"，影片增加到 5 部。

5.2.4　科普剧

云南省科学技术馆自 2008 年开始，筹备了第一部科普剧《美丽的泡泡》，并免费为公众演出，得到了昆明市中小学生的热烈追捧，科普效果显著。随后，陆续自主开发了《猪坚强和他的朋友》《仰望星空》《光的奥秘》《飞天幻梦》等科普剧，在日常的演出活动中，得到了学校师生的一致好评。其中 2017 年自编自导自演的原创科普剧《玩具店奇妙夜》，获第五届全国科技馆辅导员大赛西部赛区一等奖、全国总决赛三等奖，第五届全国科学表演大赛一等奖的佳绩。所有的比赛平台促进了业内的交流学习。

我们会把科普剧带进幼儿园、中小学，根据不同年龄段学生特点，制作不同的科普剧（彩页图 76 ～图 80）。

5.2.5　科学秀（科学实验）

带领学生们进行科学实验，偶尔加上几句诙谐的语句，或是几个很炫的实验表演，能迅速拉近学生和辅导员之间的距离，让学生在轻松的环境下接受科学知识的教育，更利于学生对科学知识的吸收和理解，从而达到科学普及的目的。

5.2.6　动手制作

利用安全可靠的设备，让学生在给自己制作小纪念品的同时，能亲身体会机床的基本工作原理、各种机床之间的工作流程。

5.2.7　馆社合作

我们与地州科协、社区科协都建立良好的合作关系机制，为体现省馆的引领带动作用，我们经常参与各地科普活动及昆明市区重大科普活动，如"科技周""科普日"及各地区主题科普活动。

6. 调查问卷

2017 年，我不定期对前来参观或参与科普活动的观众进行随机调查，共收回调查问卷 218 份，该项调查问卷便于了解观众的年龄结构、文化程度及感兴趣的展品等，为未来发展方向研究提供支撑。

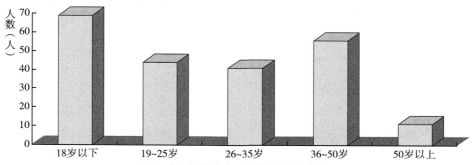

图 1　参与科普活动观众的年龄结构柱形图

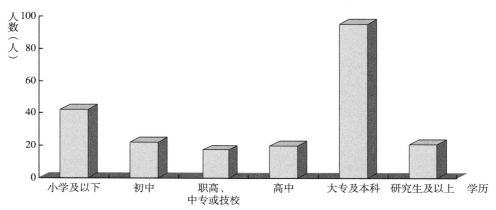

图 2　参与科普活动观众的文化水平柱形图

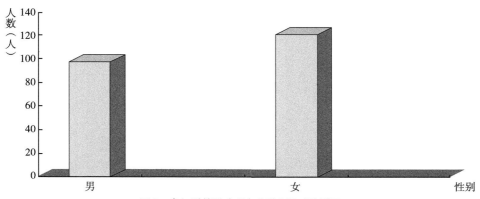

图 3　参与科普活动观众的性别组成柱形图

7. 结语

云南省科学技术馆以"弘扬科学精神、传播科学思想、倡导科学方法、普及科学知识、提升科学素质"为目的，用丰富多彩的科普活动，扩大科学普及的效果，利用好常设展览"科学的探索"和"体验科学"中国流动科技馆云南巡展两大平台，加之馆校、馆社、馆企等合作形式，最大限度地为昆明市区孩子能参与、体验更多的科普项目而努力，同时把科技与欢乐带到边疆、山区及少数民族地区，让那里的学生感受"体验科学"带来的无穷魅力。

参考文献

［1］景佳. 科普活动的策划与组织实施［M］. 北京：华中科技大学出版社，2001：5.
［2］曹健萍. 科普类场馆中科普活动方式的探讨［J］. 科学中国人，2017（15）.

中国地质博物馆的百年历史与创新发展

贾跃明 [①]　刘树臣　陈嫒嫒　徐翠香　何哲峰　阮佳萍

摘 要　中国地质博物馆始建于 1916 年，是一家具有百年历史的国家级博物馆，也是中国人自己建造的第一座公立自然科学博物馆。自 1916 年正式开馆至今，中国地质博物馆薪火相传、英才辈出、成就斐然，以典藏系统、成果丰硕、陈列精美，称雄于亚洲同类博物馆，并在世界范围内享有盛誉。近年来，中国地质博物馆坚持"典藏立馆，人才强馆，科技兴馆"的发展方略，培育"博物、博学、博爱"核心价值理念，在服务支撑国土资源工作、开展社会服务和科普教育活动、推进全国地学博物馆发展等方面起到了重要作用，然而同时也面临大楼建筑功能单一、空间局促、展览科研典藏条件受限、国际合作尚显不足等挑战，亟需通过规划建设新馆、国际联合办展等形式实现新时期的创新发展。

关键词　中国地质博物馆　百年历史　创新发展

中国地质博物馆成立于 1916 年 7 月，是中国人自己创建的第一座公立自然科学博物馆。以典藏系统、成果丰硕、陈列精美称雄于亚洲同类博物馆，并在世界范围内享有盛誉。始终坚持"典藏立馆，人才强馆，科技兴馆"的发展方略，培育"博物、博学、博爱"核心价值理念，在服务支撑国土资源工作、开展社会服务和科普教育活动、推进全国地学博物馆发展等方面起到了重要作用。2016 年 7 月 20 日，在建馆 100 周年之际，习近平主席致信祝贺，充分肯定了中国地质博物馆在地球科学研究、地学知识传播等方面取得的显著成绩，以及在发展我国地质事业、提升全民科学素质方面起到的重要作用，并对今后建设发展提出了明确要求，寄予了殷切希望，指明了前进方向和奋斗目标。

---

①　贾跃明：中国地质博物馆馆长、党委书记，研究员。主要研究方向：国土资源科技管理和战略研究，目前主要从事地学博物馆管理和科学普及工作。地址：北京市羊肉胡同 15 号，邮编：100034，邮箱：522191845@qq.com。

1. 中国地质博物馆百年辉煌历史

1.1 百年历程

中国地质博物馆始建于 1916 年，历经百年风雨，却未有一日中断，始终作为独立的科学机构存在且不断发展，是中国近代科学发端的见证者、中国地质事业的传承者及中国科学普及工作的开创者，是中国科学史、地学史、博物馆史上的传奇。1916 年 7 月 14 日，中国人自己培养的首批 18 名地质学子毕业，在丰盛胡同 3 号举行"学生成绩展览会"，让中国地质博物馆的雏形——地质矿产陈列馆登上了历史舞台，精选展出的 917 件标本也成为中国地质博物馆馆藏的源头。

1935 年，地质矿产陈列馆因日寇侵华搬迁南京，在珠江路 942 号重新建馆，陈列面积 1500 平方米，设 12 个陈列室，于 1937 年 2 月建成开放。抗日战争全面爆发后，地质矿产陈列馆再度迁徙，辗转陆路、水路，经长沙、武汉、宜昌，于 1938 年落户重庆北碚，作为中国西部科学博物馆的重要组成部分，重新设计开放，分矿物岩石、地层古生物、脊椎动物化石及土壤 4 室。抗战胜利后再迁南京，中华人民共和国成立后重回北京。

目前的馆大楼位于北京西四，是由周恩来总理 1956 年亲自批准兴建，建成于 1958 年，更名为"地质部地质博物馆"，建筑面积 11500 平方米，展陈面积 5000 多平方米，1959 年对社会开放，1986 年再次更名为中国地质博物馆。21 世纪初大楼修缮一新后于 2004 年 7 月重新对外开放，建有地球厅、矿物岩石厅、宝石厅、史前生物厅、国土资源厅（目前改为"百年历程厅"）5 个基本陈列展厅和 2 个临时展厅，年接待量约 50 万人次。

1.2 科学巨匠

中国地质博物馆厚重的历史既记载、见证了我国地球科学和地质事业的曲折发展，又汇聚、积淀了丰富的自然精华和无形资产。这里走出了章鸿钊、丁文江、翁文灏、谢家荣、侯德封、黄汲清、杨钟健、裴文中、程裕淇、高振西、刘东生等中国地质事业奠基者和学界翘楚。

1.3 自然精华

现有馆藏地质标本 20 万余件（套），涵盖地学各个领域。其中有蜚声海内外的东方神州龙、巨型山东龙、中华龙鸟等恐龙系列化石，北京人、元谋人、山顶洞人等古人类化石。有毛泽东主席赠送的"水晶王"、世界最大的方解石晶洞、巨型萤石方解石晶簇标本，精美的蓝铜矿、辰砂、雄黄、雌黄、白钨矿、辉锑矿、绿柱石、黑柱石、磷氯铅矿等中国特色矿物标本，以及包括世界已知最大的方柱石猫眼和全球罕见的沙弗莱石在内的种类繁多的宝石、玉石等一大批世界级、国宝级珍品。

2. 中国地质博物馆发展现状与取得成效

伴随着中国近现代科学的同步发展，中国地质博物馆传承了厚重的光辉历史和科学精神，不忘初心，与时俱进，在服务支撑国土资源工作、开展社会服务和科普教育活动、推进全国地学博物馆发展等方面起到了重要作用。

2.1 围绕中心工作，服务支撑国土资源事业发展

中国地质博物馆作为国土资源部直属单位，自 2009 年起承担国土资源科普基地管理、化石保护管理、史志编纂等新的支撑与服务职能，并以此先后成立国土资源科普基地管理办公室、国家古生物化石专家委员会办公室和国土资源史志办公室。

科普基地管理办公室成立以来，建立了一整套国土资源科普基地命名、评估、管理办法，分类命名科普基地 176 家，充分发挥了国土资源领域科技场馆、科研实验基地、资源保护区的科普作用，积极推动了我国国土资源科普事业的发展。化石专家委员会办公室在制定法规制度、建设技术支撑体系、编制保护规划、加强产地保护、强化收藏管理、科普宣传教育、建立化石村、人才培养、推动化石科考、追缴走私化石等方面成效显著，为我国的化石保护事业做出了重要贡献。史志办公室负责国土资源系统史志年鉴工作的规划指导及编纂审校工作，每年一卷连续出版《中国国土资源年鉴》，全面、系统、客观地记述了我国国土资源事业改革与发展情况，成为最重要的历史档案，为各级政府部门提供了资政依据。

2.2 依托专项经费，显著提升馆藏水平

藏品是博物馆的立身之本，是博物馆的生命线，也是博物馆区别于图书馆、展览馆、科技馆等其他社会公共文化机构的重要标志。丰富的馆藏对于博物馆建设发展具有极其重要的意义。自 2011 年以来，中国地质博物馆依托《地质遗迹标本采集与购置及综合研究》专项经费，采集、购置地质标本 3000 余件，其中包含沙弗莱石在内的大量精品地质标本和白钨矿、辉锑矿等特色地质标本，极大地丰富了馆藏精品，显著提升馆藏水平，为更好地开展藏品科学研究、展览展示、科普宣传提供了基础材料和宝贵资源。

在标本采集购置过程中，经过 5 年的探索实践，中国地质博物馆在标本采集、购置与征集工作上已形成一套行之有效的管理办法和运行机制，两个"三分离"（调研、谈判、决策相分离，询价、议价、定价相分离）原则和三个"机制"（纪律监察制约、专家咨询评价、集体参与民主决策）日渐成熟，为同行业和博物馆开展相关工作提供了参考与借鉴。

2.3 立足公益科普，传播地球科学知识

中国地质博物馆始终以传播地学知识，提高全民科学素质为己任，奉行让公众"认知地球、亲近自然，珍惜资源、保护环境，愉悦身心、陶冶情操，崇尚科学、热爱祖国"的科普教育宗旨，不断推出精品展览、创新科普教育形式、加强科普团队力量、扩大科普教育范围。

2.3.1 不断推出精品展览，服务广大社会公众

陈列是博物馆实现其社会功能的主要方式，也是博物馆特有的语言。基本陈列是博物馆的灵魂，而临时展览是陈列展览的有效补充，它能紧扣时代脉搏，吸引更多观众。近年来，中国地质博物馆每年都推出临时展览（表 1）。这些主题突出、特色鲜明的精品展览，在社会上引起了强烈的反响和广泛关注。如 2016 年为纪念中国地质博物馆建馆 100 周年推出的《百年历程——中国地质博物馆建馆 100 周年成就与精品展》，向社会大众免费开放，截至 2017 年年初，该展览省部级专场已达 30 多场，各单位团体近 400 次，展厅及巡展总参观人数约 90 万人次，展览及其附属活动得到包括中央电视台新闻联播、《人民日报》《光明日报》等各级媒体报道 1077 次，成为中国地质博物馆自 21 世纪以来规模最大、观众最多、最具吸引力并深受喜爱和赞赏的展览，荣获了"2016 年度中国古生物科普进展十大新闻""孔雀石杯"地矿科普优秀成果一等奖。此外，《首届中国矿物精品展》《世界矿物精品（2017）展》等展览展出了国内特色矿物晶体、世界宝石矿物晶体精品数百件，成为国内业界展示高质量矿物标本的经典展览，为广大观众献上了一场场视觉与精神的饕餮盛宴。

表 1　举办临时展览数量（单位：场）

2013 年	2014 年	2015 年	2016 年	2017 年（1～11 月）
11	6	7	7	6

2.3.2 组织开展主题新颖、形式活泼的"请进来"科普教育活动，受到观众喜爱

2015 年发布实施的《博物馆条例》，将博物馆的教育职能提到了首位。近年来，

中国地质博物馆科普教育活动的数量（表 2）和质量都在不断提升。"石上"Party 主题科普活动、一笔生辉——绘制绿色地球科普活动、化石"模"法师、点石成画——矿物颜料绘扇面等一系列活动题材新颖、形式活泼，极大地调动了观众，尤其是青少年了解、学习地学知识的兴趣。

表 2　科普教育活动统计表（单位：场）

2013 年	2014 年	2015 年	2016 年	2017 年（1～11 月）
10	15	13	28	25

2.3.3　通过"走出去"扩大科普教育活动辐射范围，惠及更多受众

近年来，中国地质博物馆主动承担社会责任，真诚地服务广大社会公众，走出博物馆、走出北京，面向全国公众传播地学知识，满足公众对地学知识的广泛需求。通过"走出去"赴湖南、江苏、湖北、辽宁、内蒙、西藏等地举办大型科普活动（表3），在社会上引起了广泛关注，深受好评。从 2011 年开始，中国地质博物馆坚持走基层、接地气，将地质博物馆进校园科普活动范围从北京市区，逐渐拓展到远郊区县，截至 2015 年，已经达到北京 16 区县全覆盖。从 2014 年开始，又将科普活动目光面向全国，将活动范围拓展到江西、辽宁、内蒙古、西藏、河北、湖北、湖南等地，行程数万千米，惠及公众数十万人。

此外，中国地质博物馆科普专家团队力量不断加强，从最初的依靠本馆专家力量，延伸到邀请国土资源部系统专业技术人员。从 2015 年开始，邀请李廷栋、欧阳自远、刘嘉麒、周忠和等热心地学科普的院士为青少年讲授科学知识。让青少年在聆听讲座的同时，感受科学的魅力，培养其对自然科学的兴趣。

表 3　地质博物馆科普教育活动统计表

活动名称	举办地点	时间	活动内容	参加人次
大型科普活动	湖南省长沙市、郴州市	2013 年—2017 年	中国（湖南）国际矿物宝石博览会博物馆论坛、科普论坛、联合展览	＞50 万人
	北京市	2015 年、2016 年	地学科普高峰论坛、讲座、展览	＞20 万人
	江苏省东海市	2015 年	水晶大王回归展暨中国地质博物馆藏品展	＞30 万人
	湖北省黄石市	2017 年	中国（黄石）地矿科普大会论坛、展览	＞15 万人

续表

活动名称	举办地点	时间	活动内容	参加人次
地质博物馆进校园	北京市门头沟区、密云区、朝阳区、西城区4所学校	2011年	4场科普标本展示、展览，4场讲座	约1万人
地质博物馆进校园	北京市延庆区、昌平区2所学校	2012年	2场科普标本展示、展览，2场讲座	数千人
	北京市平谷区、顺义区、门头沟区、大兴区、朝阳区5所学校	2013年	5场科普标本展示、展览，5场讲座	约1万人
	北京市丰台区、房山区、怀柔区、河北省怀来县6所学校	2014年	6场科普标本展示、展览，6场讲座	数千人
	北京市石景山区、通州区、西城区3所学校	2015年	4场科普标本展示、展览，3场讲座	数千人
	北京市顺义区、房山区、延庆区、门头沟区、西城区5所学校	2016年	5场科普标本展示、展览，5场讲座	数千人
	北京市通州区、河北省河间市4所学校	2017年	4场科普标本展示、展览，18场讲座	约1万人
科技列车行科普活动	江西省赣州市	2014年	6场科普标本展示、展览，2场讲座，1次免费鉴定	＞1万人
	辽宁省丹东市	2015年	4场科普标本展示、展览，2场讲座	约1万人
	内蒙古自治区赤峰市	2016年	2场科普标本展示、展览，1场讲座	约1万人
	西藏自治区拉萨、那曲、日喀则	2017年	3场科普标本展示、展览，5场讲座	约1万人

2.4 发挥领头作用，指导全国地学博物馆建设和发展

中国地质博物馆充分发挥在地学博物馆行业领头羊的影响力，切实加强对全国地学博物馆建设和管理方面的业务协调与技术指导。

2.4.1 成立12家分馆，加强业务指导和示范引领

为充分发挥中国地质博物馆在典藏、科研、科普方面的引领示范，以及专家团队作用，发挥地方博物馆在当地文化建设和经济建设中的作用，中国地质博物馆自1996年起在全国设立分馆，目前共建有中华恐龙馆、烟台自然博物馆、黄果树奇石馆、嘉荫神州恐龙博物馆、东海水晶博物馆等12家分馆。

2.4.2 依托协会、学会等交流平台，增强馆际交流，推动场馆建设

中国地质博物馆挂靠有中国自然科学博物馆协会国土资源博物馆专业委员会、中

国博物馆协会地质博物馆专业委员会、中国地质学会地质科普工作委员会和化石保护研究分会。依托这些平台，中国地质博物馆发挥行业领头作用，更好地凝聚了全国地质博物馆与科普场馆的力量，对全国一些地质博物馆的改建、新建工作给予了指导性的意见和建议，极大地推动了我国地质博物馆的发展建设。特别是依托国土资源博物馆专业委员会，组织、举办、参加国际国内业务交流活动10余次，举办联合展览8次。在促进地质博物馆业务发展的同时，向公众普及地学知识，弘扬科学精神，从而更好地服务经济社会发展，推动生态文明建设。

3. 中国地质博物馆面临的问题与思考

过去100年，中国地质博物馆在科研、科普和典藏等方面取得了显著成绩，然而，随着全球化的持续深入、经济社会的高速发展，公众对以科普场馆为代表的科学文化设施的期待越来越高，对科学知识的渴求越来越强烈，对知识普及的手段和全方位参观服务体验的要求也越来越具体。特别是以习近平主席为核心的党中央审时度势，提出建设世界科技强国和文化强国战略，并首次提出"科技创新、科学普及是实现创新发展的两翼"，将科学普及提高到与科技创新同等重要的地位。中国地质博物馆在科学研究、知识普及、合作办展等方面迎来了新的机遇和挑战。

3.1 面临空间局促的重大挑战，亟需谋划建设新馆

中国地质博物馆现有大楼使用已近60年，总建筑面积仅有1.1万平方米，展陈面积5000多平方米，日益凸显建筑功能单一、空间局促等一系列问题，严重制约了中国地质博物馆功能发挥和长远发展，无法与国家、社会发展进步的步伐相适应，难以达到广大公众的预期，陈列展览、科普活动、观众服务、标本典藏、藏品研究均受到严重限制。

21世纪以来，众多专家学者、领导和社会公众大力呼吁中国地质博物馆谋划和筹建新馆，解决目前馆舍老旧、空间局促、基础设施落后等问题。自2016年起，中国地质博物馆在国土资源部的大力支持下，正抓紧推进新馆建设，力争建设一座设施更为先进、功能更为齐全、展品更为丰富、气势更为恢宏，能够彰显国家实力和形象的中国地质博物馆新馆。

3.2 国际合作尚显不足，应立足特色馆藏开展联合办展等国际交流与合作

随着全球化、国际化的深入，同时在国家"一带一路"重大战略的机遇框架内，

中国地质博物馆在开展国际合作方面还有很大的发展空间。未来将大力开展馆际交流与合作，通过"走出去""请进来"与"一带一路"沿线国家博物馆开展藏品科学研究、地学知识普及和联合办展、业务培训等各个方面、各个层次以及多种形式的交流与合作，同时借助政府、学会、协会等多平台、多渠道获得政策支持和资金支持，推动地学类博物馆拓展国际化视野，加强国际间馆际交流与合作。

"希望你们以建馆百年为新起点，不忘初心，与时俱进，以提高全民科学素质为己任，以真诚服务青少年为重点，更好发挥地学研究基地、科普殿堂的作用，努力把中国地质博物馆办得更好、更有特色，为建设世界科技强国、实现中华民族伟大复兴的中国梦再立新功"，习近平主席的谆谆教诲是激励和鞭策我们不忘初心、牢记使命，恪守宗旨、真抓实干，开拓创新、砥砺奋进，为建设世界科技强国、实现中华民族伟大复兴的中国梦，做出新的更大贡献的行动指南和精神动力。

参考文献

［1］贾跃明. 百年薪火相传科学精神不朽——庆祝中国地质博物馆建馆 100 周年［J］. 地球学报，2017，38（2）：133–134.

［2］宗苏琴. 举社会之力助推藏品利用——从扬州博物馆实践看藏品利用［J］. 艺术品鉴，2016(10).

［3］王宏钧. 中国博物馆学基础［M］. 上海：上海古籍出版社，2001：246.

［4］中国自然科学博物馆协会. 中国科普场馆年鉴 2014 卷［M］. 北京：中国科学技术出版社，2014.

［5］中国自然科学博物馆协会. 中国科普场馆年鉴 2015 卷［M］. 北京：中国科学技术出版社，2016.

［6］中国自然科学博物馆协会. 中国科普场馆年鉴 2016 卷［M］. 北京：中国科学技术出版社，2017.

中国航海博物馆"馆校合作"项目助推青少年学生航海教育

吴春霞 ①

摘 要 中国航海博物馆积极响应国家"一带一路"倡议,在上海市教委的支持下,开展了"利用场馆资源提升科技教师和学生能力的馆校合作"项目,努力与上海市中小学共同建设具有一定影响力的新型科学教育研究基地,培养一批热爱科学、善于研究、乐于奉献的复合型、创新型青少年学生人才,促进中小学科技教师的专业能力提升。该项目通过面向学校、教师、学生的 5 个形式多样、内容丰富的子项目,在推进青少年航海教育方面取得了明显成效,并将进一步优化和完善,形成可复制可推广的教育模式,为我国航海事业和"一带一路"的建设努力做出新的贡献。

关键词 馆校合作 学生 航海教育

1. 航海博物馆的由来和概况

1405—1433 年,明代航海家郑和率船队七下西洋,实现了我国航海事业的壮举。2005 年 7 月 11 日,为纪念郑和下西洋 600 周年,我国决定将每年的 7 月 11 日定为"中国航海日",中华人民共和国国务院同时批复由中华人民共和国交通部和上海市人民政府共同筹建中国航海博物馆。2010 年 7 月 5 日,中国航海博物馆建成并对外开放,填补了我国没有国家级航海博物馆的空白。

中国航海博物馆地处东海边,位于上海浦东临港新城滴水湖畔,距离上海市中心城区约 80 千米(彩页图 81)。博物馆以弘扬中华民族灿烂的航海文明为己任,构建国际航海交流平台,促进青少年对航海事业的热爱,是国家打造海洋强国的重要文化宣

---

① 吴春霞,上海中国航海博物馆社会教育部主任,硕士,馆员。研究方向:博物馆公众教育和科普教育产品创新,青少年群体课外实践教育等。地址:上海市浦东新区临港新城申港大道 197 号,邮编:201306,电话:13918064191,E-mail:ruoxi_wcx@126.com。

传载体。博物馆建筑面积 46434 平方米，室内展示面积 21000 平方米，室外展示面积 6000 平方米，分设航海历史、船舶、航海与港口、海事与海上安全、海员、军事航海 6 大展馆，渔船与捕鱼、航海体育与休闲 2 个专题展区，并建有天象馆、4D 影院、儿童活动中心等。开馆至今每年的观众接待量约为 32 万人，其中青少年占 50% 左右。除常设展外，每年举办符合时代主题和办馆宗旨的特展，如《辛亥海军》《亚丁湾护航 5 周年展》《中国古代航海文物展》《馆藏西方航海文物精品展》《甲午海鉴——纪念甲午海战 120 周年》《航路 1600——中荷交流展》《钓鱼岛的历史与主权》《航向世界——中英荷交流展》《上海国际航运中心建设成果展》《18、19 世纪外销艺术品展》等。

开馆以来，在国家实施"海洋强国"战略的背景下，航海博物馆始终致力于航海知识传播和航海科技文化的普及，紧扣航海主题，策划了一系列面向不同受众的教育活动，如航海生活节、航海主题家庭日、航母彩虹服、四季品牌活动、多彩科普秀、郑和情景剧、创艺工坊等。注重与周围社区、学校的互动，并形成了以"三公里文化服务圈"为特点的系列活动和服务对象，开展深度合作。博物馆十分注重对青少年的航海教育，尤以与上海市教委联合开展的"馆校合作"项目最为系统全面，覆盖面最广，成效也最明显。

2. 航海博物馆馆校合作项目

"利用场馆资源提升科技教师和学生能力的馆校合作"项目，设定受众群体为上海市中小学校、教师和学生。项目共有 5 个子项目，包括校本课程开发、博老师研习会、微课题研究员、科学诠释者、文化服务包。各子项目立足于航海博物馆的常设展览、馆藏和现有的教育资源，充分利用馆内、馆外的教育力量，并实行全市公开招募。

2.1 "校本课程开发"（面向老师）子项目缓解了航海课程缺失的难题

该子项目通过与上海部分中小学一线教师合作，开发基于航海历史文化传承和科普知识传播的校本课程，征集基于场馆资源和校本课程相结合的学生课程项目。针对不同年级学生开发不同类型、不同难易程度、馆内和课堂相结合的研究型和探究型课程，并与定点学校联动实施和试点。

项目形成了适用于中小学的 60 门左右的校本课程，涉及航海历史、人文、科技等多方面。课程充分结合航海博物馆的陈列展览和教育资源。所有课程经过专家评审、试点后正式实施和推广，给全市中小学校无偿使用，并且在博物馆网站等平台公开发布。

2.2 博老师研习会（面向老师）子项目弥补了航海科普师资缺乏的问题

研习会成员在博物馆了解教育资源，体验教育活动，依托场馆丰富的藏品和展览教育资源，通过观摩陈展、与专家对话交流、教研员培训指导、组员间的学术讨论等，提高了教师利用场馆资源开展教学的能力，建立了教师与航海博物馆、教师与教师之间长期合作的沟通平台。

经过研习，每位博老师都制定了一份适于青少年课外实践拓展教育的活动策划案，并形成自己的参观体验方法，让孩子融入博物馆，爱上博物馆。第一阶段共有来自全市 10 个区 23 所中小学的老师报名参加，辐射宣传面达 200 余人，经综合统筹筛选，实际参与人数 48 人。研习班课程涵盖博物馆、航海、教育等多方面，开设近 20 场专家授课与交流，并策划了 40 多个航海主题活动案。

2.3 微课题研究员（面向学生）子项目逐步培养学生开展研究和动手能力

该子项目主要引导中学生尤其是高中生积极参与探究型场馆实践活动，进一步提升青少年探究和研究能力。项目利用博物馆在展示、教育、研究和文物等方面的资源，由科研人员、业务骨干和专家等担任指导老师，学生在导师的指导下全程参与微课题的研究。

为了提高项目主题的丰富性，确保微课题子项目选题与服务对象需求的匹配度，分为理论研究型与观察实验型两类。理论研究型课题是针对博物馆馆藏资源和教育主题，以知识探究和疑点难点分析为主要导向的课题研究，如清洁航海、快船船体结构研究、铁甲舰技术的发展演变等。观察研究型课题侧重于动手、实验、观察的微课题项目，趣味性较强，如船模制作、陶瓷修复、陶瓷制作、桌面水族馆等主题。此项目第一阶段参与的学生有 50 余人，所有参与学生的出勤时间将计入其社会实践学时中。

2.4 科学诠释者（面向学生）子项目给学生创造了锻炼和自我成长的平台

该子项目全市共有 120 余名学生报名，经筛选后有来自全市 7 个区 21 所学校 87 名学生参加，针对学生建立"了解—学习—思考—诠释"四个阶段的培养系统，整个活动包括认识"诠"概念、探索"诠"主题、优化"诠"细节、展示"诠"风采和精品"诠"演绎等环节，已形成较为全面的科学诠释培训文案和流程。经过培训后的学生可以通过情景式的故事讲解、互动式的科学演示等，面向公众传播航海科普知识，同时提升了自己的科学素养、创新思维、历史使命感和社会责任感。

在培训后，举行了第一届"小小科学诠释者"比赛，获奖选手可以成为航海主题活动的提前参与者，并在其中挑选优秀学员重点培养，树立典型和品牌。学生所有诠释的文案被整理成册，统一印制，以供分享及后期教育活动使用。第一阶段参与的学生有 60 余人，所有参与学生的出勤时间将计入其社会实践学时中。

2.5 文化服务包（面向学校）子项目努力缓和博物馆地处偏远的不便

受地理位置等客观因素的影响，航海博物馆必须主动走出场馆，将展览、讲座、活动及课程教具等送入学校，更大范围地为青少年普及航海知识和文化。针对学校的普适性活动内容包括但不仅限于一场展览、一堂讲座、一种体验等。其中展览分为航海科普、历史普及、艺术文化等多个类别；科普讲座涵盖航海历史科技等方面；体验则是开展一些与航海相关的动手体验活动或实地参观考察活动。该项目还于暑期开设了 6 个主题共 10 场航海夏令营活动，效果非常好。全市共有 1500 多学生报名，最后筛选了 12 个区 133 所学校 447 名学生参加。

中国航海博物馆的"馆校合作"已经完成了第一阶段的目标，即设计出了系列航海科普课程，开发了基于场馆资源的教育实践活动，逐步组建青少年学生航海教育科技教师队伍，形成一定规模的航海科普专家库，展出青少年航海研究及实践成果，与部分中小学校建立起项目试点合作机制，并在逐步探索建立科普场馆与学校间可复制可推广的馆校合作模式。同时，使中小学生初步了解了中国的航海的历史和著名的航海人物，懂得中国航海技术对世界航海发展的贡献，学习并掌握现代航海的一些基本技能，了解海洋权益对国家和人民的重要性等知识，激发学生主动探知航海知识的学习欲望。后续还将不断优化和完善第二阶段、第三阶段的馆校合作项目，努力形成可面向全市，甚至全国推广的馆校合作模式。

3. 航海博物馆发展对策及合作建议

在博物馆的发展过程中，最大的困难就是距离上海市中心 80 千米的路程，周边常住人口少，对行业博物馆的认可度等导致全年的参观量以及活动参与度不稳定。同时，年龄结构合理的人才梯队尚未形成，社会宣传和行业知名度有待提升，精品文物和品牌教育项目影响力不足，特展交流展水准还需加强等。如何完成博物馆综合实力的赶超，实现从"新馆""大馆"到"强馆"的转变，打造成集"教育娱乐化"与"专业实力派"于一体的现代化航海专题类博物馆，航海博物馆面临着诸多新挑战。

因此，在"十三五"发展规划制定时，我们重点在以下几方面寻求突破，即冲刺国家一级博物馆；打造陈列展览精品；提升学术研究水平；搭建航海科研平台；增强藏品典藏能力；提升社会教育影响力；优化场馆运营管理；拓展交流合作领域；公共文化服务人群覆盖率明显提高，年参观人数保持 10% 以上的增长率；打造一批社教品牌活动，促进航海科普产品的品牌化设计、项目化运作、社会化推动。同时做好组织、制度、人才、经费等各项保障工作。

"一带一路"已经成为国家发展主旋律，"一带一路"科普场馆应加强交流合作，如中国博物馆协会增设"一带一路"博物馆专委会；联合开展全国青少年活动赛事；合作举办主题教育项目；原创展览交流引进；文创产品联合开发销售等，共同为"一带一路"国家战略的实施做出更多的尝试和努力。

4. 结语

在当前条件下，博物馆与学校的"馆校合作"仍然处于探索阶段，实践经验并不丰富，同时受到国家正规教育的政策所限，在一段时间内无法进行大范围的实施，其效果体现也同样受到限制。然而，随着文化的繁荣发展以及对青少年综合素质教育的诉求，馆校合作必然是一个趋势，博物馆作为非正规教育的重要教育场所，也将越来越引起学校和教师的重视，引起家庭教育的兴趣，其价值和意义终有一天会真正得到实现和升级。

参考文献

[1] 宋娴. 博物馆与学校的合作机制研究 [M]. 上海：上海科技教育出版社，2016：8.

1	2
3	4
5	

1 位于尼泊尔遥远西部的丹加迪飞机博物馆
2 加德满都航空博物馆
3 维克多·阿姆巴楚米扬故居博物馆外貌
4 维克多·阿姆巴楚米扬
5 维克多·阿姆巴楚米扬的卧室

6　巴基斯坦盐岭中纳玛尔峡谷的视图，二叠系到始新世的岩系出露于峡谷中
7　巴基斯坦盐岭中的克乌拉峡谷视图，前寒武系和寒武系层序出露良好
8　赫拉克莱冬博物馆外景
9　学生在教室里上课

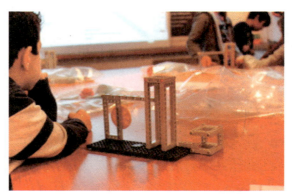

10
11
12
13

10　澳大利亚国家恐龙博物馆的入口
11　国家恐龙博物馆户外恐龙园
12　在博物馆商店中陈列各种标本
13　哈萨克斯坦共和国国家博物馆外景

```
14 │ 15
─────────
16 │ 17
─────────
    18
```

14　古代与中世纪历史展厅

15　毡质居所—kyiz ui

16　金人

17　海亚姆天文馆建筑的三维模型

18　马什哈德天文学会组织的天文科普活动

19
20
21
22

19　位于马什哈德市菲尔多西大学校园内
　　直径 10 米的天文馆
20　北京汽车博物馆
21　车辆类藏品
22　车辆类外的构成、文献、模型等藏品

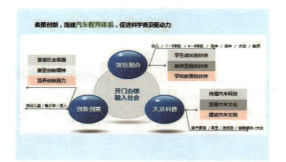

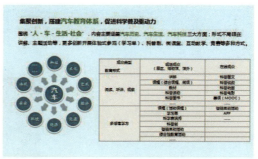

23	24
25	26
	27

23、24、25、26、27 汽车科普体系

212

相传，早在5000年前黄帝就制造了指南车。
It is said that 5,000 years ago, the Yellow Emperor invented south-pointing chariot.

中国古代伟大的思想家、教育家孔子曾乘车周游列国，传播文化思想。

Over 2,500 years ago, the great thinker and educator Confucius made a tour of numerous kingdoms in a carriage, spreading Chinese culture and ideas.

2500 years ago

2200 years ago

秦始皇统一中国后，实行了"车同轨"制度，车辆制造进入了标准化阶段。

China's first emperor, the First Emperor of Qin, unified China under one rule for the first time, including the rule "the same rutting". He standardized certain aspects of the tracks and vehicles throughout the empire.

多元文化促进国与国、城与城、人与人间的互动交流。
Multicultural events have promoted interactive exchanges between countries, cities, and people.

一路同行世界梦！
我们期待让车载着人、人载着思想，共赴"人-车-社会"的美好与远方
We look forward to continually working together toward the World Dream! We also look forward to the continuation of cars carrying people and people carrying ideas. All of these things will work together in propelling our progress toward the beautiful and far-sighted goal of integrating people, cars, and society.

31
32
33

31　多元文化

32　世界梦

33　鸟瞰科苏梅尔天文馆

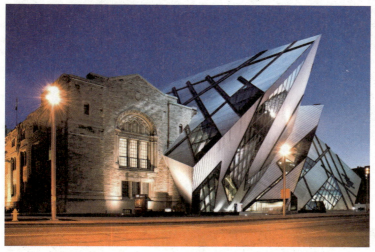

34　　34　正在放映的球幕影片《揭秘
　　35　　　玛雅先民：宇宙的观察家》
　　36　35　皇家安大略博物馆修建于
　　　　　　1933 年的女王公园侧翼和
　　　　　　翻新于 2007 年的水晶宫
　　　　36　皇家安大略博物馆在 2014
　　　　　　年展出从纽芬兰打捞出来的
　　　　　　26 米长的蓝鲸骨架

40	41	42	43
44	45	46	
	47		

40、41、42、43　科技出版物

44、45、46　国家博物馆馆长奖

47　辽宁省科技馆白垩纪之旅

48	
49	50
51	

48　VR 骑乘效果图

49　"地球号"方舟

50　动物灭绝多米诺骨牌

51　麋鹿传奇展

57

58 | 59

60

61

57　位于江岸区赵家条 104 号的武汉科学技术馆

58、59　改扩建后的展厅一角

60　武汉科技馆新馆全景

61　观众排队等待进馆参观的场景

62	63
	64
65	66

62、63、64　武汉科技馆新馆内景
65、66　丰富多彩的科普活动

67	68	69
70	71	72
73	74	75

67、68、69　少数民族双语科普读物（中文和傣文）

70、71、72　少数民族双语科普读物（中文和藏文）

73、74、75　少数民族双语科普读物（中文和彝文）

76、77、78、79、80　配套科普活动现场
81　中国航海博物馆